Petrogenesis of Metamorphic Rocks

Petrogenesis of Metamorphic Rocks

Editor

Jason Silver

scitus
academics

Petrogenesis of Metamorphic Rocks
Edited by **Jason Silver**

ISBN: 978-1-68117-593-5
Library of Congress Control Number: 2016946448

SCITUS Academics LLC,
616, Corporate Way, Suite 2, 4766,
Valley Cottage, NY 10989

www.scitusacademics.com

Printed in United States of America on Acid Free Paper

Preface

Rock metamorphism is a geological process that changes the mineralogical and chemical composition, as well as the structure of rocks. Metamorphism is typically associated with elevated temperature and pressure, thus it affects rocks within the earth's crust and mantle. Metamorphic rocks are made by either heating up or squashing the earth's crust. They are often found in mountainous regions. A metamorphic rock is a result of a transformation of a pre-existing rock. The most typical metamorphism transforms sedimentary rocks to metamorphic rocks by addition of heat during mountain building or by a large volume of magma in the crust. The original rock is subjected to very high heat and pressure, which cause obvious physical and/or chemical changes. One example is slate. Slate was originally a black mud laid down on the bottom of the sea or lake. Fossils Andreas can sometimes be found in it but they are often squashed. Due to the action of plate tectonics, compression, stress and shearing forces over long periods of time, rocks can be essentially warped and deformed, causing them to be compacted into a smaller volume of space. As a consequence, metamorphic rocks are always more dense than their original material, and also much less susceptible to erosional breakdown. As the Earth's plates move over geologic time, a plate containing igneous or sedimentary rock may become subducted under another plate. The sheer weight of the material above it can cause the rock to undergo metamorphism.

Petrogenesis of Metamorphic Rocks discusses the most common metamorphic rock types, structure and graphical representation of their mineral assemblages. It defines basic principles of metamorphism and metamorphic processes.

Table of Contents

1

CHAPTER

Geochemistry of Major Oxides in Host Rocks in Vizianagarm Manganese Ores Belt (A.P.), India

Farhat Nasim Siddiquie, Mohd Shaif*

Department of Geology, Aligarh Muslim University, Aligarh

ABSTRACT

The abundance, distribution trends and significance of the major oxides in the host rocks in Vizianagarm Manganese Ores Belt (A.P.) (between N latitude 18°12' and 18°30' and E longitudes 83°20' and 83°45'), 15 samples of host rocks from different localities of the area under study were collected and analyzed for major oxides. We describe here in major oxides geochemistry of host rocks and manganese ore deposits associated with Precambrian Khondalite and Charnockite in Vizianagarm Manganese Ores Belt (A.P.): 1) Preponderance of SiO_2 over Al_2O_3; 2) Dominance of K_2O and CaO over Na_2O; 3) Abnormally high concentration of phosphorus and a positive relationship of P_2O_5 with CaO and Ti contents; 4) Manganese increases with increases of iron, lime and soda and vice versa, 5) CaO increases with the increases of Al_2O_3, Ti, K_2O and vice versa. High P_2O_5 content in these manganese ores appears to be the result of precipitation from secondary manganese rich solutions containing dissolved phosphorus from the P_2O_5enriched host rocks. Another source of P_2O_5 may be the associated granitic and pegmatitic intrusions. Elements like K, Na, Ca, Mg, Co, Ni, Pb and Zn etc. appear to be mostly concentrated in the Mn-minerals viz. psilomelane, cryptomelane, hollandite and pyrolusite and related secondary phases [1] and [2]. Stratigraphically, the study area includes within a thick succession of Precambrian Group belonging to the Khondalite and Charnockite Groups of Dharwar Supergroup, that form a part of Eastern Ghat Complex of India. The manganiferous rocks that have been encountered in the Vizianagarm Manganese Ores Belt (A.P.) India are known as Kodurites.

Keywords: Vizianagarm, Manganese Ores, Major Oxides, Khondalite, Charnockite, A.P., Geochemistry, Eastern Ghat, Dharwar Supergroup

1. INTRODUCTION

A large number of manganese deposits occur in the Precambrian Eastern Ghats complex of Andhra Pradesh, India. These are dominantly low to medium grade metamorphic deposits with high iron, phosphorus and silica. These manganese bands (deposits) are generally syngenetic and syntectonic with metasedimentry country rock of Khondalite, quartzite and calc-granulites [3].

Manganese ores are widely distributed throughout the Peninsular India. Stratigraphically, these are mainly confined to the Precambrian terrain and are encountered in the Sausar Group and Aravalli Supergroup of central and western India, Iron ore Group and Gangpur Group of eastern India. In Andhra Pradesh, there are major occurrences of manganese ores in Srikakulam-Visakhapatnam Manganese ores belt. The deposit is in this belt over an area of 50 kms in length and 20 km in width. The study area lies 150 km NE of Visakhapatnam and includes Garividi, Garbham and Chipurupalle areas (Figure 1). Systematic geochemical study of these manganese ores in the present study area was not carried out by earlier workers. Fifteen samples of host rocks were selected from different localities of the study area. They were analysed for number of major oxides and trace elements.

2. GENERAL GEOLOGY AND STRATIGRAPHY

The manganiferous rocks that have been encountered in the Srikakulam-Visakhapatnam manganese belt of Andhra Pradesh belongs to Kodurite Series which is typically exposed at Koduru (18°16': 83°36') a locality about 80 km north of Visakhapatnam, the rock series which constitutes an integral part of the Eastern Ghat complex of India. The rock series is composed of an intimate assembalges of metamorphosed argillaceous (garnet-sillimanite gneisses) as at Chipurupalle, Rayagada and Garbham, Arenaceous (Quartzite and garnetiferous quratzite) as at Garbham, Koduru,

Perapi and Avagudem and calcareous sediments (calc-granulites and gneisses) as at Koduru and Gadabavalasa areas, this sequences belonging to granulite facies.

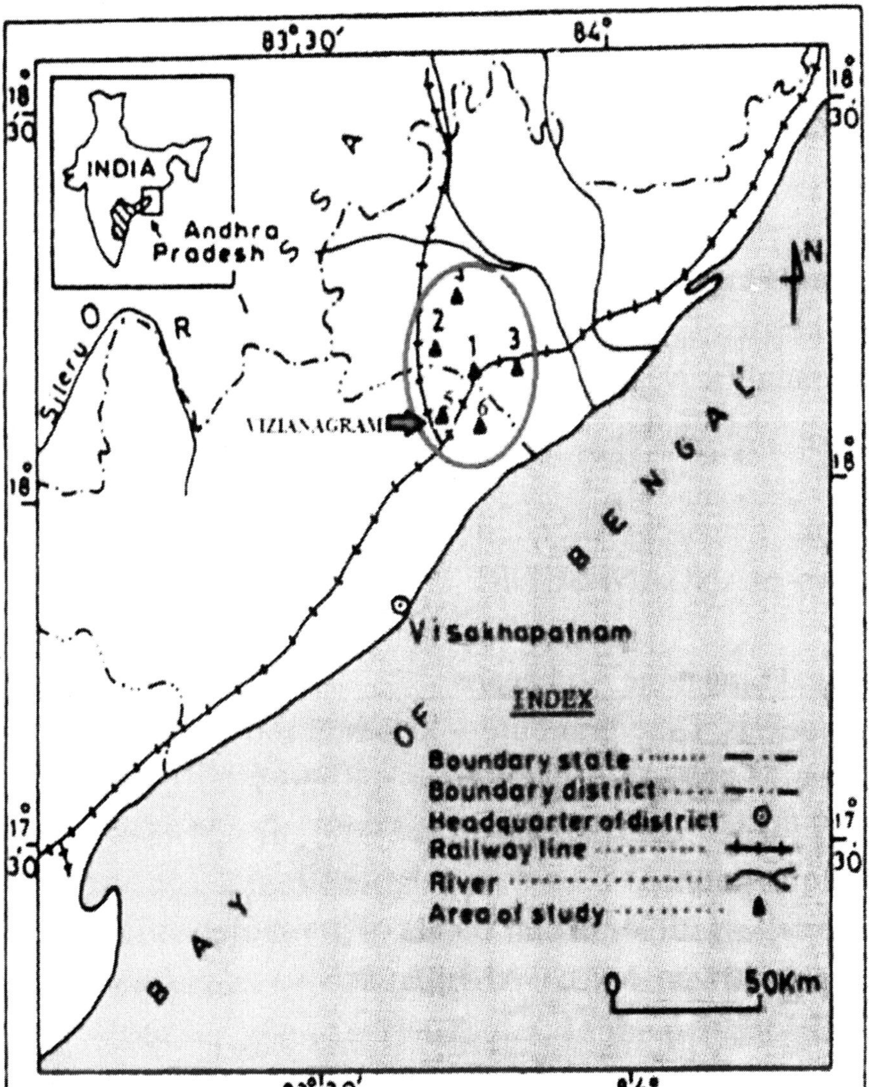

Figure 1. Location map of (1) Garividi, (2) Garbham (3) Chipurupalle manganese deposits, Vizianagram Manganese Ores Belt, (A.P.), India.

The manganese ores in the area occur primarily in association with calc-granulites, less often with quartzite and still less with the quartz garnet ± sillimanite graphite gneiss. According to [4] who made radiometric determination of the Precambrain rocks of India, the Eastern Ghat Complex was included in the Precambrain-II (2500 - 3000 m.y.) of the classification of Indian Precambrain rocks. The calc-granulite, quartzite and Khondalites being syn-sedimentary rock bodies, are conformable and co-folded with the associated lithic units such as garnetiferous quartzite, gneisses and garnet-sillimanite gneisses, etc. [5]. Subsequently, granitic and migmatitic activities of different intensities played their respective roles on these metasediments transmuting their earlier textural and mineralogical characters. However, their impact on quartzite and manganese ore is minimal, probably because these units are considered to be the resisters during migmatisation [6]. At places hypersthene gneisses (Charnockite) occur as intrusive bodies in the synsedimentary rock units. All these rocks have in turn been intruded by granite, pegmatite and quartz veins [7].

[8] recongized several rock types ranging from ultrabasic to acidic in the mineralogical composition of Kodurite and ascribed them to differentiation prior to intrusion. [9] indicated that the Kodurite series might be due to hybridism on a large scale in which a granitic intrusion bodily assimilated the manganese of the Khondalites. [10] observed that by and large the manganese ore occurs associated either with quartzite (East Garbham) or with calc-granulite (Sadanandapuram).

However, the geological association in Saibaba main, Saibaba Central, Jai Bhavani-Kottakarra quaries is not clear owing to the highly weathered nature of the rocks. [11] suggested the following stratigraphic succession of the area under study and consider the granites to be intrusive bodies into the Khondalite and Kodurite series.

4	Granites	Peninsular and Crystalline Complex
3	Charnockites	
2	Kodurites···Hybrid	

1	Calc-granulites, crystalline limestone and associated manganese ores Garnet-Sillimanite gneisses Quartzite-Garnetiferous, Feldspathic and graphite-bearing	Khondalite Series

These rocks at places exhibit relict banding. The Garbham main quarry and the Srinavasa quarry at Garbham show weathered calc-granulite-quatzo-feldslpathic rock underlying quartzite. The rock formation and associated manganese ore deposits belonging to the Khondalite and Charnockite suit of metasediments are of the Archaean age [7]. The generalised stratigraphic sequence of the rock formations of Srikakulam area which has been suggested by [7] and supported by [12] is presented as follows:

Granite, pegmatitie and quartz vein

Hypersthese-gneisses (Charnockites?)

Khondalite Series {
- Garnet-sillimanite gneisses (khondalites)
- Garnet-granulites – Ore horizon I
- Calc-granulites
- Garnet-granulites – Ore horizon II
- Garnet-sillimanite gneissess (khondalites)
- Coarse grained quartzites)
- Feldspathic quartzite
- Garnetiferous quartzite – Ore horizon III
- Feldspathic quartzites
- Coarse grained quartzites
- Garnet-sillimanite gneisses (Khondalites)
- Biotite-gneisses

Crystalline algal limestone belonging to Knondalite Group (Archaean) [13] - [17] and gritty sandstone (Miocene?) in the form of rubble and laterite capping are also observed.

The present authors finds no evidence to differ from the suggestion made by [7] and [18], regarding the stratigraphic succession in Srikakulam except that the occurrence of crystalline algal limestone and shales (red and green) which did not find place in their classification. The modified sequence is presented [13] - [17] as follows:

Some of the earlier workers have also made an attempts to classify and establish the sequence of rock formations encountered at Garividi, Garbham and Chipurupalle areas (Table 1 and Figure 2). They are as follows:

Garividi area [10]

Soil

Laterite-pisolites and concretions of

Mn-ore and calc-tufa

Pegmatite and quartz veins

Granites/garnetiferous biottie gneiss

Garnet-sillimanite gneiss (with or without graphite)

Calc-granulite and lenticles of crystalline limestone

Manganese ore and garnet granulites

Quartz-feldspar rock with occasional incidence of apatite

Garnet-sillimanite gneiss interlayered with quartzite and occasional lenticles of crystalline limestone

Garham Area [10]

Soil

Laterite-calc tufa calcreta

Pegmatite {
- pegmatite and quartz veins
- Garnetranite gneiss
- Garnetiferous and non-garnetiferous quartzo-feldspathic rock
- Biotite-gneiss

Charnockite

Garnet-sillimanite gneiss with or without graphite

Calc-granulite with manganese ore associated with bands and pockets of lithomarge and highly kaolinised quartzo-feldspathic apatite rock, the latter in the majority of cases seen underlying the Mn-ore zone.

Table 1. Geological succession of the Garividi area, Srikakulam-Visakhapatnam belt (A.P.).

Age	Group	Rock types
		Alluvium and soil
Recent		Laterite-limonite and ochre
Pleistocene (?)		Sandstone
Miocene (?)		Pegmatite and quartz vein
		Granite and granite gneisses
Archaeans ←	Charnockite Group ←	Hypersthene gneisses (porphyro-blastic) (Charnockite) granulitic hypersthene diorite and granite gneisses (Charnockite). Garnet sillimanite gneisses (Khondalite)
	Khondalite Group ←	Garnet granulite—MAGANESE ORE HORIZON – I
		* Crystalline algal limestone Archaeans
		* Green shale pockets, calc-granulite + lenses of red shale
		Garnet granulites—MANGANESE ORE HORIZON II
		Coarse grained quartzite
		Feldspethic quartzite with lenticular bands of garnetiferous quartzite—MANGANESE ORE HORIZON III
		Coarse grained quartzite
		Garnet-sillimanite-gneisses (Khondalite)
		Biotite gneisses

Two horizons were not reported by earlier workers, reported for the first time [15] and [19].

Calc-granulite highly weathered with intermixture of quartz-feldspar-apatite rock

Manganese ore with chert and lemonite quartzite

Garnet-sillimanite gneiss (with or without graphite) interlayered with quartzite

Chirpurupalle area [20]

Archaean:
- Recent laterite and loamy alluvium
- Granite gneisses
- Manganese rich rocks, "Kodurites"
- Khondalite series:
- Calc-granulites
- Quartz-garnet-sillimanite gneisses
- Quartzites

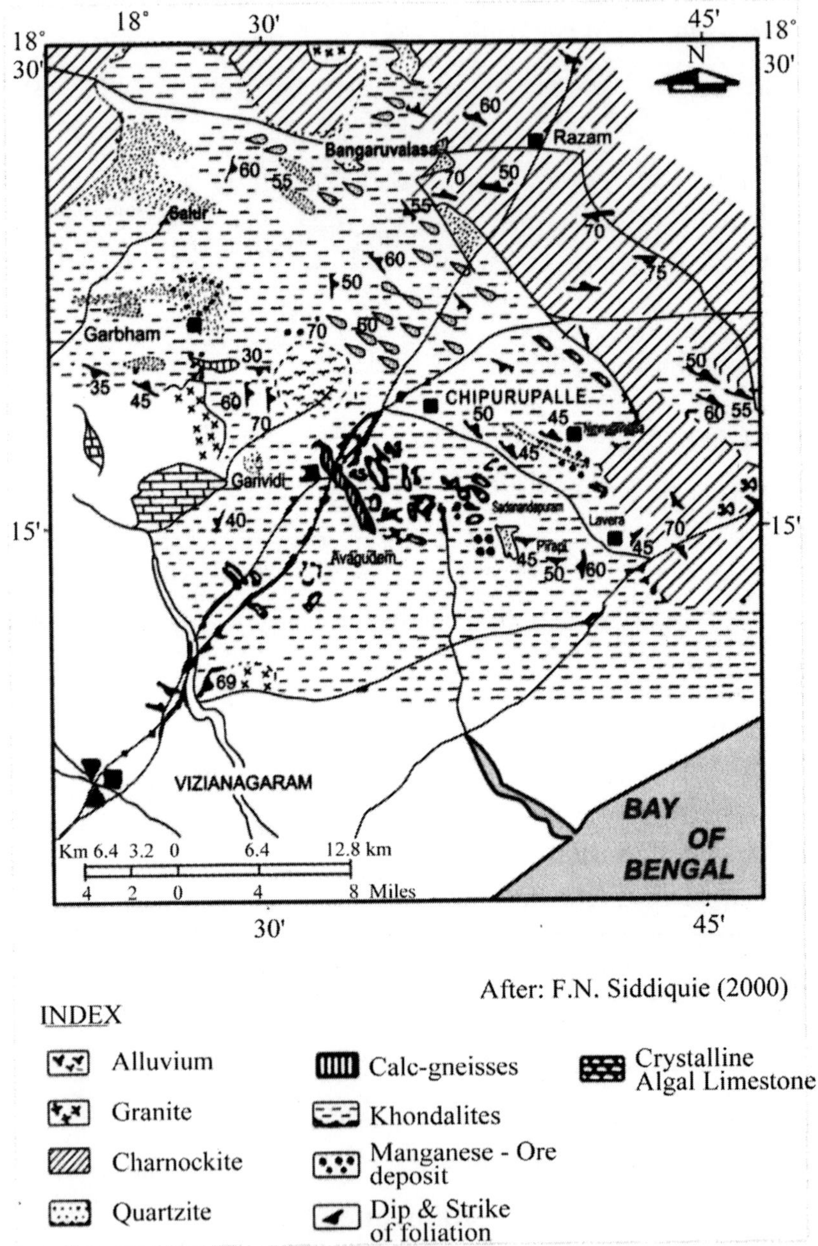

Figure 2. A generalised geological map of Garividi, Garbham and Chipurupalle manganese deposits, Vizianagram district (A.P.), India.

3. MAJOR OXIDES IN HOST ROCKS

General Statement

In the area of study, the host rocks for manganese mineralization, are mainly Khondalite and Charnockite Group of rocks.

Fifteen samples of host rocks were selected out of a large number of samples collected for their chemical analysis out of which ten samples were of Khondalites and remaining five of Charnockites. The samples were analysed chemically for the determination of their major oxides—SiO_2, Al_2O_3, TiO_2, MnO, Fe_2O_3, FeO, P_2O_5, MgO, CaO, Na_2O, K_2O and H_2O^+ and some trace elements which are discussed [21] . The relative variation trends of various major oxides are shown in Figure 3. The analytical data are presented in Table 2.

4. DISTRIBUTION OF THE MAJOR OXIDES

The abundance and variation range with average (wt%) of different major oxides of Khondalites and Charnockites are given in (Table 2 and Table 3) and the distribution of various major oxides are discussed as below:

4.1. Silica (SiO_2)

The quantitative variation trend of silica in the host rocks from 47.84 percent to 70.45 percent and gives an average value of 61.81 percent (Table 3, Figure 3). Khondalite, Charnockite rocks have wide variation of silica (Table 3). [22] reported 58.04% to 74.38% SiO_2 from Srikakulam area, [11] reported 58.4% from Garividi, [23] reported 67.10% and 69.90% respectively from Charnockites of Tamil Nadu and Karnataka. The histogram (Figure 4(a)) represents frequency percent distribution, showing polymodal nature of SiO_2 in the host rocks. The maxima is at 65 - 70 percent class. The typical mode of SiO_2 falls between 65 - 70 percent.

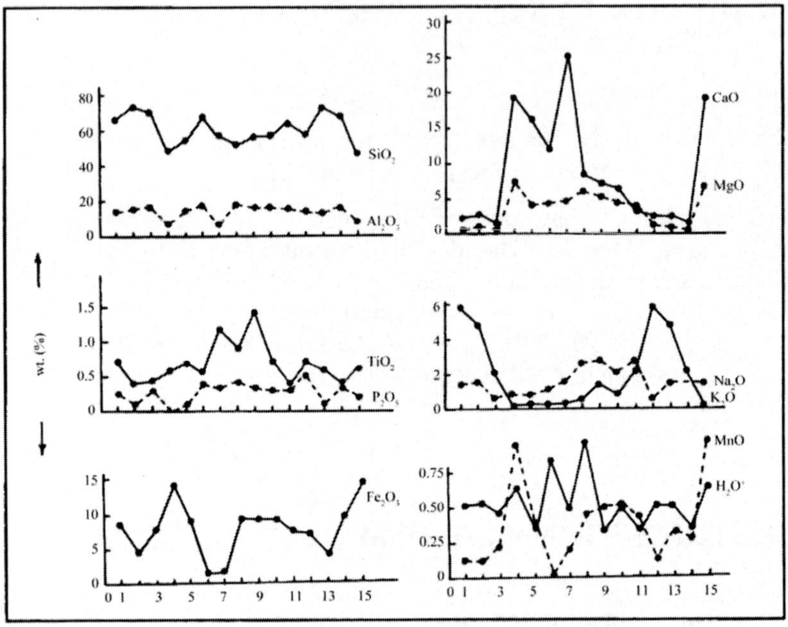

Figure 3. Variation diagrams of major oxides in Khondalite and Charnockite Groups of Garividi, Garbham and Chipurupalle areas, Srikakulam-Visakhapatnam Belt (A.P.).

Table 2. Distribution of major oxides of Khondalite and Charnockite Groups of Garividi, Garbham and Chipurupalle areas, Srikakulam-Visakhapatnam Belt (A.P.).

	KD1	KD2	KD3	KD4	KD5	KD6	KD7	KD8	KD9	KD10	CK11	CK12	CK13	CK14	CK15
SiO_2	69.35	70.3	70	48.84	56.1	68.46	58.29	52.24	56.1	59	64.16	69.09	70.45	68	47.84
Al_2O_3	13.09	13.45	16.75	6.42	13.4	10.94	6.32	17.42	16.34	16.02	15.66	13.35	13.12	16.4	7.42
MnO	0.12	0.12	0.21	0.94	0.36	0.08	0.21	0.4	0.5	0.52	0.38	0.12	0.3	0.26	0.9
Fe_2O_3	1.21	1.16	3.36	1.3	1.06	0.44	0.54	2.2	1.48	2.22	2.07	1.21	1.64	3.36	2.3
FeO	4.52	3.64	4.6	13.11	7.19	1.2	1.45	7.48	6.7	5.74	4.25	3.16	4.59	12.11	7.2
P_2O_5	0.24	0.1	0.3	n.d.	0.1	0.42	0.36	0.45	0.35	0.33	0.32	0.51	0.1	0.35	0.14
TiO_2	0.73	0.4	0.47	0.6	0.79	0.58	1.18	0.9	1.45	0.72	0.44	0.72	0.6	0.48	0.6
CaO	2.37	2.42	1.25	19.83	16.2	12.44	25.61	8.86	7.32	6.77	3.2	2.33	2.4	1.25	19.83
MgO	0.86	1.09	0.86	7.9	4.14	4.17	4.58	5.8	5.03	4.6	3.12	1.9	1.11	0.86	6.9
Na_2O	1.52	1.6	0.34	0.67	0.7	1.2	1.6	2.48	2.56	2.08	2.84	0.56	1.4	2.16	1.67
K_2O	5.94	4.96	2.16	0.02	0.24	0.16	0.2	0.64	1.42	0.94	2.2	5.9	4.97	2.09	0.02
H_2O^+	0.51	0.53	0.49	0.62	0.35	0.8	0.5	0.96	0.3	0.5	0.3	0.52	0.52	0.34	0.62
Total	100.46	99.77	100.79	100.25	100.63	100.89	100.76	99.55	100.33	100.4	100.43	100.46	99.77	100.14	100.35

KD = Khondalite; CK = Charnokite; n.d. = Not Detectable.

Table 3. Range of variation and average oxide % of Khondalite and Charnockite Groups of Garividi, Garbham and Chipurupalle areas, Srikakulam-Visakhapatnam Belt (A.P.).

Oxides	Minimum	Maximum	Average
SiO_2	47.84	70.45	61.81
Al_2O_3	6.32	17.42	13.07
TiO_2	0.12	0.94	0.36
MnO	0.44	3.36	1.45
Fe_2O_3	1.20	13.11	5.79
FeO	0.10	0.51	0.27
P_2O_5	0.40	1.45	0.70
CaO	1.25	25.61	8.80
MgO	0.86	7.90	3.52
Na_2O	0.34	2.84	1.55
K_2O	0.02	5.94	2.12
H_2O^+	0.30	0.96	0.52

4.2. Alumina (Al_2O_3)

The quantitative variation trend of Alumina in the host rocks of the study area ranges 6.32 percent to 17.42 percent and gives an average value of 13.07 percent (Table 3).

Khondalite and charnockite rocks show wide variation of Al_2O_3 (Table 3). [22] reported 9.65% to 20.64% Al_2O_3 from Srikakulam area. [24] reported 14.08% from feldspar garnetiferous granite-gneiss of Peddapantu- palli, and [23] reported 14.10% and 15.10% from Charnockites of Tamil Nadu and Karnataka. The histogram (Figure 4(b)) represents frequency percent distribution, showing polymodal nature of Al_2O_3. The maxima is at 12 - 14 and 16 - 18 percent classes.

4.3. Manganese (MnO)

The quantitative variation trend of manganese monoxide in the Khondalite and Charnockite rocks ranges from 0.12 percent to 0.94 percent and gives an average value of 0.36 percent (Table 3).

[22] reported 0.09% to 3.70% from the Srikakulam area whereas Khondalites of Sri Lanka consist 0.18% MnO [25] , [23] reported 0.05% and 0.02% from Charnockites of Tamil Nadu and Karnatka.

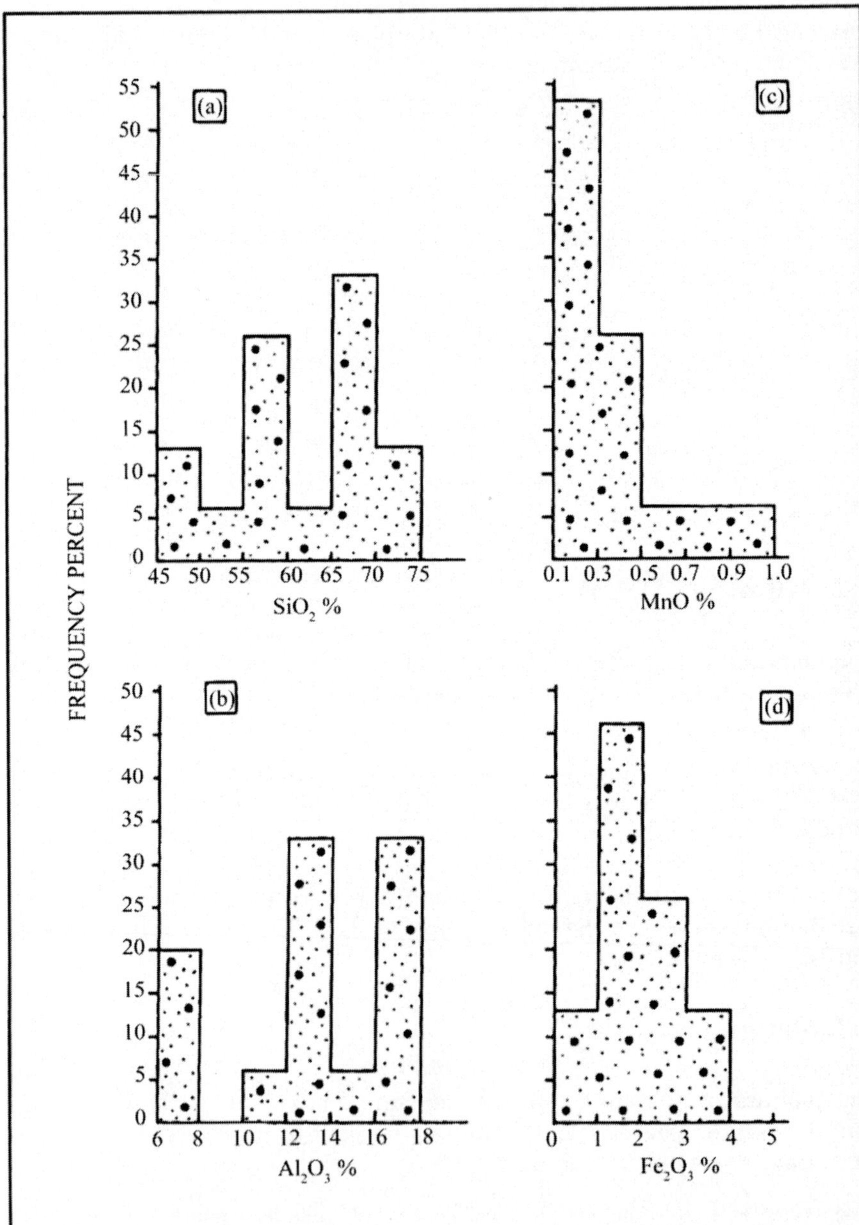

Figure 4. Histograms showing frequency percent distribution of SiO_2, MnO, Al_2O_3 and Fe_2O_3 in Khondalite and Charnockite Groups of Garividi, Garbham and Chipurupalle areas, Srikakulam-Visakhapatnam Belt (A.P.).

The histogram (Figure 4(c)) represents frequency percent distribution, showing unimodal positively skewed nature of MnO. The maxima is at 0.1 to 0.3 percent class and the typical mode of MnO also lies in the same class.

4.4. Ferric Oxide (Fe_2O_3)

The quantitative variation trend of Fe_2O_3 in the host rocks ranges from 0.44 percent to 3.36 percent and gives an average value of 1.45 percent (Table 3).

[22] reported 0.86% to 3.76% from Srikakulam area. [24] reported 1.22% from feldspar garnetiferous granite-gneiss of Peddapantupalli area, [23] reported 4.36% and 2.36% from the Charnockite of Tamil Nadu and Karnataka (Table 3 and Table 4).

The histogram (Figure 4(d)) represents frequency percent distribution, showing unimodal positively skewed nature of Fe_2O_3. The maxima is at 1 to 2 percent class and the typical mode also lies in the same class.

4.5. Ferrous Oxide (FeO)

The quantitative variation trend of FeO in the host rocks ranges from 1.20 percent to 13.11 percent and gives and average value of 5.79 percent (Table 3 and Table 4).

The Khondalites and Charnockites having higher concentration of FeO as compared to Fe_2O_3 (Table 3 and Table 4), [22] reported 1.12% to 8.04% from same area, [26] reported 9.98% from hypersthene-granulite of Narasimhunipeta, [11] reported 8.04% from Khondalites of Garividi and [23] reported 3.60% and 1.30% from Charnockites of Tamil Nadu and Karnataka (Table 3 andTable 4).

The histogram (Figure 5(a)) represents frequency percent distribution, showing polymodal nature of FeO. The maxima is at 3 to 5 percent class and typical mode also lies in the same class.

4.6. Phosphorus (P_2O_5)

The quantitative variation trend of P_2O_5 in the host rocks ranges from 0.10% to 0.51% and gives an average value of 0.27 percent (Table 3, and Table 4).

The concentration of P_2O_5 is abnormally on the higher side in the manganese ores as well as in the host rock of the study area, whereas [22] reported 0.18% to 1.01% from the various areas of Srikakulam district, [11] reported 0.75% from Khondalites of Garividi, and [24] reported 0.24% from feldspar garnetiferous granite-gneiss of peddapantupalli (Table 3 and Table 4).

The histogram (Figure 5(b)) represents frequency percent distribution, showing bimodal nature of P_2O_5. The maxima is at 0.3% to 0.4% class. The typical mode of P_2O_5 is also lies between 0.3 percent to 0.4 percent class.

4.7. Titania (TiO_2)

The quantitative variation trend of TiO_2 in the host rocks ranges 0.40 percent to 1.45 percent and gives an average value of 0.70 percent (Table 3).

Table 4. Average major oxide (wt%) of the host rocks (Khondalite and charnokite) compared with other Khondalites and Charnockites.

	I	II	III	IV	V	VI	VII	VIII
SiO_2	60.86	63.90	68.36	60.75	60.08	58.04	67.10	69.90
Al_2O_3	13.01	13.19	14.08	14.49	12.38	17.30	14.10	15.10
TiO_2	0.77	0.56	0.73	0.36	0.65	1.65	0.80	0.41
Fe_2O_3	1.49	2.10	1.22	1.75	3.28	1.20	4.36	2.36
FeO	5.70	5.97	4.51	9.98	4.20	8.04	3.60	1.30
MnO	0.35	0.39	0.11	0.10	0.18	2.27	0.05	0.02
MgO	3.90	2.77	0.85	3.26	1.95	5.90	2.31	1.09
CaO	10.30	5.80	2.37	5.10	9.43	1.40	2.82	3.20
Na_2O	1.47	1.72	1.52	2.62	2.59	0.88	3.91	4.79
K_2O	1.66	3.03	5.93	1.20	3.06	3.24	2.23	1.49
P_2O_5	0.26	0.28	0.24	0.05	-	0.75	0.02	0.08
H_2O	0.55	0.46	0.44	0.51	1.80	0.15	0.47	0.16

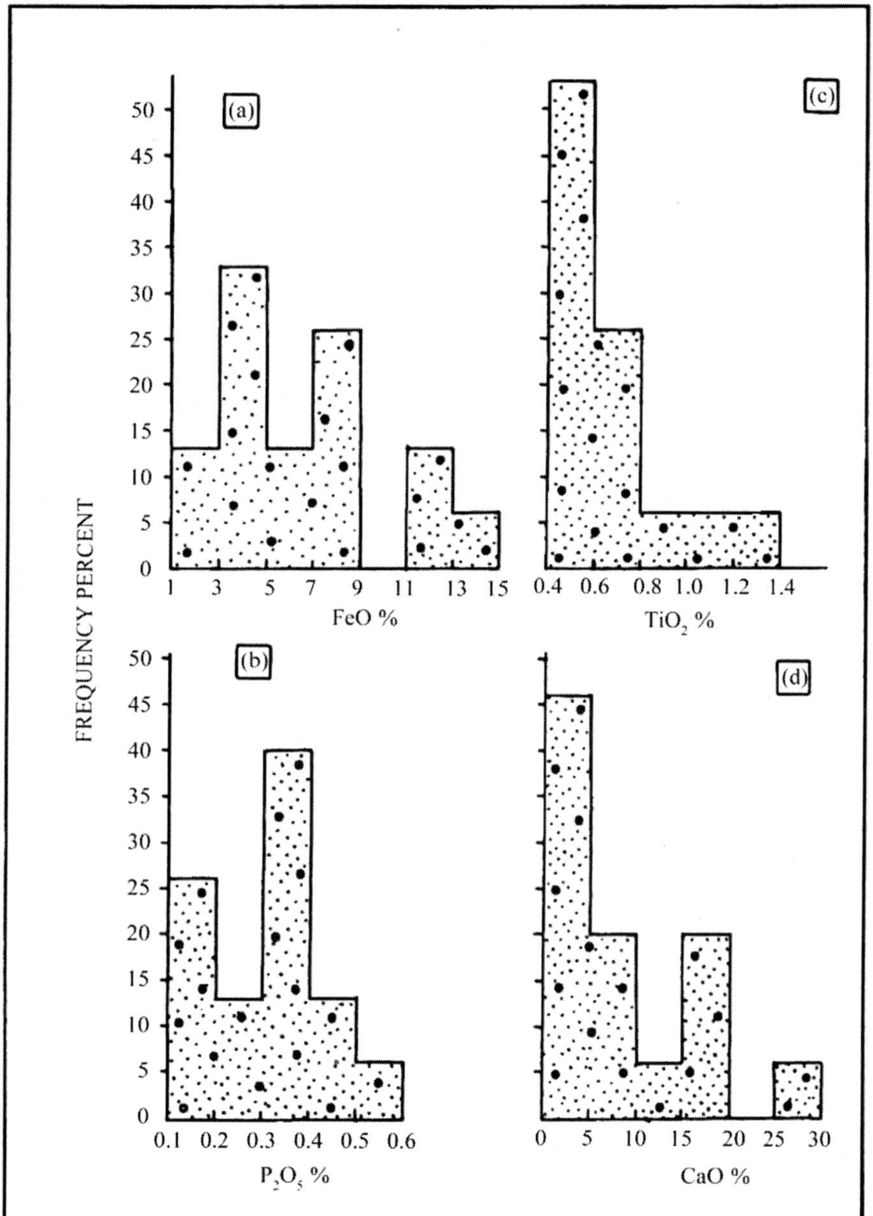

Figure 5. Histograms showing frequency percent Distribution of FeO, TiO_2, P_2O_5 and CaO in Khondalite and Charnockite Groups of Garividi, Garbham, and Chipurupalle areas, Srikakulam-Visakhapatnam Belt (A.P.).

[24] reported 0.73% from feldspar garnetiferous granite-gneiss of Peddapantupalli, Khondalite of Sri Lanka shows 0.65% TiO_2 [25] . [11] reported 1.65% from Khondalite of Garividi and [23] reported 0.80% and 0.41% from Charnockites of Tamil Nadu and Karnataka.

The histogram (Figure 5(c)) represents frequency percent distribution showing unimodal, positively skewed nature of TiO_2. The maxima is at 0.4 to 0.6 percent class and the typical mode also lies between 0.4 to 0.6 percent class.

4.8. Lime (CaO)

The quantitative variation trend of calcium oxide in the host rocks of the study area ranges from 1.25 percent to 25.61 and gives and average value of 8.80 percent (Table 3 and Table 4).

The CaO in the Khondalites and Charnockites of the study area shows a wide variation range from 1.25% to 25.61% (Table 3 and Table 4), Rao, K.V.R. (1955) reported 5.10% from hypersthene-granulite from the Narasimhunipeta and Krishnan (1934) reported 9.43% CaO in Khondalites of Sri Lanka (Table 3 and Table 4).

The histogram (Figure 5(d)) represents frequency percent distribution, showing bimodal nature of CaO. The maxima is at 1 percent to 5 percent class. The typical mode lies between 1 percent to 5 percent class.

4.9. Magnesia (MgO)

The quantitative variation trend of MgO in the host rocks ranges from 0.86 percent to 7.90 percent and gives and average value of 3.52 percent (Table 3 and Table 4).

Rao, K.V.R. (1955) reported 3.26% from hypersthene-granulite of narasimhunipeta, [11] reported 5.90% from Khondalites of Garividi and [23] reported 2.31% from Charnockites of Tamil Nadu (Table 3 and Table 4).

The histogram (Figure 6(a)) represents frequency percent distribution, showing bimodal nature of MgO. The typical mode of MgO and maxima lie between 0 to 2 and 4 to 6 percent classes.

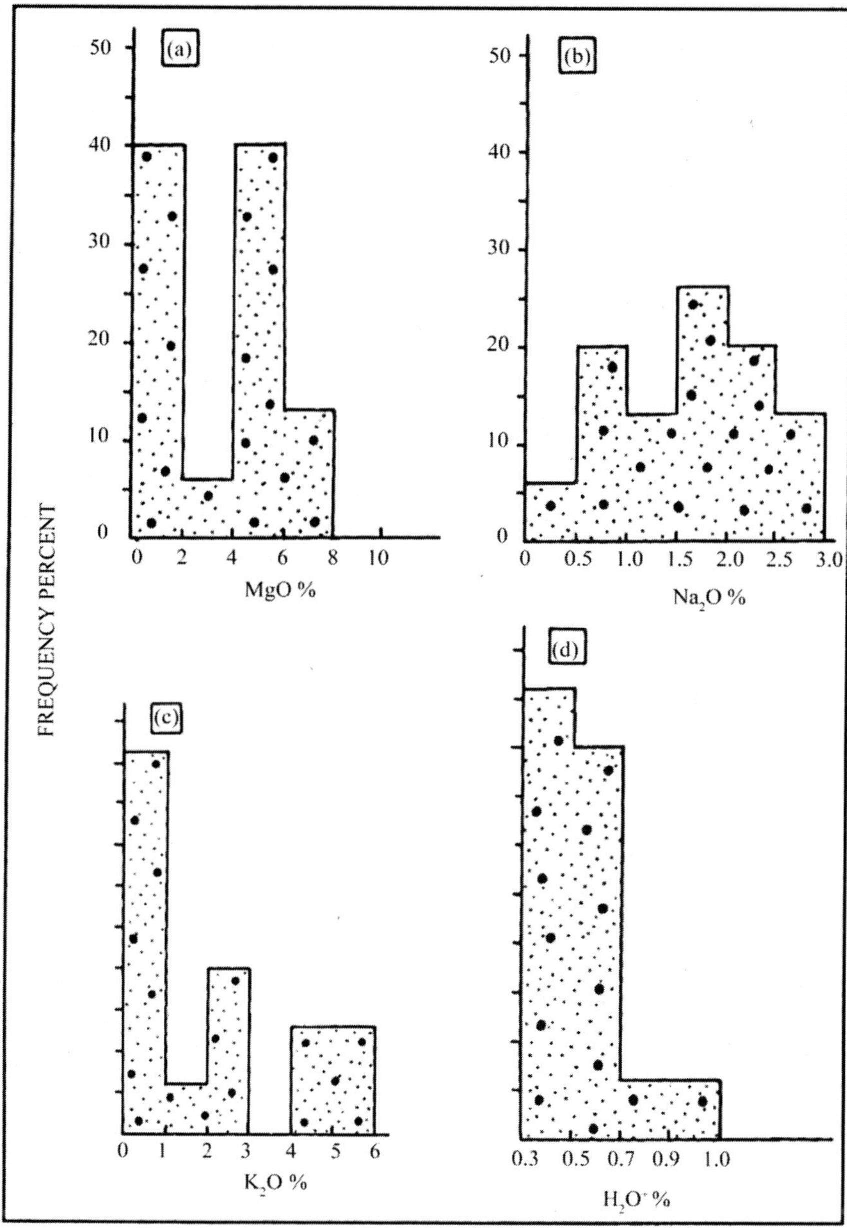

Figure 6. Histograms showing frequency percent distribution of MgO, K_2O, Na_2O and H_2O^+ in Khondalite and Charnockite Groups of Garividi, Garbham and Chipurupalle areas, Srikakulam-Visakhapatnam Belt (A.P.).

4.10. Soda (Na_2O)

The quantitative variation trend of Na_2O in the host rocks ranges from 0.34 percent to 2.84 percent and gives an average value of 1.55 percent (Table 3 and Table 4).

Rao, K.V.R. (1955) reported 2.62% from hypersthene granulite of Narasimhunipeta, 2.59% reported in Khondalites of Sri Lanka Krishnan (1934), and [23] reported 3.91% and 4.79% from Charnockites of Tamil Nadu and Karnataka (Table 4).

The histogram (Figure 6(b)) represents frequency percent distribution, showing polymodal nature of Na_2O. The maxima is at 1.5 percent to 2.0 percent class, and the typical mode also lies in the same class.

4.11. Potash (K_2O)

The quantitative variation trend of K_2O in the host rock ranges from 0.02 percent to 5.49 percent and gives an average value of 2.12 percent. [24] reported 5.93% from feldspar garnetiferous granite-gneiss of peddapantupalli, 3.06% reported from Khondalites of Sri Lanka [25] .

[11] reported 3.24% from Khondalites of Garividi, and [23] reported 2.23% from Charnockites of Tamil Nadu (Table 4).

The histogram (Figure 6(c)) represents frequency percent distribution, showing polymodal nature of K_2O. The maxima is at 0 to 1 percent class and typical mode also lies in the same class.

The Khondalites and Charnockites have wide variation of potassium oxide. This higher amount of K_2O is due to the presence of potash feldspar and micaceous mineral in these rocks. The high potash contents in Khondalites and Charnockites are possibly due to the migmatisation, which took place by the injection of feldspathic (K-feldspar) fluid. The megacrystic K-feldspar in them crystallised from the injected fluid and also by the segregation of the neo-mobilised feldspathic components during migmatisation. According to [27], the main supply of alkalies came probably from the intrusive igneous rocks, another probable source might be the quartz with its numerous fluid inclusion.

4.12. Water (H_2O^+)

The quantitative variation trend of H_2O^+ in the host rocks ranges from 0.30% to 0.96% and gives an average value of 0.52 percent.

[24] reported 0.44% from feldsdpar garnetiferous granite-gneiss of peddapantupalli, [26] reported 0.51% from hypersthene-granulite of Narasihunipeta, and [23] reported 0.47% from Charnockites of Tamil Nadu.

The histogram (Figure 6(d)) represents frequency percent distribution, showing unimodal positively skewed nature of H_2O^+. The maxima is at 0.3 to 0.5 percent and typical mode also lies in the same class.

5. MUTUAL RELATIONSHIP

An attempt has been made to determine the geochemical relationship of MnO with SiO, Al_2O_3, FeO, K_2O, Na_2O, CaO (Figures 7(a)-(f)), and Fe_2O_3 in Khondalite and Charnockite Groups of Garividi, Garbham and Chipurupalle areas, Srikakulam-Visakhapatnam Belt (A.P.). The mutual relationships of significant oxides are plotted in various diagrams and discussed as follows:

5.1. Silica (SiO_2)

The computed correlation coefficient of silica shows strongly antipathetic relationship with MnO (Figure 7(a)) and FeO (relationship of SiO_2 with Fe_2O_3 and FeO and Al_2O_3 with Fe_2O_3 in Khondalite and Charnockite Groups of Garividi, Garbham and Chipurupalle areas, Srikakulam-Visakhapatnam Belt (A.P.)) and MgO and shows slightly negative relationship with TiO_2, Na_2O and H_2O^+ (Table 5).

These relationships reveal that with the increase of SiO_2 there is correspondingly decrease in MnO, FeO, TiO_2, CaO, MgO, Na_2O and H_2O^+ and vice versa. The negative relationship of silica with manganese and FeO probably indicates that manganese and FeO are the substitutes of silica. It appears that the negative relationship of MnO with silica is possibly due to the alteration of manganese bearing minerals into silicate minerals. The highly variable silica/alumina ratio (Table 6) and sympathetic relationship of silica with Al_2O_3, Fe_2O_3 and K_2O favours the coexistence of quartz and aluminosilicate minerals in the Khondalite and Charnockite rocks. The higher concentration of silica is due to the presence of quartz and other siliceous minerals a characteristic feature of the original pellitic sedimentary rocks. Furthermore the high concentration of silica in the rocks of study area is possibly due to the presence of scapolite.

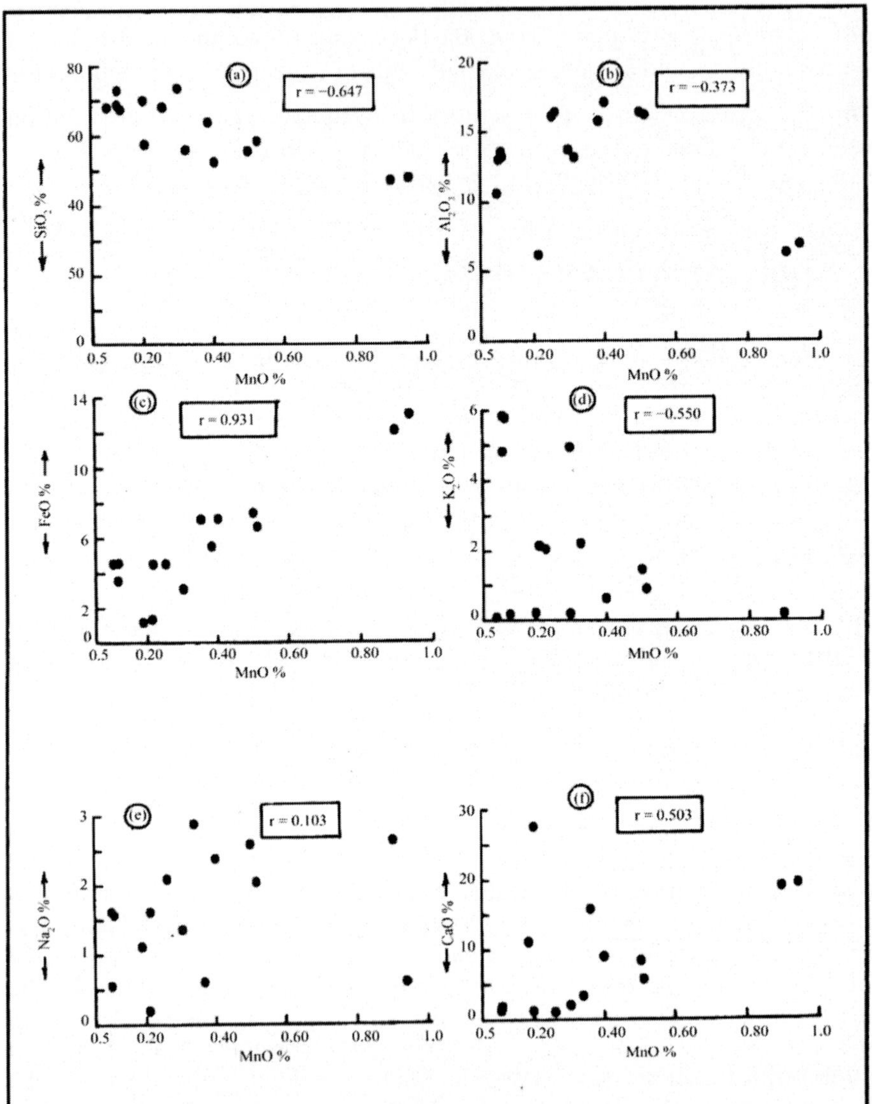

Figure 7. Relationship of MnO with SiO2, Al2O3, FeO, K2O, Na2O and CaO in Khondalite and Charnockite Groups of Garividi, Garbham and Chipurupalle areas, Srikakulam-Visakhapatnam Belt (A.P.).

Table 5. Correlation coefficient between the various constituents of Khondalite and Charnockite Groups of Garividi, Garbham and Chipurupalle areas, Srikakulam-Visakhapatnam Belt (A.P.).

S.No.	SiO_2	Al_2O_3	MnO	Fe_2O_3	FeO	P_2O_5	TiO_2	CaO	MgO	Na_2O	K_2O	H_2O^+	Ni	Co	Pb	Zn	Cr	Sr	V	Li
SiO_2	1	0.3863	−0.0348	0.0348	−0.7988	0.2349	−0.4119	−0.7385	−0.9348	−0.1741	0.7329	−0.2199	0.1747	0.0957	0.6072	−0.2634	0.0899	0.0178	−0.2974	0.0779
Al_2O_3		1	−0.3746	0.5374	−0.2285	0.4139	−0.0116	−0.7976	−0.4863	0.3681	0.2813	−0.2005	0.3542	0.2514	0.0668	0.0683	0.239	−0.0308	−0.1491	−0.3063
MnO			1	0.1993	0.9319	−0.474	0.0923	0.5032	0.7941	0.1035	−0.5504	0.0573	−0.0163	0.2032	−0.249	0.0689	−0.2299	−0.219	0.1018	0.1688
Fe_2O_3				1	0.2476	0.0803	−0.3351	−0.4456	−0.2154	0.1929	−0.0402	−0.1686	0.3964	0.6076	0.2426	−0.3531	0.1611	−0.3805	−0.1415	0.091
FeO					1	−0.467	0.0407	0.3514	0.6977	0.0139	−0.4058	0.0484	−0.1092	0.1699	−0.2559	0.0429	−0.0546	−0.0091	0.1808	0.0044
P_2O_5						1	0.298	−0.2433	−0.1491	0.2804	0.0272	0.1629	0.2422	0.1302	−0.1324	0.0787	0.2954	−0.3134	0.3457	−0.3347
TiO_2							1	0.3417	0.3732	0.2686	−0.2434	−0.0778	−0.1786	−0.4448	−0.5717	0.4041	0.2491	0.0493	0.5239	−0.1731
CaO								1	0.7822	−0.1503	−0.7267	0.2387	−0.225	−0.3526	−0.5382	0.1953	−0.2272	0.0612	0.238	0.0786
MgO									1	0.1214	−0.7766	0.3612	−0.1259	−0.1526	−0.6402	0.3623	−0.2005	−0.0782	0.2634	−0.0911
Na_2O										1	−0.1265	−0.1089	0.0251	0.3904	−0.0253	0.0677	0.1261	−0.3141	−0.0978	−0.4564
K_2O											1	−0.1963	−0.1984	0.0819	0.6583	−0.2417	0.2086	0.0984	−0.0662	0.1648
H_2O–												1	−0.4218	−0.365	−0.2954	0.2541	0.1423	0.1041	−0.005	−0.0424
Ni													1	0.429	0.0073	−0.0018	−0.5629	−0.4311	−0.0747	0.2016
Co														1	0.5297	−0.347	−0.1564	−0.5655	−0.2778	0.0288
Pb															1	−0.6955	0.0855	−0.338	−0.2282	0.3454
Zn																1	−0.2832	0.3346	−0.2687	−0.125
Cr																	1	0.2135	0.3407	−0.4597
Sr																		1	−0.1375	−0.378
V																			1	−0.1025
Li																				1

Silica shows a sympathetic relationship with Al_2O_3, Fe_2O_3, P_2O_5 and K_2O (r = 0.732) (Table 6). The positive relationship of silica with these elements is possibly due to the presence of potash feldspar and other alumino-silicate minerals.

5.2. Alumina (Al_2O_3)

The alumina is negatively but insignificantly related with MnO (Figure 7(b)), FeO (Figure 7(c)), TiO_2, MgO, and H_2O^+ and strongly antipathetic with CaO (r = 0.79. As evident from the (Table 6), alumina shows positive relationship with SiO_2, Fe_2O_3 (r = 0.73, P_2O_5, Na_2O and K_2O). However, a strong positive relationship appears between Al_2O_3 & Fe_2O_3 (r = 0.73). By correlation coefficient it therefore appears that like other few oxides, e.g., FeO, TiO_2, MgO, H_2O^+, the alumina is also not significantly related with MnO. Further the positive relationship of alumina with silica, Fe_2O_3, Na_2O and K_2O indicates the presence of sillimanite, kyanite, andalusite, orthoclase, microcline. The high Al_2O_3/Fe (total iron) ratio (Table 6) in both Khondalite and Charnockite is identical to that found in alumino-silicate minerals.

The abnormally high Al_2O_3 content is some samples of host rocks may be due to the presence of white specks of kaoline, found in cleavages and cracks of garnet [11] . These specks appear to be the cause of these abnormally high values. The abundance of alumina and silica can be assigned to the argillaceous nature of the parent rocks.

Table 6. Fe_2O_3/FeO, SiO_2/Al_2O_3, Na_2O/K_2O, CaO/MgO, MnO/P_2O_5, MnO/CaO, Al_2O_3/TiO_2, MnO/Fe (total iron) Al_2O_3/Fe (total iron) ratios of Khondalite and Charnockite Groups of Garividi, Garbham and Chipurupalle areas, Srikakulam-Visakha- patnam Belt (A.P.).

Sample Nos.	Fe_2O_3/FeO	SiO_2/Al_2O_3	Na_2O/K_2O	CaO/MgO	MnO/P_2O_5	MnO/CaO	Al_2O_3/TiO_2	MnO/Fe (Total Iron)	Al_2O_3/Fe (Total Iron)
1	**0.26**	**5.29**	**0.255**	**2.75**	0.5	0.050	17.93	0.020	2.284
2	0.318	5.22	0.322	2.22	1.2	0.049	33.62	0.031	3.539
3	0.730	4.17	0.157	1.45	0.7	0.168	35.63	0.026	2.104
4	0.099	7.60	33.5	2.51	0.94	0.047	10.7	0.065	0.445
5	0.147	4.18	2.91	3.91	3.6	0.022	16.9	0.043	1.625
6	0.366	6.25	7.5	2.98	0.42	0.014	18.86	0.109	6.67
7	0.372	9.22	8.0	5.59	0.58	0.008	5.74	0.105	3.175
8	0.305	2.99	3.87	1.52	0.88	0.045	18.35	0.042	1.853
9	0.197	3.43	1.80	1.45	1.42	0.068	11.26	0.056	1.823
10	0.331	3.68	2.21	1.47	1.57	0.076	22.25	0.058	1.795
11	0.360	4.09	1.29	1.02	1.18	0.118	35.59	0.048	2.005
12	0.284	5.17	0.094	1.22	0.23	0.051	18.50	0.021	2.445
13	0.518	5.36	0.281	2.16	3.0	0.125	21.86	0.062	2.733
14	0.732	4.14	1.03	1.45	0.74	0.208	34.16	0.032	2.062
15	0.189	6.44	83.5	2.87	6.42	0.045	12.36	0.062	0.514

The generally high concentration of alumina is possible due to the presence of alumina-silicate minerals, i.e., feldspar, micaceous minerals and sillimanite (in garnet-sillimanite gneisses). In feldspar Si^{4+} (radius 0.42 Å) is replaced by Al^{3+} (radius 0.50 Å) in Si-O tetrahedra [28].

5.3. Manganese (MnO)

The MnO shows strong antipathetic relationship with SiO_2 ($r = -0.84$) and K_2O (Figure 7(d)), whereas it shows feebly negative relationship with Al_2O_3 and P_2O_5.

As evident from Table 4, the MnO, in the Khondalite and Charnockites of the present area have strong positive relationship with FeO ($r = 0.93$) and MgO ($r = 0.79$) and slightly positive relationship with Fe_2O_3, TiO_2, Na_2O (Figure

7(e)), Cao (Figure 7(f)) and H_2O^+, MnO/CaO ratio is constant. The positive relationship of MnO with Fe_2O_3, FeO and the constant MnO/Fe (total iron) (Table 6) ratio indicate that iron and manganese were delivered in the same proportion. The constant MnO/Fe ratio suggests coprecipitation of Mn and Fe under fairly oxidizing condition. The small quantity of manganese determined from the country rocks has most probably been derived from the manganese silicate minerals like manganese bearing garnets, pyroxenes, and other Mn bearing minerals presents in the host rocks ([14] [29] - [31]) possibly in association of ferruginous magnesium rich calc-granulite and garnet sillimanite gneiss and hypersthene gneisses.

5.4. Iron Oxides

The relationship of Fe_2O_3 with TiO_2, CaO, MgO, K_2O and H_2O^+ is negative as Fe_2O_3 increase with the decrease of these elements or vice versa.

The antipathetic relationship of Fe_2O_3 with MgO in Khondalite and Charnockite rocks may be explained by the replacement of Fe^{3+} (0.67 Å) by Mg^{2+} (0.66 Å) [32] . The higher FeO/Fe_2O_3values is Khondalite from the Ea- stern Ghat granulite belt of India indicate lower oxidation conditions in Early Proterozoic [33] . In Eastern Ghat granulite belt of India the variation in FeO/Fe_2O_3 ratio from one area to the other suggests that these sediments were deposited in different basins with different pH. Eh conditions of deposition [33]. The Khondalite and char- nockite rocks of the area under study, exhibit composition variation in iron (total iron), ranging from 1.64% to 14.41% and exhibit high MgO values. Random variation is also observed in K_2O, Na_2O and iron in ternary diagram showing high and random distribution of MgO in Khondalites (Δ) and Charnockites (•) of Garividi, Garb- ham and Chipurupalle areas, Srikakulam-Visakhapatnam Belt (A.P.) Fe_2O_3 (Total Iron), (Table 7 and Table 8).

The significant relationship between various other oxides are also taken into consideration for example relationships of Al_2O_3 with TiO_2 (Figure 8(a)), CaO with MgO (Figure 8(b)) and Fe_2O_3(Figure 8(c)) and FeO (Figure 8(d)).

5.5. Titania (TiO_2)

Titanium oxide shows positive correlation with MnO, FeO, P_2O_5, CaO, MgO and Na_2O whereas SiO_2, Al_2O_3 (Figure 8(a)), Fe_2O_3, K_2O and H_2O^+ are negatively correlated with TiO_2. The positive correlation of TiO_2 with CaO (r = 0.34), MgO (r = 0.37) and Na_2O (r = 0.26) clearly indicates that much of the TiO_2 contains in Khondalite and Charnockite rocks were possibly derived from the titanium bearing minerals like sphene.

5.6. Lime (CaO)

The graph and computed correlation coefficient of calcium show positive relationship with MnO in relationship of MnO with SiO_2, Al_2O_3, FeO, K_2O, Na_2O and CaO in Khondalite and Charnockite Groups of Garividi, Garbham and Chipurupalle areas, Srikakulam-Visakhapatnam Belt (A.P.), FeO, TiO_2 and H_2O^+ and strongly positive relationship with MgO (r = 0.78), (Figure 8(b)). CaO show antipathetic relationship with Fe_2O_3, P_2O_5 and Na_2O and strongly antipathetic relationship with SiO_2 (r = 0.78), Al_2O_3 (r = −0.79) and K_2O (r = 0.726). The ratio of CaO and MnO is constant. The Khondalite and Charnockite rocks of the study area having fairly high concentration of CaO (range from 1.25% to 25.61%). Furthermore diopside, wollastonite, anorthite, scapolite, piedmontite and calcite also contributed varying proportion of CaO to the Khondalite and Charnockite rocks. The lime content seems to be of migmatitic origin [38] . The relationship between various other oxides are also taken into consideration for example relationships of Fe_2O_3(Figure 9(a)), SiO_2 with FeO (Figure 9(b) and Figure 9(c)) and in these host rocks of Garividi, Garbham and Chirpurupalle areas.

5.7. Phosphorus (P_2O_5)

The concentration of P_2O_5 is abnormally on the higher side in the manganese ores as well as in the host rocks of the study area. However, P_2O_5 shows insignificant negative relationship with MnO, FeO, CaO and MgO.

The ratio of MnO and P_2O_5 is variably low and high. Whereas P_2O_5 shows positive relationship with SiO_2, Al_2O_3, Fe_2O_3, TiO_2, Na_2O, K_2O and H_2O^+. The positive relationship of P_2O_5 with SiO_2, $Al_2O_3$and K_2O appears to be due to the presence of alumino-phosphate minerals probably leucophosphate formed by the reaction of phosphate rich solutions with clay minerals. The chemical reaction which indicates the formation of leucophosphate as given by [34] is as follows:

$$Al_2Si_2O_5(OH)_4 + 2HPO_4^{2-} \rightarrow 2AlPO_4 \cdot 2H_2O + 2SiO_2 + H_2O$$

$$2AlPO_4 \cdot 2H_2O + KOH \rightarrow KAl_2(PO_4)_2 OH \cdot 2H_2O + 2H_2O$$

Table 7. Recalculted to 100 wt% of bulk components (Fe_2O_3 + MgO), Na_2O and K_2O in Khondalite rocks of Garividi, Garbham and Chipurupalle areas, Srikakulam-Visakhapatnam Belt (A.P.).

Sample No	(Fe_2O_3 + MgO)	Na_2O	K_2O
1	46.9	10.8	42.2
2	42.7	13.9	43.3
3	77.9	3.0	19.0
4	97.0	2.9	0.08
5	92.9	5.2	1.8
6	81.0	16.7	2.2
7	78.4	19.1	2.3
8	82.9	13.5	3.4
9	77.8	14.2	7.9
10	81.7	12.5	5.6

Table 8. Recalculated to 100 wt% of bulk components (Fe_2O_3 + MgO), Na_2O and K_2O in Charnockite rocks of Garividi, Garbham and Chipurupalle areas, Srikakulam-Visakhapatnam Belt (A.P.).

Fe_2O_3 total iron.

Sample No	(Fe_2O_3 + MgO)	Na_2O	K_2O
1	68.4	17.7	13.7
2	53.2	4.0	42.6
3	48.1	11.4	40.4
4	67.4	16.5	16.0
5	92.6	7.2	0.08

Further the positive relation of P_2O_5 with Fe_2O_3 clearly indicates the ferruginous nature of phosphate minerals. According in [1] [27] [31] [35] , the source of phosphorus in the original manganiferous rocks is apatite and spessartite which are the common constituent of the rocks associated with gondite and Kodurite types.

[36] reported up to 4.10 percent P_2O_5 in spessartite. The presence of apatite in the host rocks indicates that they were subjected to metamorphism in a more hydrous environment as also suggested by [37].

The manganese silicate-carbonate, clay and silica. In few mines as in Avagudem the manganese carbonate was either deposited in the form of carbonate or formed through diagenesis of manganese oxide [39].

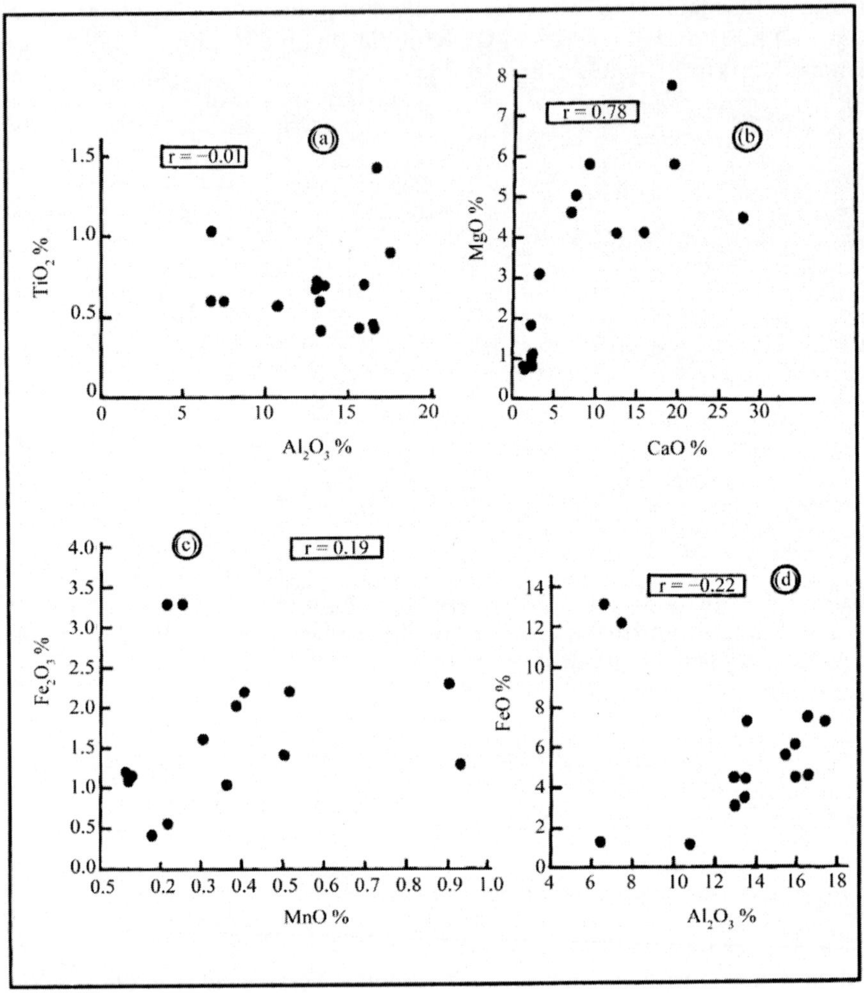

Figure 8. Relationship of Al_2O_3 with TiO_2 and FeO, and CaO with MgO and MnO with Fe_2O_3 in Khondalite and Charnockite Groups of Garividi, Garbham and Chipurupalle areas, Srikakulam-Visakhapatnam Belt (A.P.).

5.8. Magnesia (MgO)

The present study reveals that MgO is positively correlated with FeO, CaO, MnO, TiO_2, Na_2O and H_2O^+ whe- reas it is negatively correlated with Al_2O_3, Fe_2O_3, P_2O_5, SiO_2 and K_2O (Figures 9(a)-(c)).

The Khondalites and Charnockites of the study area show higher concentration of MgO (ranges from 0.86% to 7.90%). [33] also reported higher concentration of MgO in Khondalites from the Eastern Ghat granulite belt of India.

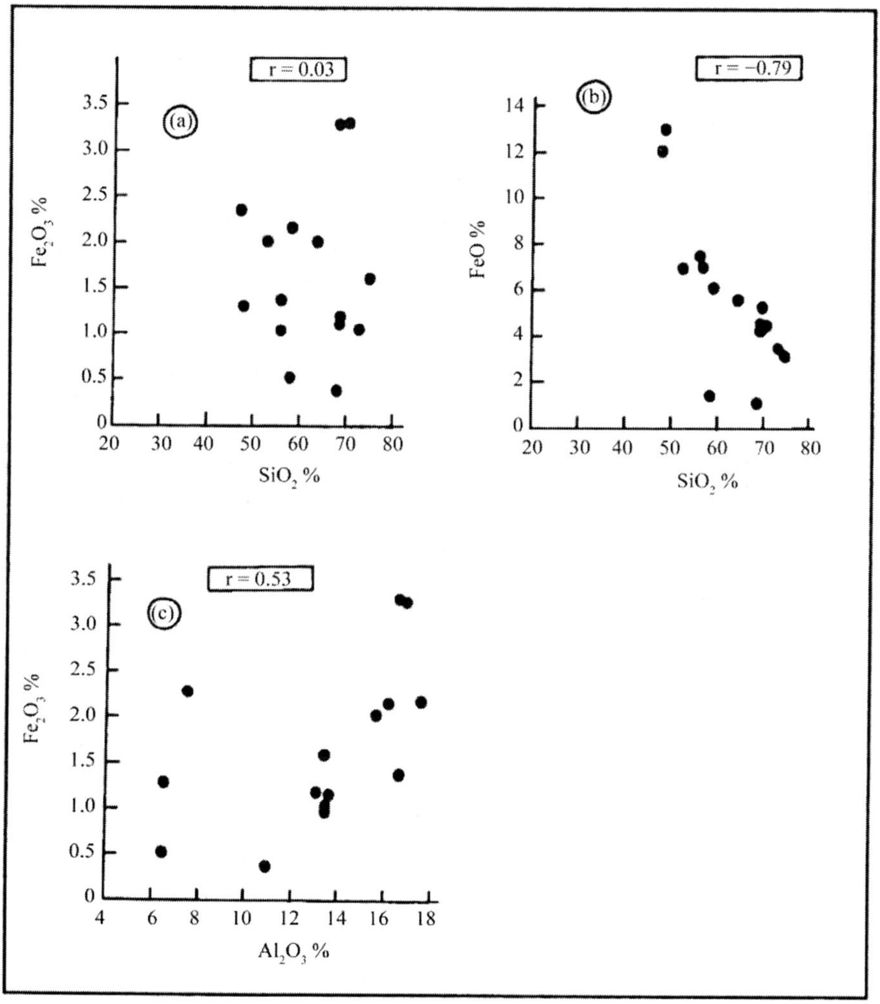

Figure 9. Relationship of SiO_2 with Fe_2O_3 and FeO and Al_2O_3 with Fe_2O_3in Khondalite and Charnockite Groups of Garividi, Garbham and Chipurupalle areas, Srikakulam-Visakha- patnam Belt (A.P.).

Table 9. Recalculated to 100 wt% of bulk components MgO, Na_2O and K_2O in Khondalite rocks of Garividi, Garbham and Chipurupalle areas, Srikakulam-Visakhapatnam Belt (A.P.).

Sample No	MgO	Na_2O	K_2O
1	10.3	18.2	71.3
2	14.2	20.9	64.8
3	25.5	10.1	64.2
4	92.9	7.8	.2
5	81.4	13.7	4.7
6	75.4	21.6	2.8
7	71.7	25.0	3.1
8	65.0	27.8	7.1
9	55.8	28.4	15.7
10	60.3	27.2	12.3

These Khondalite with high MgO content can be compared with Late-Archaean-Early Proterozoic metapelites. In the area under study the Khondalite and Charnockite rocks exhibit compositional variations (Table 9 and Table 10) in MgO (ranges from 0.86% to 7.90%), K_2O (ranges from 0.02% to 5.94%) and Na_2O (ranges from 0.34% to 2.84%). The ternary diagram (Figure 10) shows wide scatter distribution of MgO, K_2O and Na_2O.

The concentration of magnesium oxide is possibly due to the presence of garnet, spinel, hornblende, magnesium rich pyroxenes and biotite in the rocks of the study area. All these minerals pointing to the argillaceous nature of the original sediments which gave rise to these rocks.

5.9. Soda (Na_2O)

In the Khondalite and Charnockite rocks soda shows positive relationship with Al_2O_3, MnO in relationship of MnO with SiO_2, Al_2O_3, FeO, K_2O, Na_2O and CaO in Khondalite and Charnockite Groups of Garividi, Garbham and Chipurupalle areas, Srikakulam-Visakhapatnam Belt (A.P.), Fe_2O_3, FeO, P_2O_5 and TiO_2.

These rocks in few samples show exceptionally high Na_2O/K_2O ratio. The ternary diagram (Figure 11) shows wide scatter distribution of K_2O and Na_2O.The presence of soda is due to the presence of feldspar and also due to the presence of scapolite in the Khondalites of study area [11] . The main supply of alkalies probably came from the igneous intrusion which cut across the Khondalite and Charnockite rocks. The high soda concentration in these rocks suggests the existence of predominantly tonalite/trondhjemite rocks in the source area. A study of granitic/gneissic episodes of the South Indian shield [40] - [42] indicates that the Early Archaean.

Post-greenstone granitic and gneissic episodes of the stable craton are mostly tonalitic/trondhjemitic in nature and K-rich granites appeared only in Late Proterozoic which explains better the high soda content in Proterozoic Khondalites of the Eastern Ghats. Predominance of soda rich tonalites and trondhjemites in Early Archean crust is also reported from other shield areas [43] [44] . Wide distribution in Na_2O from area to area may indicate to different temperatures of Na-rich and mafic source material indifferent basins [33] .

5.10. Potash (K_2O)

K_2O shows insignificant negative relationship with CaO, MgO, MnO in relationship of MnO with SiO_2, Al_2O_3, FeO, K_2O, Na_2O and CaO in Khondalite and Charnockite Groups of Garividi, Garbham and Chipurupalle areas, Srikakulam-Visakhapatnam Belt, Fe_2O_3, FeO, TiO_2, Na_2O and H_2O^+. However, it shows strongly positive relationship with SiO_2 and slightly positive relationship with Al_2O_3 and P_2O_5 (Table 4). The dominance of K_2O over Na_2O is characteristic of parametamorphic rocks [20] and these rocks have exceptionally high and variable Na_2O/K_2O ratio. Potassium and sodium also show close coherence with aluminium, which is possibly due to the presence of soda and potash feldspar in these rocks.

5.11. Water (H_2O^+)

Water shows positive relationship with MnO, FeO, P_2O_5, CaO and MgO and negative relationship with SiO_2, Al_2O_3, Fe_2O_3, TiO_2, Na_2O and K_2O.

Table 10. Recalculated to 100 wt% of bulk components MgO, Na_2O and K_2O in Charnockite rocks of Garividi, Garbham and Chipurupalle areas, Srikakulam-Visakhapatnam Belt (A.P.).

Sample No	MgO	Na_2O	K_2O
1	38.2	34.8	26.9
2	22.7	6.6	70.5
3	14.8	18.7	66.4
4	16.8	42.2	40.9
5	80.3	19.4	0.2

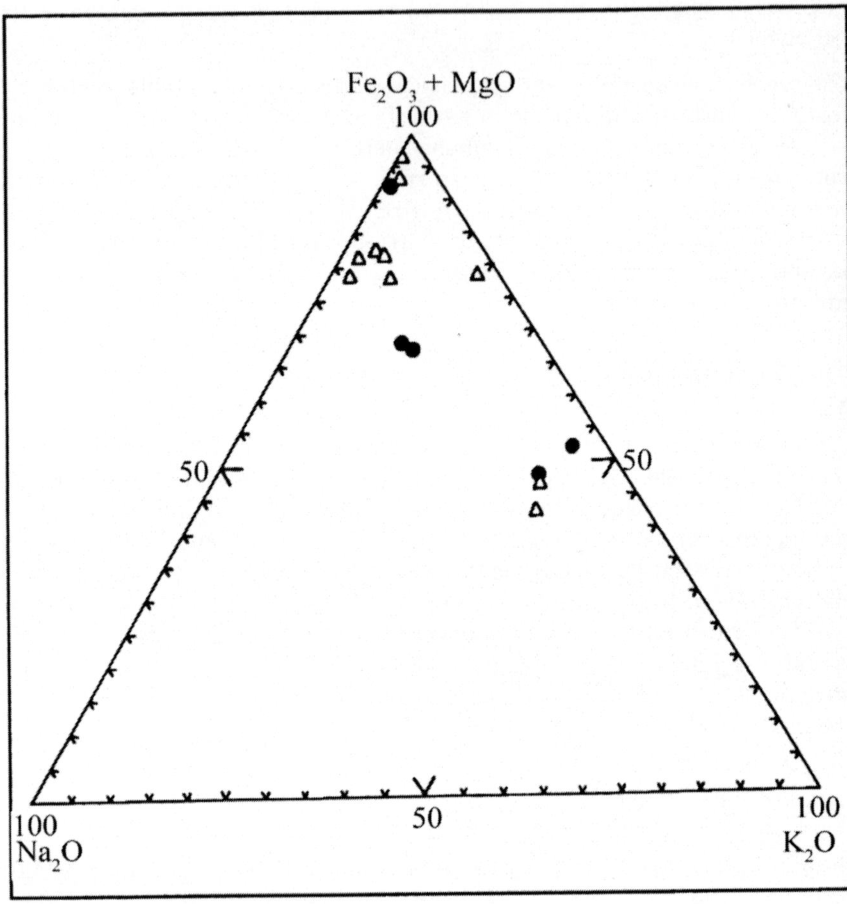

Figure 10. Ternary Diagram showing high and random distribution of MgO in Khondalites (Δ) and Charnockites (•) of Garividi, Garbham and Chipurupalle areas, Srikakulam-Vi- sakhapatnam Belt (A.P.) Fe_2O_3 (Total Iron).

6. CONCLUSIONS

The analysis of 15 host rocks samples reveals that:

- Silica increases with the increases of Al, phosphorous and potash and vice versa.

- Alumina increases with the increases of SiO_2, P_2O_5, Na_2O and K_2O and vice versa.
- Mn increases with the increases of Fe and vice versa.
- MnO also shows strong sympathetic relationship with MgO and Na_2O.
- Phosphorous decreases with increase of MnO and vice versa.
- Ti increases with the increases of Mn and vice versa.
- Lime increases with the increases of Mn and vice versa.
- Mg increases with the increases of MnO and vice versa.
- Soda increases with the increases of both Mn and Fe and vice versa.
- Potash decreases with the increases of both Mn and Fe and vice versa.
- The geochemical behavior of the host rocks indicate that the evolution of the ores took place in two different stage.

In first stage, sediments were rich in Mn, deposited in a basin during typical condition of oxidation. The chemical alteration and metamorphism of these sediments gave rise to the formation of primary metamorphic

Mn minerals. In the later stages these metasediments suffered metasomatism due to intense igneous intrusions which caused hybridization at places.

In the second stage, the previously formed Mn ores and Mn-silicates of the country rocks like Mn-rich garnets and pyroxenes which played an important role in the formation of secondary Mn minerals viz. psilomelane-cry- ptomelane, hollandite, pyrolusite and wad etc. by the alteration. The meteoric water rich in CO_2 played an important role in solution, transportation, precipitation, concentration and finally redeposition of the secondary Mn ores.

The concentration and redeposition have largely taken place from colloidal state or solution in oxidizing environment. It is assumed that Mn and Fe are in colloidal state. The anions and cations of minor and trace elements are adsorbed on the surface of negatively charged $Mn(OH)_4$ and positively charged $Fe(OH)_3$ particles. The high K_2O/Na_2O, CaO/MgO and Co/Ni ratios in these ores are due to high adsorption capacities of K_2O, CaO and Co then Na_2O, MgO and Ni:

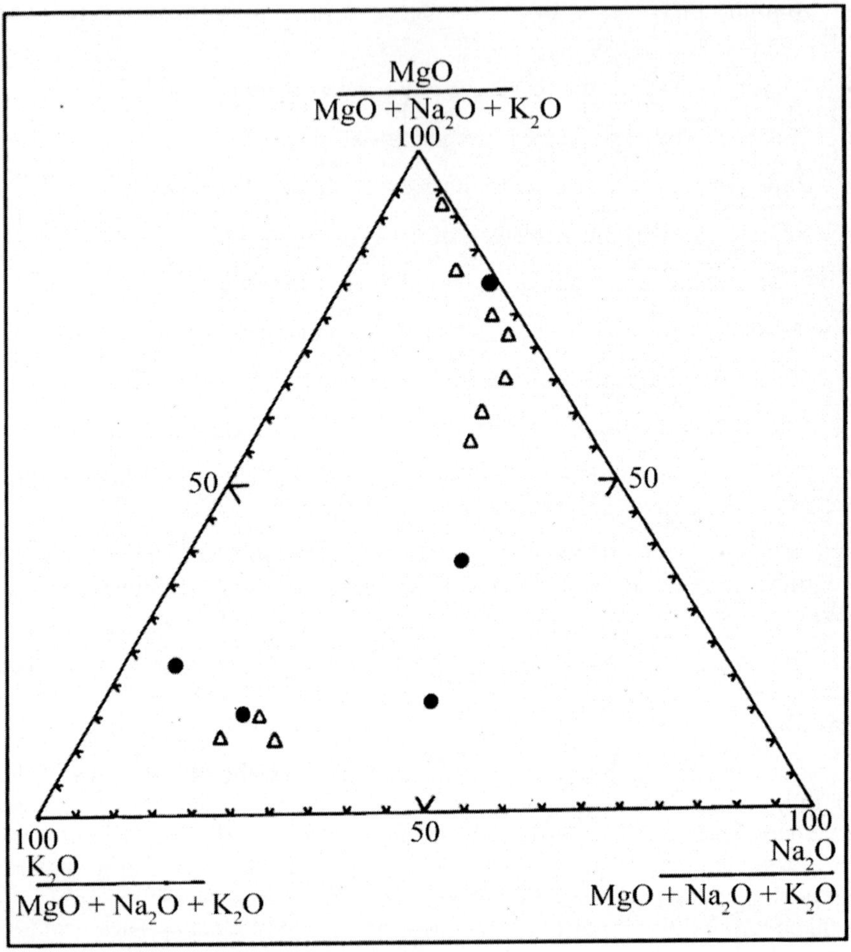

Figure 11. Ternary Diagram showing Wide Scatter of Na_2O and K_2O in Khondalites (Δ) and Charnockites (•) of Garivi- di, Garbham and Chipurupalle areas, Srikakulam-Visakha- patnam Belt (A.P.).

Presumably the Mn and Fe solutions first form in feebly acidic medium with at pH value around 7. However with an increase in alkalinity Mn^{2+} and Fe^{2+} they are converted to colloidal $Mn(OH)_4$ and $Fe(OH)_3$. The solubility of SiO_2 also increases with the increase pH (about 9), the Al_2O_3 dissolves into solution. The addition of $Al(OH)_3$ into solution causes drastic changes in character of $Mn(OH)_4$colloids which are precipitated as oxide ores. Originally the Mn deposits have been derived from sources on the bordering Cratons, which are also the probable provinces for the sediments in this part of the basin.

ACKNOWLEDGEMENTS

Farhat Nasim Siddiquie, Mohd Shaif The authors are grateful to Prof. M. Rahimullah, Chairman Department of Geology and Dean Faculty of Science, AMU Aligarh. To Director National Institute of Oceanography, Dona Paula, Goa for providing research facilities.

REFERENCES

1. Siddiquie, F.N. (2001) Source of Unusually High P_2O_5 in Manganese Ores of Srikakulam District (A.P.), India. Journal of Applied Geochemistry, 3, 78-84.
2. Siddiquie, F.N., Ali, S.A., Bhat, S.H. and Burhamuddin, Kh. (2015) Genetic Association of Trace Elements of with Special Emphasis on Cu, Ni, and Cr in the Manganese Ores of Vizianagaram District, (A.P.), India. Journal of Applied Geochemistry, 17, 66-77.
3. Acharya, D.C. and Nayak, B.K. (2002) Mineral Chemistry of Manganese Ores Associated with Precambrian Eastern Ghats Complex of Andhra Pradesh, Orissa, India. In: Siddiquie, F.N., Ed., Metallogenesis of Manganese Ores of Srikakulam-Visakhapatnam Belt, Icon Publication Ltd., New Delhi, 213 p.
4. Sarkar, S.N. (1968) Precambrian Stratigraphy and Geochronology of Peninsular India. Dhanbad Publishers, India, 33 p.
5. Roy, S. (1966) Syngenetic Manganese Formations of India. Jadavpur University Press, Calcutta, 219 p.
6. Mahnert, K.R. (1968) Migmatites and the Origin of Granitic Rocks. Els. Pub. Co., 393 p.
7. Rao, G.V. (1969) The Geology and Manganese Ore Deposits of Parts of Vizakhapatnam (Visakhampatnam) Manganese Belt, Srikakulam District, A.P., India. Bulletin of Geological Survey of India, Series A, No. 35, 129 p.
8. Fermor, L.L. (1909) Manganese Ore Deposits of India. Memoir of Geological Survey of India, 37, 272.
9. Middlemiss, C.S. (1914) General Report of Geological Survey of India. Geological Survey of India, 36, 102-105.

10. Rao, S.V.G.K. and Sarma, K.J. (1978) Report on Assessment of Manganese Ore in Srikakulam and Visakhapatnam Districts, A.P., India. Unpublished Report, Geological Survey of India.

11. Prabhakhar Rao, G. (1950) The Geology of Manganese Bearing Rocks of Garividi and Garbham, Vizagapatnam District, A.P., India. Quarterly Journal of the Geological, Mining and Metallurgical Society of India, 22, 25-42.

12. Venkataramaiah, K., Doshi, C.M. and Rao, S.V.G.K. (1978) A Report on the Assessment of the Manganese Ore Deposits in Srikakulam and Visakhapatnam District, A.P., India. Unpublished Report, Indian Bureau of Mines.

13. Siddiquie, F.N. (1982) A Geochemical Appraisal of the Manganese Ores of Garividi, Garbham and Chirpurupalle mines, Srikakulam District, A.P., India. Unpublished M.Phil. Thesis, Geology Department, Aligarh Muslim University, Aligarh, 92 p.

14. Siddiquie, F.N. (1986) Geology and Geochemistry of Manganese Ore Deposits of Srikakulam District (Andhra Pradesh), India. Unpublished Ph.D. Thesis, Geology Department, Aligarh Muslim University, Aligarh, 253 p.

15. Siddiquie, F.N. (2000) On the Occurrence of Lepocinclis from Khondalites of Srikakulam District (A.P.). Bulletin of the Indian Geologist's Association, 33, 63-68.

16. Siddiquie, F.N. (2003) Some Modification in the Stratigraphy of Manganese Bearing Formations, Srikakulam District (A.P.), India. Journal of Science and Technology, 11, 137-154.

17. Siddiquie, F.N. (2004) Metallogenesis of Manganese Ores of Srikakulam-Visakhapatnam Belt. Icon Publication Pvt. Ltd., New Delhi, 194.

18. Sharma, N.L., Ali, I. and Perraju, P. (1974) Geology of Parts of Srikakulam, Chipurupalli and Palakonda Taluks, Srikakulam District, A.P., India. Unpublished Report, Geol. Surv. India.

19. The Indian Express (Express Newsline) (2002) Geologist Discovers New Green Alga. Science Talk, 7.

20. Krishna Rao, J.S.R. (1954) The Geology of Chipurupalle Area, Vizagapatnam District, A.P., India, with Special Reference to the Origin of Manganese Ores. Quarterly Journal of the Geological, Mining and Metallurgical Society of India, 26, 36-44.

21. Siddiquie, F.N. and Bhat, S.H. (2010) Geochemistry of Trace Elements in Manganese Ores and Their Host Rocks in Vizianagaram Manganese Ores Belt (A.P.), India. Journal of Applied Geochemistry, 10, 38-72.

22. Janardhan, A.S., Newton, R.C. and Hansen, E.C. (1983) Transformation of Peninsular Gneiss to Charnockite in Southern Karnataka. Memoir 4, Geological Society of India. Proceedings of the Indo-U.S. Workshop, Hyderabad, 417-435.

23. Krishna Rao, J.S.R. (1975) Some Problems on the Manganese Ore Deposits of A.P., India. Acta Mieralogica-Petrogra- phica, 22, 133-141.

24. Gupta, B.P. (1973) Petrology of Leptynites and Garnetiferous Apatite Gneiss around Garbham, Srikakulam District, A.P., India. Journal of the Geological Society of India, 14, 113-123.

25. Krishnan, M.S. (1934) Khondalite from the Road between Banderwala and Nuwara, Eliya M.P. 1. Ceylon (Sri Lanka). Record of Geological Survey of India, 67, 395.

26. Raghava Rao, K.V. (1955) In: Rao, G.V., 1959. The Geology and Manganese Ore Deposits of Parts of Vizakhapatnam (Visakhapatnam) Manganese Belt, Srikakulam District, A.P. India. Bulletin of Geological Survey of India, Series A, No. 35, 1-129.

27. Mookherjee, A. (1961) Distribution of Minor Elements in the Gonditic Manganese Ore and Its Geochemical Significance. Economic Geology, 56, 723-729.

28. Rankama, K. and Sahama, Th.G. (1950) Geochemistry. University of Chicago Press, Chicago, 912 p.

29. Siddiquie, F.N. and Raza, M. (1990) Major and Trace Elements Variability in Manganese Ores of Srikakulam District (A.P.). Indian Geologist's Association, Chandigarh, Vol. 23, 99-109.

30. Siddiquie, F.N. and Raza, M. (1990) Trace Elements in the Manganese Ores of Srikakulam Distt. (A.P.) India (Abstract). National Seminar on Precambrian Geology, Madras, 61.

31. Siddiquie, F.N. and Raza, M. (2008) Geochemical Environment and Metallogenesis of Manganese Ores of Vizianagaram Manganese Belt Andhra Pradesh, India. Journal of the Indian Association of Sedimentologist, 27, 1-16.

32. Drever, J.I. (1971) Magnesium Iron Replacement in Clay Minerals in Anoxic Marine Sediments. Science, 172, 1334- 1336.

33. Divakara Rao, V. (1983) Khondalites from the Eastern Ghat Granulite Belt Geochemistry and Origin. Geophysical Research Bulletin, 20, 233-242.

34. Nriagu, J.O. (1972) Stability of Vivianite and Iron-Pair Formation of the System $Fe_3(PO_4)_2$-H_3PO_4-H_2O. Geochimica et Cosmochimica Acta, 36, 459-470.

35. Krishna Rao, J.S.R. and Venkataramaraju, D. (1966) A Study of Apatites from Srikakulam District, A.P., India. Quarterly Journal of the Geological, Mining and Metallurgical Society of India, 38, 177-181.

36. Mason, B. and Berggren, T. (1941) A Phosphate-Bearing Spessartite Garnet from Wodgina, Western Australia. Geologiska Föreningen i Stockholm Förhandlingar, 63, 413-418.

37. Mitchell, L., Faust, G.T., Hendricks, S.B. and Reynolds, D.S. (1943) The Mineralogy and Genesis of Hydroxyl Apatite. American Mineralogist, 28, 356.

38. Bhattacharyya, C. (1972) Granitisation in Relation to Evaluation of the Charnockite Series from the Eastern Ghat Hills, Srikakulam District, A.P., India. Neues Jahrbuch für Mineralogie Monatshefte, No. 5, 220-240.

39. Bostrom, K. (1967) The Problem of Excess Manganese in Pelagic Sediments. In: Ableson, P.H., Ed., Researches in Geochemistry, Vol. 2, John Wiley and Sons, New York, 420-452.

40. Naqvi, S.M., Divakara Rao, V. and Narain, H. (1978) The Primitive Crust: Evidence from the Indian Shield. Precambrian Research, 6, 323-345.

41. Ali, S. and Devakara Rao, V. (1980) Geochemistry and Origin of Acidic Rocks of Shimoga, Karnataka, India. Journal of Gelogical Society India, 21, 91-96.

42. Divakara Rao, V. and Rama Rao, P. (1982) Granitic Activity and Crustal Growth in India Shield. Precambrian Research, 16, 257-271.

43. Sheraton, J.W. (1980) Geochemistry of Precambrian Metapelites from East Antarctica: Secular and Metamorphic Variations. Bureau of Mineral Resources Journal of Australia Geology and Geophysics, 5, 279-288.

44. Glikson, A.Y. (1979) Early Precambrian Tonalite-Trondhjemite.

2

CHAPTER

Pressure-Temperature and Fluid Evolution of Quartzo-Feldspathic Metamorphic Rocks with a Relic High-Pressure, Granulite-Facies History from the Central Erzgebirge (Saxony, Germany)

Arne P. Willner[1,*], Kerstin Rötzler[2] and Walter V. Maresch[3,†]

[1]*Institut fur Mineralogie, Ruhr-Universitat BochumD-44780 Bochum, Germany*

[2]*Geogorschungszentrum, TelegraphenbergD-14473 Potsdam, Germany*

[3]*Institut fur Mineralogie, Westfalische Wilhelms-Universitat, Corrensstr24, D-48149 Munster, Germany*

ABSTRACT

The Gneiss-Eclogite Unit is a composite tectonometamorphic unit within the Variscan Erzgebirge mega-antiform. It comprises migmatitic para- and orthogneisses, high-temperature (HT) mylonites, kyanite-bearing granulites, eclogites and garnet peridotites. Four different quartzo-feldspathic assemblages are recognized, in which maximum conditions of up to 830°C and 21 kbar were determined. The assemblages are characterized by the nearly complete prograde breakdown of biotite, by high grossular content (23–47 mol %) of garnet in the presence of albite, and high Si contents of phengite [3.3–3.4 per formula unit (p.f.u.)]. Water activities at this stage are variable and range from <0.15 to >0.4. The maximum pressures indicated for individual rock volumes may vary considerably between 12 and 24 kbar at 700–800°C, so that non-coherency of the entire Gneiss-Eclogite Unit appears likely during the high-pressure event itself. After decompression, concomitant with penetrative HT myulonitization, hydration led to overprinting of the rocks to variable degrees, owing to channelized fluid influx. Partial equilibration at medium-pressure conditions of about 7–10 kbar and 600–700°C occurred, involving abundant retrograde migmatization. The water activity increased to 0.5–1.0. During later exhumation, deformation and re-equilibration at 2–3 kbar and 400–500°C were concentrated in local, discrete, ductile normal fault zones. The kinked geometry of the PT path is thus characterized by (1) high-pressure (HP) equilibration, followed by near-isothermal decompression at high temperatures, during which rocks from different depths were amalgamated, and (2) extensive hydration and re-equilibration at medium pressures, followed by rapid cooling during continued uplift, when the entire unit came into contact with cooler, now over- and underlying units. This scenario is attributed to continent collision, orogenic collapse and disintegration of the HP unit during continuing collision, crustal stacking and uplift controlled by extension.

Key words: continent collision; fluid evolution; high-pressure granulite facies; PT history; quartzo-feldspathic rocks

INTRODUCTION

The Erzgebirge is part of the metamorphic basement of the internal Mid-European Variscides exposed in Saxony and the northernmost Czech Republic (Fig. 1). It represents an antiformal megastructure with a large core composed of medium- to high-grade mica schists and gneisses as well as innumerable intercalations of eclogite from decimetres up to several hundred metres in length. Recently, these eclogites have been studied intensively (Klápová, 1990; Massonne, 1992, 1994; Schmädicke *et al.*, 1992), and it has been found that some record ultra-high-pressure conditions up to 27 kbar and 800°C. Even higher maximum pressures (33–40 kbar; Kotková, 1993; Massonne & Grosch, 1995) have been reported from garnet peridotite lenses that represent remnants of the lithospheric mantle. These contrasting maximum pressures point to a very heterogeneous and strongly incoherent basement. Concurrent petrological work on the quartzo-feldspathic and metapelitic rocks enclosing these eclogite and garnet peridotite lenses also points to high-pressure conditions, although the pressures appear to be considerably lower than in the eclogites and garnet peridotites (Kotková, 1993; Rötzler *et al.*, 1993;Willner *et al.*, 1993, 1994; Rötzler, 1995). In addition, these studies showed that the Erzgebirge consists of a stack of discrete units with contrasting pressure-temperature evolution. Thus the Erzgebirge may be regarded as a model region for studying the continent-continent collision process, especially with regard to the relationships between eclogites and 'normal' crustal rocks enclosing them, as well as the uplift history that resulted in the nappe stack observed today.

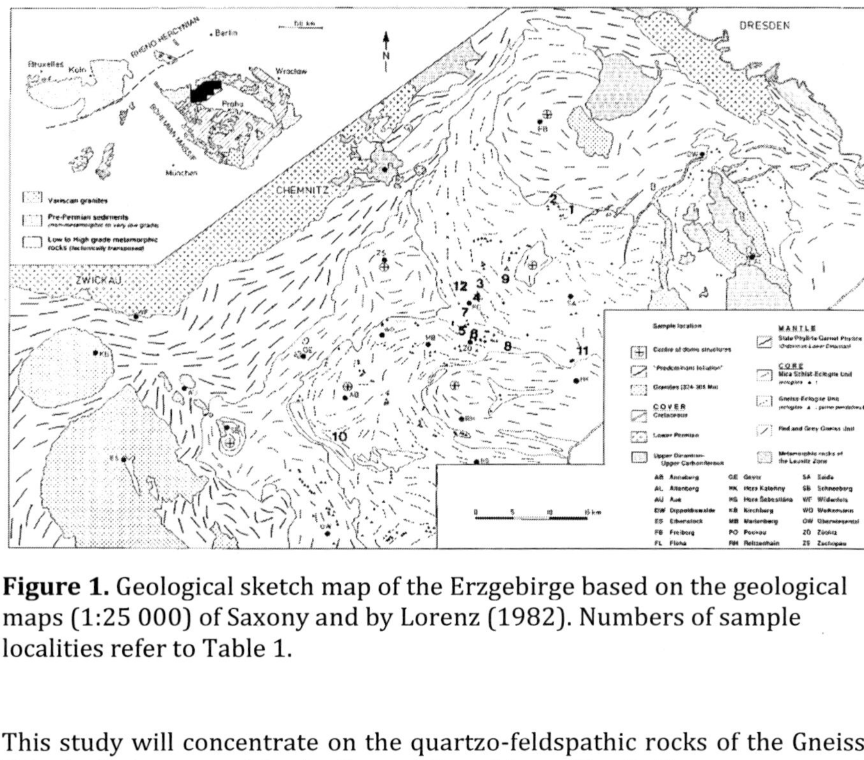

Figure 1. Geological sketch map of the Erzgebirge based on the geological maps (1:25 000) of Saxony and by Lorenz (1982). Numbers of sample localities refer to Table 1.

This study will concentrate on the quartzo-feldspathic rocks of the Gneiss-Eclogite Unit, exposed in the Central Erzgebirge (Fig. 1). Key occurrences in this unit are small areas with high-temperature (HT) mylonites of granitic composition intercalated parallel to the predominant gneissose foliation. Such rocks preserve not only a relic deformation fabric, but also mineral phases with compositions reflecting an early stage during the metamorphic evolution. Typical are kyanite granulites nearly free of hydrous minerals. These and adjacent rocks were mapped as 'granulitic gneiss' at a scale of 1:25 000 at the turn of the century (e.g. Reinisch, 1929), when initial detailed mapping was done on a purely lithological basis. Definition was primarily based on the characteristic rock fabric. This paper will deal with the petrogenesis of such rocks and the gneisses surrounding them; together they form a discrete tectonic unit in the Central Erzgebirge, the Gneiss-Eclogite Unit. Two major topics will be addressed: (1) high-pressure (HP) metamorphism of a variety of quartzo-feldspathic assemblages under granulite-facies conditions in the uppermost stability field of albite, and (2) the uplift and exhumation process of these rocks, as deduced from an improved understanding of the mechanisms of metamorphic overprinting at high temperatures. In detail, the following set of interrelated problems must be considered:

1. What are the highest *PT* conditions recorded in quartzo-feldspathic and metapelitic rocks, particularly in the vicinity of eclogites? What happens to these rocks in the extremely wide transitional field involving eclogite, granulite and amphibolite facies? To what extent do both eclogites and adjacent acid rocks have a common history?
2. Do the quartzo-feldspathic rocks themselves have a common *PT* history? What is the scale of coherence within the high-pressure Gneiss-Eclogite Unit of the Central Erzgebirge?
3. What are the relationships of the HT mylonite lenses to the surrounding gneisses? Are they relics within an otherwise overprinted coherent unit with a common *PT* history, or are they allochthonous bodies?
4. What is the role of the metamorphic fluid phase during petrogenesis of the acid metamorphic rocks? What are the specific mechanisms of retrograde overprinting in the quartzo-feldspathic rocks in general?
5. What information concerning the exhumation history may be derived from the *PT* evolution? Is this *PT* evolution typical for a specific geotectonic environment and how is it related to megascale processes during the Variscan orogenesis?

GEOLOGICAL SETTING

The Erzgebirge antiform structure (Fig. 1) is roughly oval in shape and bounded by the Elbe zone in the east and the Tertiary Eger graben in the south. It is composed of numerous subsidiary dome structures within the medium- to high-grade metamorphic rocks that deform tectonic units with mainly subhorizontal and subparallel transposition foliations. These core rocks are overlain by a phyllitic mantle. The overall 'onion-skin' structure of the entire crystalline complex, as defined by the dominant foliation planes, is a phenomenon that is accepted by all workers in the area. However, although the subhorizontal units of the Erzgebirge were already interpreted as structural units formed by tangential tectonics during the first half of this century (e.g. Kossmat, 1916, 1925;Scheumann, 1935), all later subdivisions for mapping purposes in the 1960s until the 1980s were made on the basis of lithostratigraphic units (e.g. Hoth *et al.*, 1983; Lorenz & Hoth, 1990). These lithostratigraphic units make up rock associations that may be traced along strike for considerable distances. Unfortunately, this subdivisional system was entirely based on an orthostratigraphical philosophy, disregarding clear differences in *PT* conditions between adjacent units and the omnipresent transposition structures of variable age, genesis and displacement character.

In addition, the accepted view until the end of the 1980s was that metamorphic grade gradually increased from the chlorite zone in the outer shell towards anatexis in the core, thus resulting in concentric isograds (Wienholz *et al.*, 1979; Hofmann *et al.*, 1981). Nevertheless, a remarkably reduced metamorphic profile was observed at the northwestern edge of the mega-antiform. Until the beginning of this decade, metamorphic grade was characterized only by a description of observed assemblages. A first attempt to reconstruct *PT* paths for the Erzgebirge on this basis was made by Krentz *et al.* (1990), who suggested continuously higher pressures towards the core of the mega-antiform. The proposed *PT* loops barely reach the kyanite stability field. A HT-LP imprint well within the sillimanite stability field was recognized in the easternmost part of the Erzgebirge (Bergner, 1990)

However, the first extensive geothermobarometric work revealed that the crystalline Erzgebirge complex was certainly not a coherent body with gradual *PT* variations throughout its history, but rather a stack of subhorizontal structural units with considerable discontinuities in maximum *PT* conditions and markedly different *P-T* trajectories. Hence, for the purpose of a better overview, some major units–largely identical to previous major lithostratigraphic units–were summarized and redefined byWillner *et al.* (1994)and Rötzler (1995). The pressure-temperature-deformation histories of these units differ considerably, but even the units themselves cannot be regarded as homogeneous and coherent bodies. From the bottom to the top of the stack these are as follows (Fig. 1):

1. A 'Grey and Red Gneiss Unit' consists predominantly of monotonous two-mica orthogneisses of granitic composition, with very rare intercalations of garnet amphibolite and metapelite. Eclogites are lacking. Pb-Pb results from zircons with a magmatic morphology yield consistent crystallization ages of 550 Ma for the orthogneisses (Kröner *et al.*, 1995).

2. The overlying 'Gneiss-Eclogite Unit' is composed of predominantly garnet-bearing ortho- and paragneisses with intercalations of HT eclogite and garnet peridotite. Pb-Pb ages of zircons with magmatic morphology vary around 480 Ma, those of metamorphic zircons around 340 Ma (evaporation method; Kröner & Willner, 1995;Kotková *et al.*, 1996; Willner *et al.*, 1996). Nd-Sm mineral or whole-rock isochrons in HT eclogites and garnet peridotites yield ages between 333 and 360 Ma, interpreted as dating the thermal peak of metamorphism (Schmädicke *et al.*, 1995).

3. The 'Mica-schist-Eclogite Unit' follows toward the top, where metapelites dominate with abundant intercalated lenses of eclogite, marble, calc-silicate rock and orthogneiss.

4. The 'Phyllitic Mantle' is composed of slates and phyllites with intercalated quartzite. These rocks represent low-pressure rocks. It is the only unit which contains rocks with known biostratigraphic ages, which range from Silurian to Lower Dinantian (Kurze, 1966), probably also including Ordovician and Cambrian (Freyer *et al.*, 1994). Nevertheless, transposition has also obscured orthostratigraphic relationships in this unit. The contact with the underlying units is a continuous subparallel transposition foliation.Rötzler (1995)identified a separate unit containing garnet phyllites near the contact with the underlying Mica-Schist-Eclogite Unit.

The first pronounced unconformity originated around 325 Ma, when undeformed upper Dinantian clastic sediments were deposited in the NE of the Erzgebirge (Kurze, 1966). This marks the approximate end of the exhumation of the metamorphic core. However, crustal dynamics continued in parts of the now-exposed basement within a near-surface environment, as indicated by pronounced granitic magmatism of crustal anatectic origin intruding rather late (324–305 Ma; Gerstenberger, 1989) into a shallow crustal level (Seltmann *et al.*, 1991). Hence the crustal dynamics recorded by the metamorphic rocks is mainly Variscan and must have occurred, if the geochronological results are correctly interpreted, between 340 and 325 Ma, indicating extremely rapid exhumation rates.

The Gneiss-Eclogite Unit of the Central Erzgebirge features the highest *PT*conditions so far determined, whereas the over- and underlying rock units definitely exhibit lower pressure and temperature histories (Rötzler *et al.*, 1993; Willner *et al.*, 1994, 1995; Klemm, 1995; Rötzler, 1995).

PETROGRAPHY

Small volumes of rock characterized by HT mylonites and facies-critical kyanite granulites are found within the rather homogeneously deformed stack of gneisses that make up most of the Gneiss-Eclogite Unit (e.g. Zöblitz, south of Pockau, Grundau, Saidenbach dam; see Table 1). Lenses of eclogite and garnet peridotite are associated with these occurrences on a regional scale. These rocks are crucial for an understanding of the Gneiss-Eclogite Unit, because here an older deformational fabric and mineral assemblage are preserved that have been obliterated in the surrounding, volumetrically dominant gneisses owing to recrystallization and retrograde overprinting (e.g. Krohe, 1997). A detailed map and a first thorough account of the petrography and microfabrics of such rock association were presented by Behr *et al.* (1965)in the Zöblitz area. Those workers already interpreted the area as part of a deep-seated shear horizon.

Table 1. Location of analysed samples

1	Ed5a	metablastic gneiss	Lichtenberg, 100 m W of Hornmühle
2	Ed16a	metablastic gneiss	200 m N of Mulda
3	E42g	assemblage 1	Saidenbach dam
	E42-1b	assemblage 1	
	E42i	metablastic gneiss	
4	E43c, e	migmatitic gneiss	Görsdorf Quarry
5	8216	assemblage 1	Burgberg NW of Zöblitz
	E41a	assemblage 1	
6	E3	assemblage 1	Quarry E of Zöblitz
	E726b	assemblage 2	
	E735c	assemblage 3a	
	E726d	assemblage 3a	
	E735a	assemblage 3b	
	E534	assemblage 3b	Morgensternhöhe Ansprung
7	E29a, d, e	assemblage 2	Road cut 500 m S of Pockau
	E28a	metablastic gneiss	
8	E4c	assemblage 3a	500 m NW of Grundau
	E4d	assemblage 3b	
9	E537	assemblage 3b	Mittelsaida
10	E602	assemblage 3a	1 km N of Bärenstein, road to Kühberg
	E36a	assemblage 3a	
11	E47b	metablastic gneiss	Old railway station, Niederlochmühle
12	E45	metablastic gneiss	Rauenstein castle

The HT mylonites may be almost white, owing to the almost total lack of biotite, and generally show a strong, millimetre-scale foliation. The microfabric is dominated by layers of oriented, large relic quartz and feldspar grains that exhibit marginal recrystallization to variable degrees. Nevertheless, completely homogeneous, strain-free rocks with a polygonal texture–but of comparable mineralogy–may also occur intercalated with the HT mylonites on the scale of a few metres. Also on the scale of several metres, the HT mylonites gradually pass into the surrounding overprinted regional gneisses by (1) an increasing degree of recrystallization, (2) a loss of preferred orientation in the quartz domains, and, most prominent, (3) growth of biotite at the expense of garnet and white mica.

For the overprinted regional gneisses we have chosen the term 'metablastic gneiss', as introduced by Mehnert (1968) with reference to migmatized gneiss areas. The term 'metablastesis' describes isochemical recrystallization of an assemblage of minerals at high temperatures. The mineral assemblage can remain unchanged. The emphasis is thus placed on textural rather than mineralogical re-equilibration. We use this term to distinguish overprinted rocks with visible garnet relics from those gneisses free of garnet in which there is no evidence of an equilibration stage before the one now observed and hence no evidence of an overprint or recrystallization. Such 'normal' gneisses without garnet but with abundant white mica and biotite are not considered further in this paper. The metablastic regional gneisses may in part be migmatitic. The planar fabric observed is a mimetic foliation without the relic deformational structures of the HT mylonites. It corresponds to the dominant regional transposition foliation s_2 (Krohe, 1992, 1997; Sebastian, 1995). Rare rootless intrafolial folds may be observed deforming a relic older foliation defined by quartz-feldspar layers (s_1). Kinematic indicators are very rare; they suggest predominant flattening as a major deformational component. A pronounced stretching lineation is oriented east-west to ENE-WSW.

Both the HT mylonites and the surrounding overprinted gneisses are cut by local discrete ductile shear planes with recrystallized quartz, feldspar and white mica. These s_3 planes are generally flat-lying, with normal-fault kinematics and a pronounced ENE-WSW stretching lineation (Krohe, 1997).

Three major peak *PT* assemblages can be distinguished in the HT mylonite domains in quartzo-feldspathic and metapelitic rocks, as follows.

Assemblage 1: quartz + garnet + plagioclase + K-feldspar + kyanite + rutile + (white mica) + (biotite) ± (omphacite)

On the basis of their mineral assemblage, with original prograde white mica being decomposed to potassic feldspar and kyanite, these rocks may be regarded as felsic granulites. This rock type is rarely foliated and is generally found in small low-strain domains between HT mylonites. Such rocks are characterized by an equigranular polygonal fabric (0.02–0.2 mm) of mainly quartz, K-feldspar, plagioclase and garnet. Some rare, coarser-grained quartz-feldspar veinlets may possibly represent former anatectic melts.

Quartz is the most common matrix phase and is similar in grain size to the feldspars in undeformed rocks. In HT mylonites, layers of small recrystallized quartz grains (<0.02 mm) have formed by recrystallization of a few large elongated individuals (up to 0.2–0.5 mm; 'disc' quartz) with a strong preferred lattice orientation. Quartz also forms inclusions in garnet (orientated inclusion trails), kyanite, plagioclase and potassic feldspar.

Potassic feldspar and plagioclase have variable grain size up to 0.5 mm. The cores of larger K-feldspar grains contain exsolution strings and patches (Fig. 2). In large plagioclase grains, exsolution lamellae are more rarely preserved and mostly developed as patches. However, the rims of all large feldspar grains and all small recrystallized feldspar grains are homogeneous on a microscopic scale. Large deformed augen can also be devoid of exsolution lamellae. In high-strain domains, feldspar porphyroclasts are strongly deformed. Recrystallized grains around their rims are very small (20–100 mm). Subgrain misorientation increases towards the rim of the porphyroclasts (core-mantle structure). Hence, subgrain rotation was an important mechanism of recrystallization.

Garnet is remarkably rich in inclusions, comprising numerous phases such as quartz, kyanite, plagioclase, K-feldspar, rutile, magnetite, pyrite, zircon, monazite and very rare biotite and white mica. Also typical are polycrystalline inclusions predominantly of quartz-K-feldspar-plagioclase and quartz-K-feldspar aggregates that may exhibit symplectitic intergrowth on a cryptocrystalline scale. This type of inclusion is characterized by radial cracks in the surrounding garnet, a feature resembling textures interpreted to be due to the breakdown of coesite inclusions in garnet in Erzgebirge eclogites (Schmädicke *et al.*, 1992). Some garnets show corroded grain boundaries decorated with biotite. In places, garnet forms partial coronas around large plagioclase or kyanite grains.

Kyanite is similar in grain size to all the other matrix minerals; it exhibits preferred orientation and deformation in HT mylonites. Inclusions of white mica, plagioclase, quartz, magnetite, potassic feldspar, monazite and rarely biotite are observed. Garnet, biotite and white mica may survive and replace kyanite.

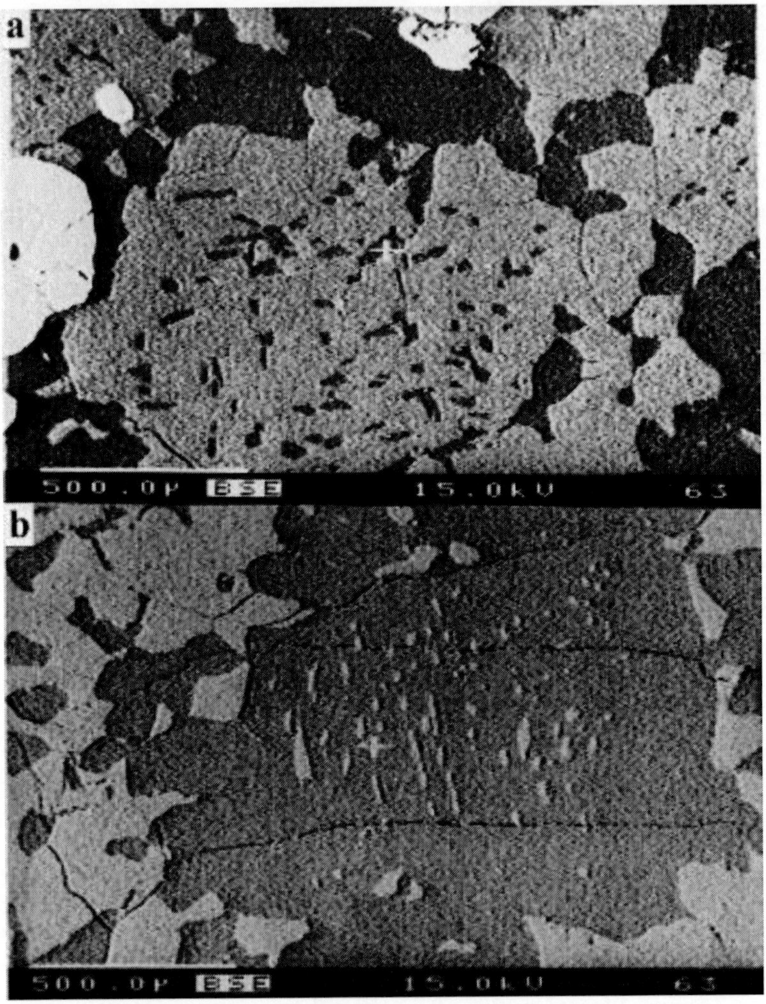

Figure 2. Back-scattered electron image of (a) perthitic and (b) antiperthitic cores of feldspar grains in a granoblastic kyanite granulite. Sample E42g.

The amount of biotite observed is generally very low in comparison with typical quartzo-feldspathic assemblages. It is usually absent in the matrix of these rocks, but can occur as 0.2 mm flakes that may in fact be primary. In one sample biotite was observed as an inclusion in garnet and in kyanite. In all other cases biotite represents a retrograde product growing around the rims of garnet and white mica. Some larger biotite grains represent symplectites intergrown with plagioclase.

White mica generally grows around kyanite at the expense of potassic feldspar and kyanite. The degree of this retrograde white mica formation varies strongly. However, rare inclusions in garnet, kyanite and large zircons also indicate the existence of prograde white mica.

Omphacite was observed in one undeformed sample (E42–1b). Fine-grained omphacite-plagioclase symplectites appear to represent pseudomorphs (Fig. 3) that correspond well in shape and size to the surrounding grains of an equigranular polygonal fabric. This suggests that the precursor mineral (clinopyroxene or amphibole) formed part of a dominant assemblage involving plagioclase.

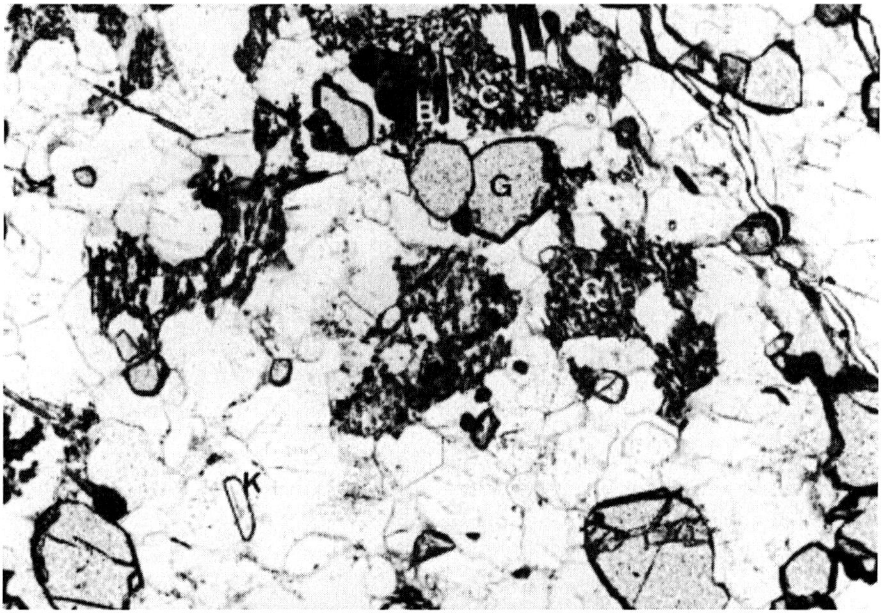

Figure 3. Clinopyroxene-plagioclase symplectites (C) in a granoblastic matrix with plagioclase, quartz, garnet (G), kyanite (K) and biotite-plagioclase symplectites (B) in a kyanite granulite. Sample E42–1b. Plane-polarized light. Width of photo is 1.75 mm.

Accessory phases are rutile, magnetite, apatite, zircon and monazite. In HT mylonites rutile and apatite may be aligned within the plane of foliation. Rutile sometimes shows narrow rims of ilmenite. Late rare alteration phases are chlorite, associated with garnet or biotite, and kaolinite replacing kyanite.

Assemblage 2: quartz + garnet + plagioclase + K-feldspar + white mica ± rutile + (biotite) ± (sphene)

Most rocks with this assemblage occur as HT mylonites with features similar to deformed kyanite-bearing granulites, but are characterized by somewhat more pronounced recrystallization. This is probably the precursor rock of most of the regional overprinted gneisses of the Gneiss-Eclogite Unit, which have a similar mineral assemblage, but high and very variable biotite/garnet ratios. Both rock types have a granitic composition, but represent different stages of microfabric development.

Again, the microfabric of the HT mylonites is characterized by layers of oriented quartz and feldspar porphyroclasts. Quartz forms augen up to 2.5 mm in size and smaller recrystallized grains of 0.05–0.25 mm. Feldspars are more strongly deformed than those of the kyanite-bearing HT mylonites. Exsolution lamellae are more rarely observed in the feldspar porphyroclasts and are more restricted to the inner cores compared with those of assemblage 1. Both types of feldspar form augen up to 1 mm in size.

Grains of garnet up to 1 mm in diameter contain inclusions of rutile, white mica, plagioclase, K-feldspar, quartz, zircon and K-feldspar-quartz aggregates. They usually have strongly corroded grain boundaries decorated with unoriented biotite. Where the grain boundaries are straight, this is due to an overgrowth of a second generation of garnet, which may grow along the grain boundaries of coarse matrix phases and enclose retrograde biotite.

White mica is invariably stable in the presence of quartz, and often occurs as inclusions in garnet. In the HT mylonites white mica is often deformed and strongly oriented parallel to the mylonitic foliation (Fig. 4) with a corona of finer-grained biotite, which may be symplectitically intergrown with K-feldspar. Locally, very fine-grained quartz-white mica symplectites also form on the grain boundaries of the white mica.

Biotite is always an unoriented breakdown product of white mica (Fig. 5) and garnet, and may be entirely lacking in some rocks. Similarly, idioblastic unoriented sphene is a retrograde matrix phase with inclusions of round rutile in some samples.

Assemblage 3a: quartz + garnet + white mica + rutile + kyanite ± (ilmenite) ± (biotite); assemblage 3b: quartz + garnet + white mica+ rutile + plagioclase ± (ilmenite) ± (biotite)

This rock type is characterized by the total absence of potassic feldspar, a markedly larger grain size compared with the previous assemblages, and is generally unfoliated or less well foliated owing to pronounced recrystallization. Plagioclase and kyanite are mutually exclusive. On the basis of their mineral content, these rocks may be regarded as metapelites.

Garnet xenoblasts occur with strongly variable grain size ranging from a few millimetres up to several centimetres. The grains are strongly poikiloblastic, containing inclusions of white mica, rutile, kyanite, quartz, pyrite, apatite and very rarely biotite. The cores may be enriched in inclusions, sometimes oriented after a relic foliation or even presenting helicitic trails as a relic internal fabric. Very small oriented needles of rutile form sharply bounded zones that may represent pseudomorphs after a former Ti-rich precursor phase such as biotite. Garnet is locally decomposed to biotite along the rims.

White mica forms conspicuous unoriented crystals up to several millimetres in diameter, or is oriented parallel to the main s_2 foliation as is kyanite. There is no sign that white mica was unstable in the presence of quartz during peak metamorphism. Large xenoblasts of kyanite may be kinked, and contain inclusions of white mica. Some kyanite, on the other hand, is strongly corroded and enclosed in matrix white mica. Both white mica and kyanite may be enclosed in garnet. Some white mica has recrystallized parallel to the local s_3 foliation.

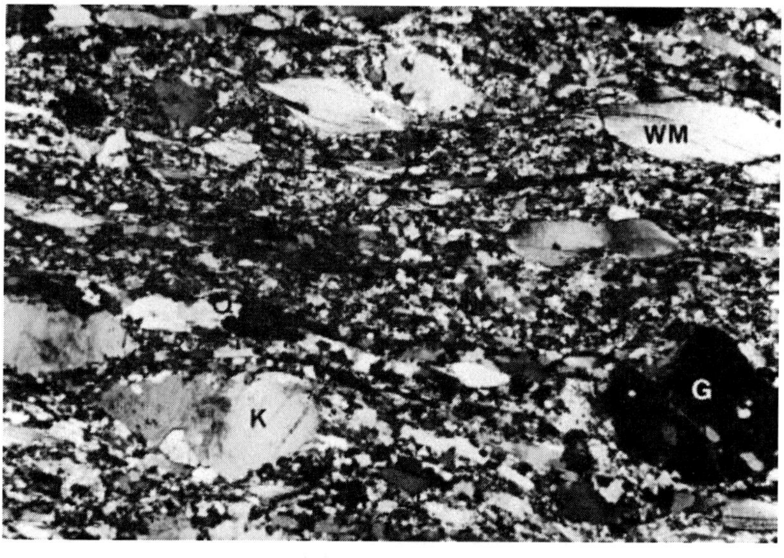

Figure 4. HT mylonite lacking biotite and with a relic deformation fabric. Deformed white mica (WM), quartz (Q), potassic feldspar (K) and garnet (G) alternate with quartz layers. Sample E726a. Crossed polarizers. Width of photo is 4.5 mm.

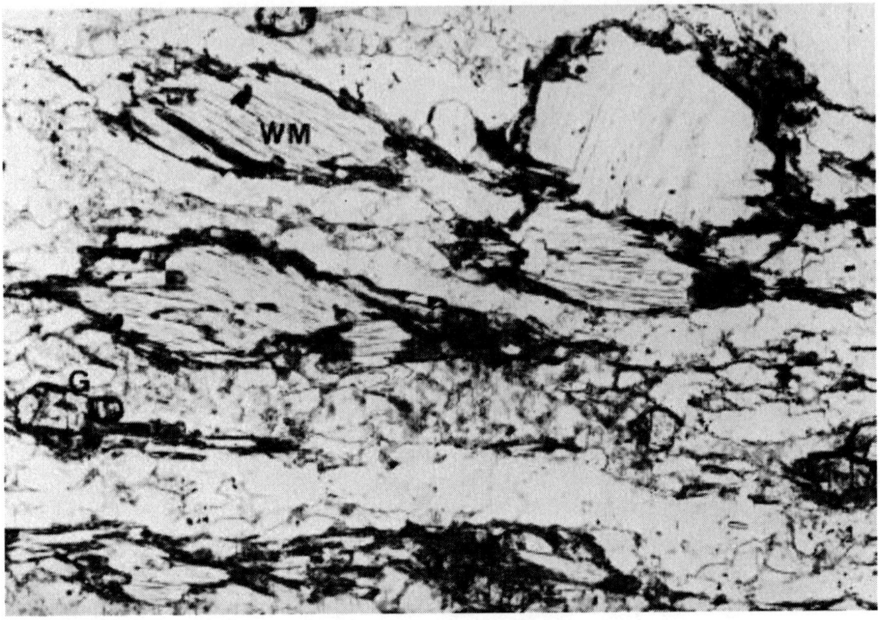

Figure 5. Fine-grained retrograde biotite grows along the margins of deformed white micas (WM) and garnet (G) in an HT mylonite. Sample E29e. Plane-polarized light. Width of photo is 1.75 mm.

When plagioclase occurs instead of kyanite, it has a grain size and shape similar to matrix quartz (up to a few millimetres), but is also found as an inclusion in garnet. Biotite is an accessory constituent forming occasional narrow rims around white mica, but is also a breakdown product of garnet. The degree of retrograde biotite formation is restricted relative to the K-feldspar-bearing assemblages 1 and 2. Ilmenite forms around abundant matrix rutile; both are enclosed in matrix white mica. Rutile is generally more abundant and larger as an inclusion in garnet. Tourmaline, monazite, apatite and large round zircons are also found among the accessories. Rare magnetite occurs as inclusions in garnet.

Overprinted metablastic gneisses

These are the most widespread rock types in the Gneiss-Eclogite Unit, along with some garnet-free gneiss containing plagioclase, K-feldspar, biotite and white mica in approximately equal amounts. These rocks show a strong increase in the amount and grain size of biotite with respect to the assemblages 1, 2 and 3. In particular, there may be a continuous transition over only a few metres towards rocks of assemblage 2 , with the biotite/garnet ratio increasing with increasing recrystallization. A strongly planar fabric is still characteristic in hand specimen; asymmetric fabric elements that may be used as shear indicators are generally lacking.

Feldspars and quartz have recrystallized completely, with a strong grain-size variation (0.1–2 mm) owing to recrystallization at high temperatures. Exsolution lamellae in feldspars are rare and observed only in the cores of very large grains. The larger grains may enclose biotite, white mica and corroded garnet. Biotite and white mica have usually recrystallized mimetically within the main s_2 foliation with similar coarse grain size. The strong parallel foliation is mesoscopically identical to the fabric of the HT mylonites. The formation of biotite at the expense of white mica is always clearly discernible and represents the most distinguishing feature of all gneisses in the Gneiss-Eclogite Unit, apart from garnet resorption.

Garnet may show two generations of growth similar to those in the HT mylonites; they are readily distinguished. The older generation is represented by strongly resorbed grains with 2–40 μm inclusions of quartz, K-feldspar, white mica and rutile. In some cases, only accessory relics within biotite flakes are preserved. Garnets of the second generation have straight grain boundaries, enclose coarse matrix quartz and biotite, and tend to form atoll garnets up to 3 mm in diameter. The growth of this garnet obviously occurred along the grain boundaries of the recrystallized biotite-bearing assemblage under static conditions. Sometimes growth of small garnet 2 grains may be observed along the boundaries of large biotites. Occasionally, garnet 1 cores rich in tiny inclusions may be distinguished from garnet 2 rims free of inclusions.

Idioblastic sphene or ilmenite grains are always present, both often containing inclusions of rutile. Small resorbed relics of kyanite may be found occasionally.

Evidence of anatectic partial melting of variable intensity is omnipresent in the metablastic gneisses of the entire Gneiss-Eclogite Unit and may be a dominant feature in some areas [e.g. near Görsdorf and Mulda; seeWienholz *et al.* (1979)]. In some cases, metatexites form, with neosomes of mainly tonalitic and more rarely granitic composition well separated from melanosomes; in other cases, partial melting is more diffuse. The presence of abundant biotite and white mica in these migmatitic gneisses–also within the neosomes–and the lack of large concentrations of anhydrous minerals such as aluminosilicate, garnet, orthopyroxene or cordierite, particularly in the neosomes, indicates anatexis under water-saturated conditions. Coproduction of anhydrous solids together with the melt would be involved.

All rocks of assemblages 1–3, as well as the metablastic gneisses, may show discrete foliation planes of a strongly non-coaxial deformation with normal fault kinematics (D3; Krohe, 1997). These discrete planes contain recrystallized fine-grained quartz, feldspar and white mica. The formation of green biotite, chlorite and epidote is locally abundant where this late ductile imprint is documented.

MINERAL CHEMISTRY AND ZONING

Analytical methods

Mineral analyses were obtained using electron microprobes (Camebax and Cameca SX50) in Bochum and Potsdam. Operating conditions were 15 kV and 14 nA, with 20 s counting time on peaks and a slightly defocused electron beam to minimize alkali loss when measuring micas or feldspars. The following minerals and synthetic materials were used as standards: andradite glass (Ca, Fe), pyrope (Si, Al, Mg), jadeite (Na), topaz (F), spessartine (Mn), NaCl (Cl), K-silicate glass and rutile (Ti). Representative point analyses are given Tables 2–4. A complete list may be obtained from the first author upon request.

Table 2. Representative compositions of garnet

	E42-1b			E29e			E43e		E736a		E735a	
	core	rim	grt2	core	rim	grt2	grt1	grt2	core	rim	core	rim
SiO_2	37·84	38·07	38·46	36·94	37·16	37·53	37·00	37·38	36·87	37·13	36·60	37·08
TiO_2	0·16	0·07	0·09	0·14	0·12	0·08	0·37	0·09	0·08	0·02	0·13	0·00
Al_2O_3	21·44	21·68	21·79	20·77	20·97	21·21	20·51	20·98	20·61	20·88	20·73	21·05
Cr_2O_3	0·00	0·02	0·04	0·03	0·04	0·01	0·00	0·00	0·00	0·00	0·03	0·01
Fe_2O_3	0·91	0·55	0·54	1·43	0·84	0·99	1·55	1·76	0·94	0·85	0·97	0·91
FeO	25·63	25·06	23·60	25·17	27·98	25·46	30·56	21·83	35·17	36·33	38·66	36·83
MnO	0·45	0·47	0·34	1·69	0·63	0·72	0·83	0·72	1·47	0·38	0·34	0·25
MgO	4·74	5·63	6·04	1·34	2·11	1·64	1·56	0·95	1·99	3·48	2·49	3·63
CaO	8·93	8·13	9·13	12·33	9·87	12·87	8·39	16·89	3·37	1·51	0·82	1·24
Sum	100·10	99·68	99·93	99·84	99·72	100·51	100·77	100·60	100·50	100·68	100·77	101·00
Cations based on 24 O												
Si	5·920	5·938	5·949	5·898	5·934	5·925	5·909	5·887	5·954	5·951	5·915	5·920
Al^{IV}	0·080	0·062	0·051	0·102	0·066	0·075	0·091	0·113	0·046	0·049	0·085	0·080
Al^{VI}	3·874	3·923	3·921	3·808	3·880	3·873	3·769	3·780	3·877	3·895	3·863	3·882
Cr	0·000	0·003	0·005	0·004	0·004	0·001	0·000	0·000	0·000	0·000	0·004	0·001
Fe^{3+}	0·107	0·065	0·063	0·172	0·101	0·117	0·187	0·209	0·113	0·103	0·118	0·117
Ti	0·019	0·009	0·011	0·016	0·015	0·009	0·044	0·011	0·010	0·002	0·015	0·000
Sum	4·000	4·000	4·000	4·000	4·000	4·000	4·000	4·000	4·000	4·000	4·000	4·000
Mg	1·106	1·310	1·394	0·319	0·502	0·385	0·372	0·223	0·480	0·831	0·601	0·864
Fe^{2+}	3·353	3·269	3·053	3·361	3·736	3·362	4·082	2·875	4·749	4·869	5·225	4·917
Mn	0·059	0·062	0·045	0·228	0·085	0·096	0·113	0·096	0·201	0·051	0·047	0·034
Ca	1·497	1·359	1·513	2·110	1·689	2·177	1·435	2·850	0·582	0·259	0·142	0·212
Sum	6·015	6·000	6·005	6·018	6·012	6·020	6·002	6·044	6·012	6·010	6·015	6·027

Table 3. Representative compositions of feldspar, pyroxene and sphene

	Plagioclase						
	E42-1b			E28e			E43e
	I*	IIC	IIR	I*	IIC	IIR	
SiO_2		65·83	66·92	65·32	65·90	67·30	64·56
Al_2O_3	20·82	21·59	22·81	20·95	20·78	22·29	21·88
Fe_2O_3	0·00	0·03	0·03	0·00	0·06	0·03	0·07
BaO	0·06	0·00	0·02	0·00	0·03	0·03	0·06
CaO	2·41	2·09	3·26	1·90	1·59	3·16	2·63
Na_2O	9·04	9·83	9·85	9·46	10·58	9·61	10·03
K_2O	2·41	0·36	0·19	1·78	0·35	0·14	0·26
Sum	100·00	100·86	101·52	100·00	100·73	100·01	100·39
Cations based on 8 O							
Si	2·914	2·909	2·837	2·910	2·933	2·846	2·871
Al	1·086	1·106	1·168	1·090	1·067	1·158	1·131
Fe^{3+}	0·000	0·001	0·001	0·000	0·002	0·001	0·003
Sum	4·000	4·016	4·006	4·000	4·002	4·005	4·005
Ba	0·001	0·000	0·000	0·000	0·001	0·001	0·001
Ca	0·087	0·097	0·152	0·090	0·074	0·149	0·124
Na	0·776	0·828	0·828	0·810	0·894	0·821	0·853
K	0·136	0·020	0·011	0·100	0·019	0·008	0·015
Sum	1·000	0·945	0·991	1·000	0·988	0·979	0·992

	Potassic feldspar						Clinopyroxene		Sphene
	E42-1b		E29e		E43e		E42–1b		
	I*	II	I*	II					
SiO_2	64·96	62·87	65·22	64·51	63·82	SiO_2	53·71	SiO_2	29·83
Al_2O_3	19·04	19·11	18·79	18·43	18·75	TiO_2	0·26	TiO_2	33·59
Fe_2O_3	0·00	0·04	0·00	0·03	0·03	Al_2O_3	11·17	Al_2O_3	3·79
BaO	0·50	0·54	0·11	0·16	0·41	FeO	5·88	Fe_2O_3	0·53
CaO	0·37	0·08	0·24	0·00	0·00	MgO	7·60	CaO	28·29
Na_2O	2·98	1·72	2·41	1·15	1·18	CaO	15·61	H_2O	0·18
K_2O	12·15	13·64	13·22	15·43	15·20	Na_2O	4·89	F	1·17
Sum	100·00	98·03	100·00	99·91	99·40	Sum	99·02	Sum	96·90
Cations based on 8 O						*Cations based on 6 O*			
Si	2·973	2·955	3·086	2·988	2·972	Si	1·949	Si	0·991
Al	1·027	1·058	1·014	1·006	1·029	Al^{IV}	0·051	Ti	0·839
Fe^{3+}	0·000	0·001	0·000	0·001	0·001			Al	0·147
Sum	4·000	4·014	4·000	3·995	4·002	Al^{VI}	0·427	Fe^{3+}	0·011
						Ti	0·007	Ca	1·007
Ba	0·008	0·010	0·002	0·003	0·008	Fe	0·179	OH	0·042
Ca	0·019	0·004	0·012	0·000	0·000	Mg	0·411	F	0·116
Na	0·264	0·157	0·214	0·104	0·107	Ca	0·607		
K	0·709	0·818	0·772	0·912	0·903	Na	0·344	O†	4·834
Sum	1·000	0·989	1·000	1·018	1·017	Sum	1·975		

*Composition integrated (see text); C, core; R, rim.
†Three cations; $OH=(Al+Fe^{3+})-F$; $O=[(\Sigma \text{ positive valencies})-OH-F]/2$.

Table 4. Representative mineral analyses of micas

	White mica												
	E42-1b	E42g in Zc	E42g in Ky	E29e in Grt	E29e core	E29e rim	E29a WM III	E736a in Grt	E736a core	E736a rim	E735a in Grt	E735a core	E43e
SiO_2	47·09	47·21	46·27	49·97	48·37	47·74	44·91	50·52	49·59	49·29	49·64	47·98	48·63
TiO_2	2·46	2·61	0·31	0·90	1·43	0·89	0·13	0·37	0·29	0·41	0·86	1·58	0·69
Al_2O_3	29·33	28·29	39·98	29·28	27·89	29·07	32·12	28·37	27·38	30·02	28·00	29·91	29·19
Cr_2O_3	0·04	0·04	0·00	0·01	0·06	0·02	0·00	0·05	0·03	0·08	0·03	0·02	0·00
FeO	2·62	1·54	0·40	3·21	3·95	3·87	4·72	2·55	2·57	1·98	2·46	2·59	4·48
MnO	0·00	0·03	0·01	0·03	0·03	0·02	0·03	0·00	0·06	0·01	0·00	0·03	0·02
MgO	2·40	3·55	0·29	2·88	2·25	2·00	1·87	2·82	2·76	2·20	2·66	1·82	2·51
CaO	0·00	0·00	0·52	0·04	0·00	0·00	0·00	0·00	0·00	0·00	0·00	0·00	0·00
BaO	0·56	0·34	0·00	0·14	0·11	0·00	0·00	0·11	0·17	0·23	0·15	0·17	0·19
Na_2O	0·31	0·45	6·49	0·19	0·20	0·23	0·19	0·56	0·29	0·59	0·38	0·47	0·15
K_2O	10·65	10·73	1·82	9·72	11·04	10·92	10·59	9·90	10·39	9·93	10·72	10·74	10·76
F	0·00	0·49	0·15	0·47	0·59	0·55	0·60	0·10	0·14	0·11	0·15	0·05	0·03
Cl	0·05	0·03	0·00	0·01	0·00	0·00	0·00	0·02	0·04	0·01	0·04	0·02	0·00
H_2O	4·43	4·19	4·63	4·33	4·16	4·16	4·10	4·45	4·32	4·42	4·37	4·43	4·47
Sum*	99·93	99·41	100·83	101·01	99·86	99·27	98·59	99·76	97·95	99·23	99·37	99·78	101·16

Ions based on 22 O

Si	6·362	6·392	5·903	6·601	6·560	6·497	6·171	6·736	6·764	6·601	6·685	6·457	6·504
Al^{IV}	1·638	1·608	2·097	1·399	1·440	1·503	1·829	1·264	1·236	1·399	1·315	1·543	1·496
Al^{VI}	3·033	3·907	3·914	3·159	3·018	3·160	3·372	3·193	3·164	3·340	3·130	3·200	3·105
Ti	0·250	0·266	0·030	0·090	0·146	0·091	0·013	0·037	0·004	0·041	0·087	0·160	0·070
Cr	0·004	0·005	0·000	0·001	0·007	0·002	0·000	0·006	0·030	0·008	0·003	0·002	0·000
Fe	0·297	0·174	0·043	0·355	0·448	0·440	0·493	0·284	0·293	0·221	0·277	0·292	0·501
Mn	0·003	0·003	0·001	0·004	0·003	0·002	0·004	0·000	0·007	0·001	0·000	0·003	0·002
Mg	0·483	0·716	0·054	0·566	0·454	0·406	0·383	0·561	0·560	0·440	0·534	0·365	0·500
Sum	4·070	4·071	4·042	4·175	4·078	4·106	4·265	4·080	4·058	4·051	4·031	4·023	4·178
Ca	0·000	0·000	0·070	0·005	0·000	0·000	0·000	0·000	0·000	0·000	0·000	0·000	0·000
Ba	0·030	0·018	0·000	0·008	0·006	0·000	0·000	0·006	0·009	0·012	0·008	0·003	0·010
Na	0·081	0·119	1·605	0·048	0·054	0·059	0·050	0·144	0·077	0·154	0·100	0·115	0·040
K	1·836	1·853	0·297	1·639	1·909	1·895	1·857	1·684	1·807	1·696	1·841	1·858	1·835
Sum	1·947	1·990	1·972	1·700	1·969	1·954	1·907	1·834	1·893	1·862	1·949	1·976	1·885
F	0·000	0·209	0·059	0·184	0·237	0·222	0·243	0·042	0·060	0·047	0·064	0·022	0·014
Cl	0·011	0·007	0·000	0·001	0·000	0·000	0·001	0·005	0·008	0·002	0·008	0·004	0·000
OH	3·989	3·784	3·941	3·815	3·763	3·778	3·756	3·954	3·932	3·951	3·928	3·974	3·987

*Sum corrected for fluorine.

	White mica			
	E42-1b	E3 in Ky	E29e in Grt2	E43e
SiO_2	36·24	38·41	36·44	35·54
TiO_2	3·77	3·10	3·47	2·05
Al_2O_3	13·78	17·05	15·80	17·62
Cr_2O_3	0·00	0·10	0·00	0·03
FeO	18·54	11·01	20·58	23·33
MnO	0·05	0·03	0·06	0·19
MgO	11·60	16·13	8·34	6·77
CaO	0·02	0·00	0·07	0·00
BaO	0·34	n.d.	0·09	0·08
Na_2O	0·10	0·13	0·04	0·07
K_2O	9·33	9·34	9·92	9·39
F	0·47	n.d.	0·88	0·21
Cl	0·48	n.d.	0·24	0·03
H_2O	3·52	4·12	3·40	3·74
Sum*	98·11	99·42	98·90	99·04
Ions based on 22 O				
Si	5·628	5·592	5·638	5·536
Al^{IV}	2·372	2·408	2·362	2·464
Al^{VI}	0·151	0·517	0·520	0·771
Ti	0·441	0·295	0·404	0·240
Cr	0·000	0·012	0·000	0·003
Fe	2·408	1·340	2·663	3·039
Mn	0·006	0·003	0·007	0·025
Mg	2·684	3·502	1·923	1·572
Sum	5·690	5·669	5·519	5·650
Ca	0·004	0·000	0·011	0·000
Ba	0·020	0·000	0·006	0·005
Na	0·030	0·036	0·013	0·021
K	1·848	1·735	1·957	1·865
Sum	1·902	1·771	1·987	1·891
F	0·232	n.d.	0·433	0·009
Cl	0·126	n.d.	0·062	0·103
OH	3·643	4·000	3·505	3·888

Before turning to geothermobarometric calculations, extensive studies of mineral inhomogeneities must be undertaken to allow the correlation of small domains with particular stages of equilibration. For this purpose, qualitative digital element distribution maps were used extensively. Such maps were produced by measuring X-ray impulses for 2 s per point during a two-dimensional scan across a grain. The data were subsequently processed with appropriate software (Bernhardt *et al.*, 1995). Three elements were usually measured simultaneously. Quantitative line scans were also used. Individual grains of all minerals within most samples were studied in this manner. With this procedure it was possible to select core compositions unequivocally, despite the fact that zoning is asymmetrical in all phases. Particular care was taken to analyse inclusion phases.

Garnet

Garnet composition varies strongly among the different assemblages (Fig. 6; Table 2): unusually high grossular contents characterize garnet core compositions of assemblages 1 (22–33 mol %), 2 (10–35 mol %) and 3b (10–20 mol %), and the metablastic gneisses (25–45mol %), in contrast to those of the feldspar-free assemblage 3a (<5 mol %).

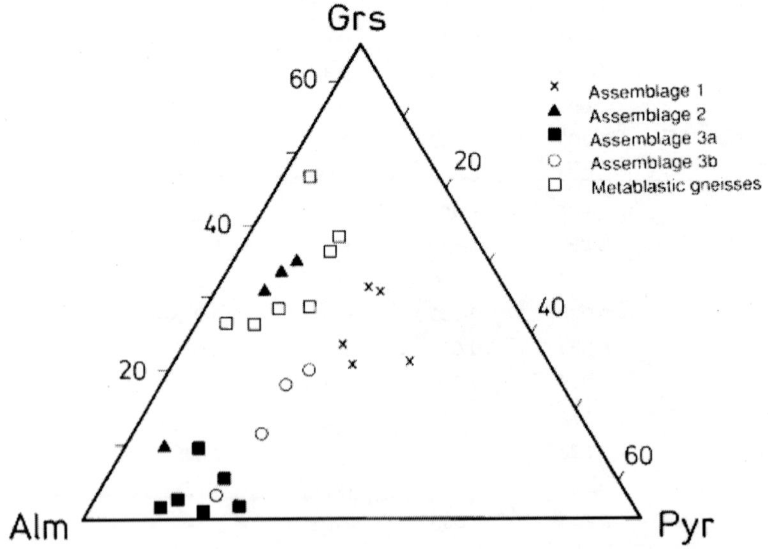

Figure 6. Grossular-almandine-pyrope ratios of garnet core compositions.

Assemblage 1 has particularly pyrope-rich garnets (20–33 mol %), in contrast to all other white-mica-bearing assemblages (5–15 mol %). Whereas almandine contents are generally 50–70 mol % in most assemblages, the feldspar-free assemblage 3a in particular has very almandine-rich garnets (80–92 mol %), owing to an Fe- and Al-rich whole-rock chemistry. Spessartine contents are minor in all assemblages, generally below 2 mol %, but rarely up to 7 mol %.

Conspicuous and very irregular zoning of garnet is a consistent feature with similar characteristics in all assemblages (Fig. 7). Most typical is a strong decrease of the grossular component from the core towards the corroded rim. Zoning roughly follows the shape of the corrosion embayments, indicating that it cannot represent a prograde growth feature. The second generation of garnet has idioblastic grain boundaries and discontinuously overgrows the irregular rims of the resorbed garnets of the first generation. It is characterized by a strong increase in Ca content. This garnet generation is present in all rocks, but in differing amounts. In weakly overprinted rocks it is only detectable by two-dimensional element distribution mapping. The boundaries between the two garnets are generally sharp, but may be somewhat transitional in some cases. There may also be a slight decrease of the grossular component towards the rim of the second garnet generation. In strongly overprinted gneisses this generally atoll-shaped idioblastic second generation dominates; small, irregular relics of the corroded first generation (Fig. 7b) may be distinguished only by careful element distribution mapping.

Fe contents generally decrease and Mg contents increase toward the garnet rims. Fe/(Fe + Mg) always decreases continuously from core to rim, more or less concentrically with respect to the grain boundaries, even across the boundaries between the two garnet generations. This may be seen in all assemblages. Only in the outermost rims, especially of the second-generation garnet in the metablastic gneisses, does the trend reverse again. Mn is either unzoned or decreases slightly from core to rim. In general, compositions of small garnets are equivalent to rim compositions of large grains.

Plagioclase

In fresh, unaltered samples of all plagioclase-bearing assemblages 1, 2 and 3b (Table 3) the core composition is albite (An_{7-11}). The Ca content increases continuously towards the rim (An_{11-15}). The zoning is very asymmetrical (Fig. 8a). Patches in the outermost rim or in very small matrix

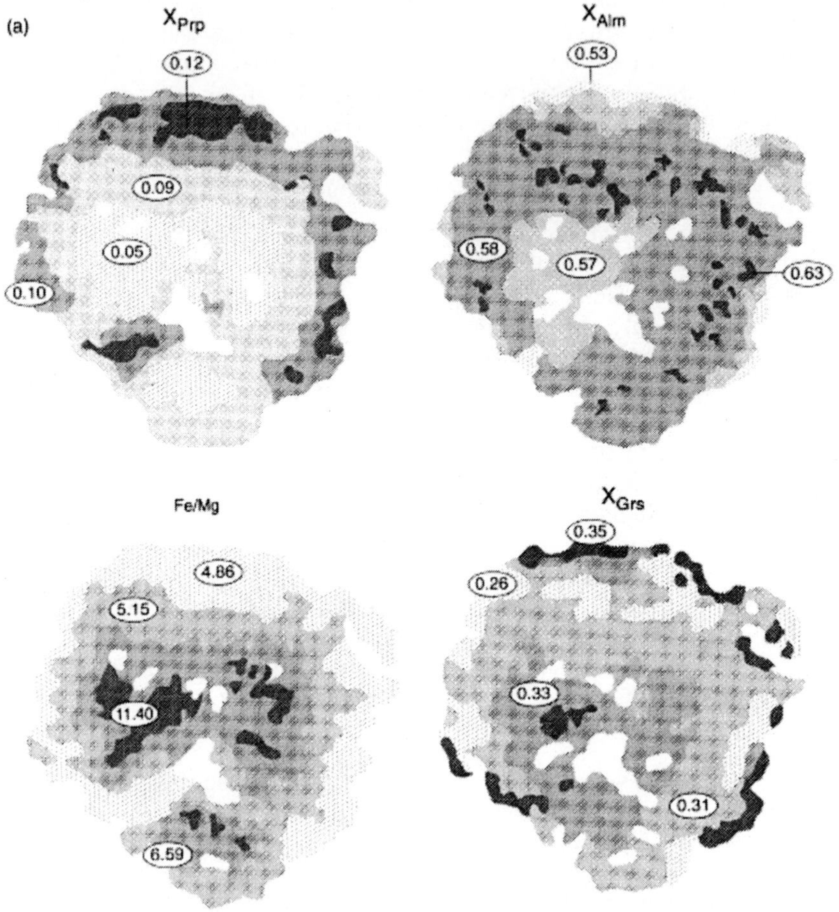

Figure 7. Characteristic element distribution maps of garnet in (a) assemblage 2 (E29e) and (b) overprinted metablastic gneiss (E43e).

grains may contain up to 20–36 mol % anorthite component. In metablastic gneisses, anorthite contents may increase from 10 to 30 mol % from the cores toward the rims. Rare antiperthitic cores in some larger grains, especially in assemblages 1 and 2, allow reconstruction of an earliest plagioclase composition (Table 3). Normally, inclusions of plagioclase in garnet are also albite with up to 10 mol % anorthite component. Rare exceptions are oligoclase (An_{20}) inclusions in garnet of assemblages 1 and 2.

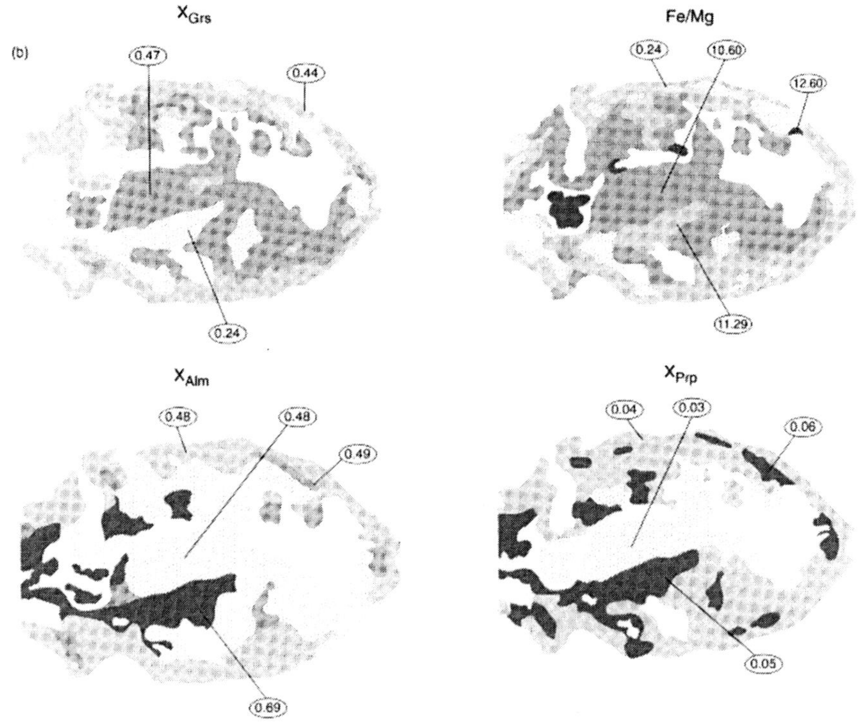

Figure 8. Characteristic element distribution maps of (a) plagioclase; (b) white mica from assemblage 2 (E29e).

K-feldspar

K-feldspar has a remarkably uniform composition in all assemblages (88–92 mol % orthoclase component; Table 3), both in homogeneous as well as in perthitic grains. However, the integrated composition of the perthitic cores in larger grains reveals an original composition near the temperature peak with ~71–77 mol % orthoclase component (Table 3).

White mica

The composition of white mica may vary slightly from assemblage to assemblage, but most strongly from one generation to the next (Table 4). In general, contents of Si [>3.2 per formula unit (p.f.u.)] and Ti (>0.03 p.f.u.) are enhanced; Mg contents always exceed those of Fe, whereas the paragonite content is <5 mol %. Al contents are generally somewhat lower than expected according to the Tschermak substitution, indicating some ferrimuscovite component.

Zoning is omnipresent, strongly asymmetric and irregular, with Si and Ti contents decreasing near the rim, whereas Al increases (Fig. 8b). Mg/(Mg + Fe) (0.55–0.75) and Na/(Na + K) (0.04–0.08) generally show no systematic variation from core to rim, but decreasing Mg/(Mg + Fe) and increasing Na/(Na + K) are occasionally observed.

Two generations are clearly detectable in assemblage 1. Relic prograde white mica can be found in garnet, kyanite or zircon, occasionally with high Si contents (3.2–3.35 p.f.u.). The Na contents may also be high (6–9 mol %), and paragonite was also found. Mg/(Mg + Fe) tends to be higher than in the matrix white mica. White mica inclusions are more abundant in assemblages 2 and 3 as well as in metablastic gneisses. In assemblage 2 their composition is comparable with that of the cores of matrix white mica. As in the metablastic gneisses, Si contents in assemblage 2 micas vary from 3.2 to 3.3 p.f.u. with a strong decrease to 3.15 Si p.f.u. near the rims, and Ti contents range from 0.06 to 0.12 p.f.u.

In white micas of assemblages 3a and 3b, differences between included and matrix white mica are always evident. The included white mica consistently shows lower Ti but higher Si and Na contents. The composition of the included and matrix white mica varies considerably (Si 3.15–3.4 p.f.u.; Ti 0.02–0.1 p.f.u.) among different samples.

In all the various assemblages, fine-grained white mica of a late third generation grew parallel to discrete, ductile normal shear planes (Krohe, 1997). This generation exhibits significantly lower Si and Ti contents (3.07–3.14 Si p.f.u.; 0.02–0.04 Ti p.f.u.) with identical compositions in all assemblages.

Omphacite

Omphacite is only known from symplectites in one sample of assemblage 1; it has a continuously variable composition from 20 to 34 mol % jadeite component. Plagioclase in these symplectites is richer in An (14 mol %) than the cores of the matrix plagioclase.

Biotite

Biotite compositions (Table 4) vary considerably, even within single grains. However, variations within single grains are not systematic. Only in strongly overprinted gneisses do biotite compositions tend to be more uniform. Mg/(Mg + Fe) varies between assemblages, being more magnesian in assemblage 1 (0.45–0.65) than in assemblage 2 (0.25–0.45), whereas biotites from assemblages 3a and 3b cover the complete spectrum. However, the latter are considerably higher in Al content (>0.4 p.f.u). Very rare inclusions of biotite in garnet 1 have lower Mg/(Mg + Fe) values than the respective matrix biotite.

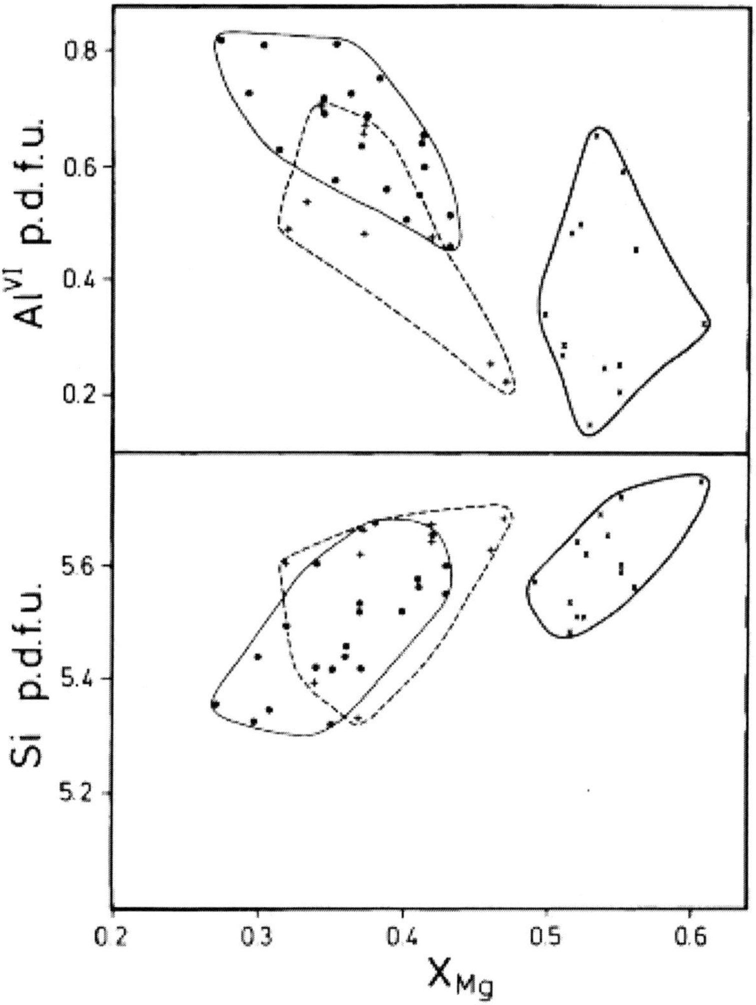

Figure 9. Al^{VI} and Si vs X_{Mg} = Mg/(Mg + Fe) variations in biotite of selected rocks of assemblage 1 (× E42–1b) and assemblage 2 (•, E29a; +, E29d).

In three samples a noticeable statistical tendency for a decrease in Si and an increase in AlVI with decreasing Mg/(Mg + Fe) was found (Fig. 9). Both effects should reflect consistent continuous changes of *PT* conditions. Ti contents are relatively high (0.05–0.2 p.f.u.), similar to those found in other HT rocks (Schreurs, 1985), but do not show any regular variation. Remarkably high chlorine contents of 0.01–0.08 p.f.u. were detected in all relict, non-overprinted assemblages, which strongly exceed those of biotites from amphibolite-facies gneisses (<0.005 p.f.u.; Willner, 1994). Biotites from overprinted gneisses always have low Cl contents.

Sphene

Sphene particularly occurs in rocks of assemblage 1, showing considerable contents of F (0.08–0.17 p.f.u.) that correspond to increased Al contents according to the substitution $Ti^{4+} + O^{2-} = Al^{3+} + F^{-}$ (Table 3).

PHASE RELATIONSHIPS

A notable feature of all quartzo-feldspathic rocks studied is the total absence of epidote, even as an inclusion mineral, except as a late retrograde alteration product. Epidote was observed by Le Goff & Ballèvre (1990), who studied high-pressure granitic orthogneisses characterized by the breakdown of the anorthite component in plagioclase to form Ca-rich garnets as in the quartzo-feldspathic rocks considered here. Those workers suggested that epidote should be stable in such assemblages below ~600°C. Hence our samples must have formed well above the stability field of epidote-biotite-quartz as delineated by Le Goff & Ballèvre (1990).

Furthermore, the evolution of the present quartzo-feldspathic rocks is dominated by the nearly complete breakdown of prograde biotite toward peak metamorphic conditions, and retrograde growth of new biotite. The HT mylonitization is bracketed by an early biotite-absent assemblage containing plagioclase, K-feldspar, quartz and kyanite or white mica, and a retrograde assemblage always containing biotite and white mica. Phase relationships at such HP-HT conditions may be illustrated by petrogenetic grids for the KMASH and KCMASH systems, involving the phases phlogopite, pyrope, potassic feldspar, quartz, kyanite, muscovite, Al-celadonite, grossular and anorthite (Figs 10 and 11). These phases exist as components of solid solutions in the rocks considered and the reactions shown are therefore multivariant and continuous over a very large *PT* field. Prominent is the continuous breakdown of biotite in the presence of quartz and kyanite or white mica towards high temperature and high pressure.

The three major biotite-absent assemblages may represent different water activities at constant pressure and temperature (Fig. 10). The water-dependent reactions affecting the fields of the three assemblages are

$$Ms_{WM} + Qtz = Or_{Kfs} + Ky + H_2O \tag{1}$$

and

$$3Ms_{WM} + Ace_{WM} = 4Or_{Kfs} + Py_{Grt} + 4H_2O. \tag{2}$$

Both reactions may be used as hygrometers. At given *PT* conditions and comparable whole-rock compositions, water activity decreases from assemblage 3 towards 1 in the KMASH system (Fig. 10). The AFM plot inFig. 12ashows the alternative assemblages 1 and 2 for rocks with very similar bulk rock composition that may be an effect of variable water activity. Notable is the strong shift in garnet composition owing to complete breakdown of phengite in assemblage 1.

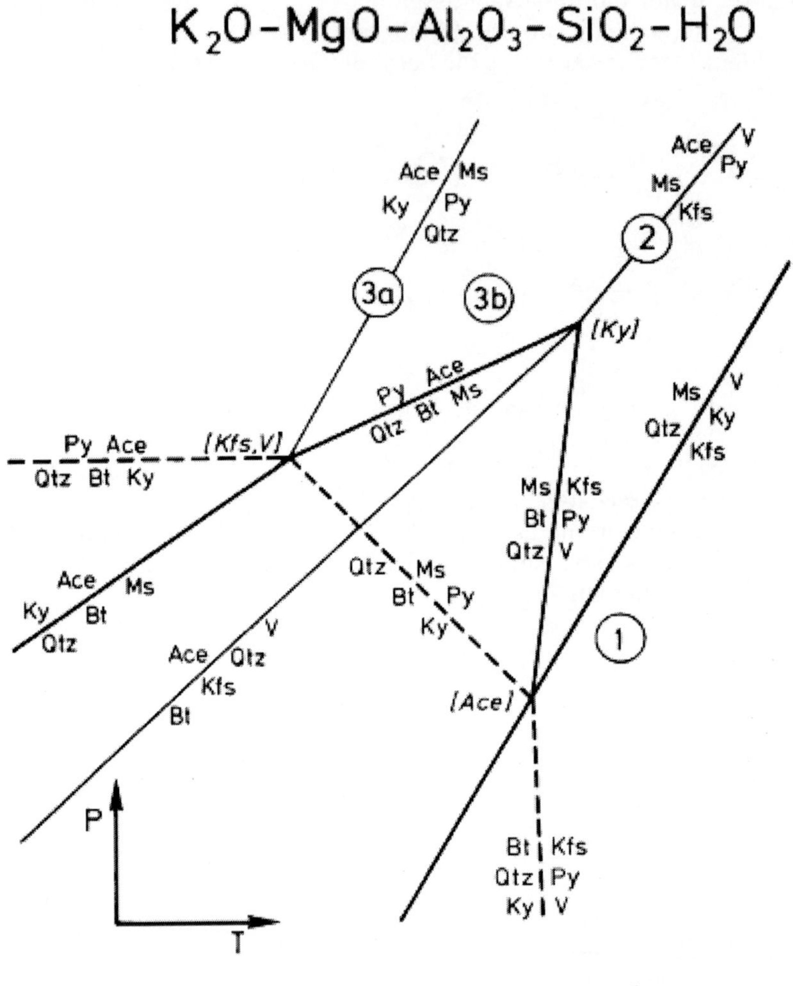

Figure 10. Schematic partial *PT* grid involving the mineral components Al-celadonite, muscovite, pyrope, quartz, phlogopite, kyanite and K-feldspar in the system K_2O-MgO-Al_2O_3-SiO_2-H_2O. Numbers refer to assemblages 1–3 (see text). Bold lines mark the upper stability of biotite + muscovite, dashed lines that of biotite + kyanite.

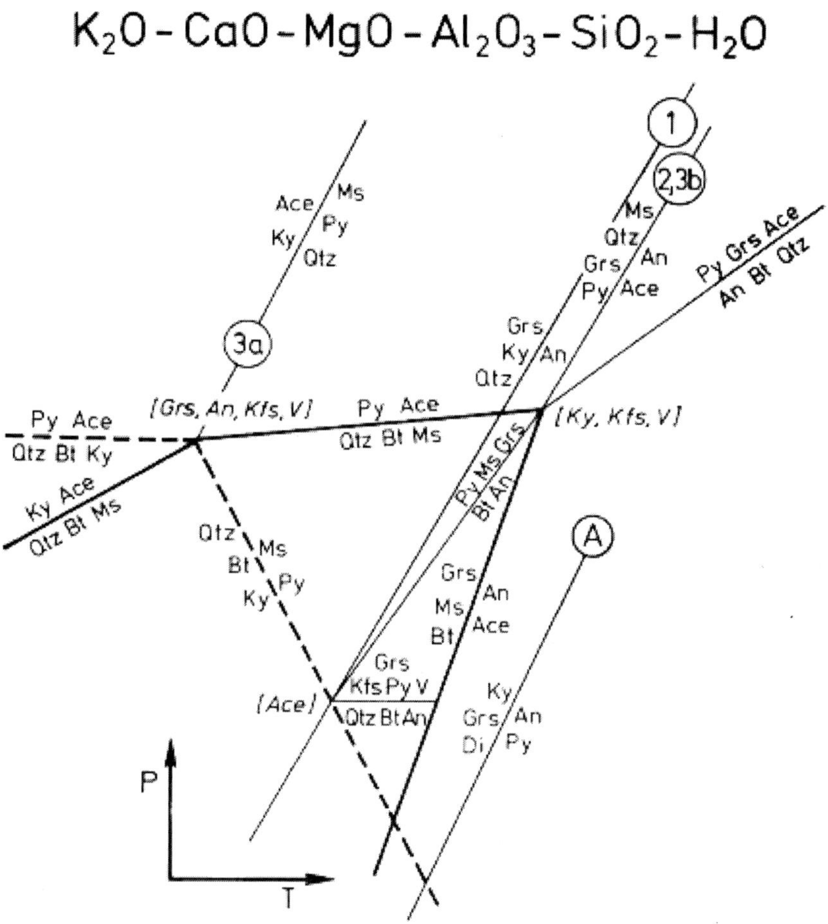

Figure 11. Schematic partial *PT* grid involving the mineral components Al-celadonite, muscovite, pyrope, grossular, anorthite, quartz, phlogopite, kyanite and K-feldspar in the system K_2O-CaO-MgO-Al_2O_3-SiO_2-H_2O. Numbers refer to assemblages 1–3 (see text). Additional reaction A including diopside is placed relatively to the proposed grid. Bold lines mark the upper stability of biotite + muscovite, dashed lines that of biotite + kyanite.

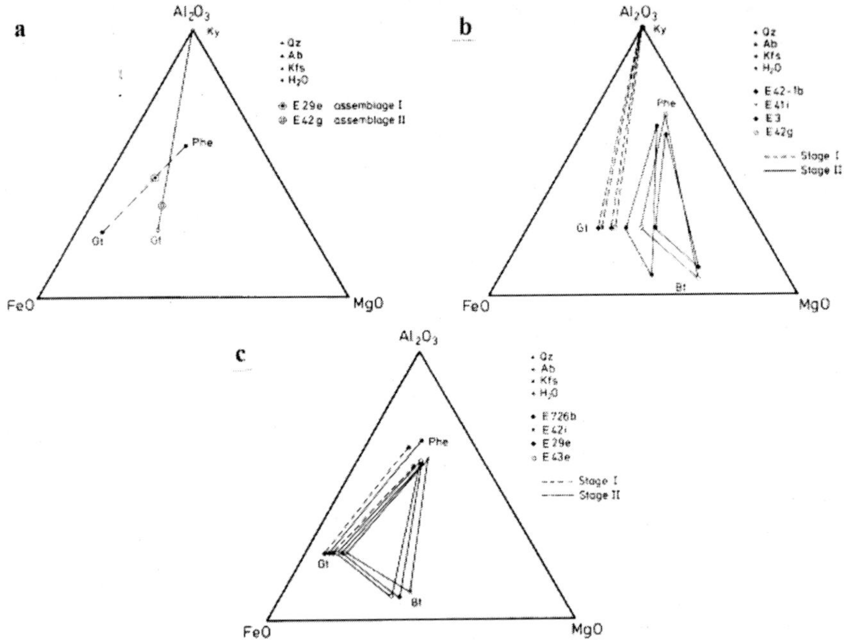

Figure 12. Projections of the phase compositions of assemblages 1 and 2 in AFM diagrams from quartz, K-feldspar, albite and H_2O for two successive equilibration stages: (a) the alternative assemblages 1 and 2 for HP-HT stage I caused by variable water activities; (b) assemblage 1 compositions for the HP-HT stage I and a retrograde stage II in local rock domains; (c) as in (b), for assemblage 2.

The K-feldspar-free assemblage 3 may involve either kyanite or plagioclase (assemblage 3a or 3b, respectively). In the KCMASH system (Fig. 11) both assemblages are separated by the multivariant reaction

$$\mathrm{Grs_{Grt}} + 2\mathrm{Ky} + \mathrm{Qtz} = 3\mathrm{An_{Plag}} \quad (3)$$

which may serve as a barometer for assemblage 1. Assemblages 2 and 3b are defined by the continuous barometer reaction

$$2\mathrm{Grs_{Grt}} + \mathrm{Py_{Grt}} + 6\mathrm{Qtz} + 3\mathrm{Ms_{WM}} = \quad (4)$$

$$6\mathrm{An_{Plag}} + 3\mathrm{Ace_{WM}}$$

and assemblage 3a by

$$3Ace_{WM} + 4Ky = 3Ms_{WM} + Py_{Grt} + 4Qtz. \quad (5)$$

Jadeitic clinopyroxene is a retrograde phase in only one sample of assemblage 1. The principal occurrence of jadeitic clinopyroxene below the pure end-member reaction

$$Jad + Qtz = Ab \quad (6)$$

depends on the stability of a diopside component in clinopyroxene. Where anorthite and grossular coexist as Ca-bearing phase components, coexistence of an additional diopside component at equilibrium is governed by the multivariant reaction

$$3Di_{Cpx} + Grs_{Grt} + 6Ky = Py_{Grt} + 6An_{Plag}. \quad (7)$$

Such a reaction occurs only in assemblage 1. However, this assemblage contains additional quartz. Hence, to overlap with the multivariant reaction (3), the pyrope component in garnet must be low relative to the grossular component (Fig. 11). Otherwise, the anorthite component in plagioclase would have to be absent in rocks containing grossular-rich garnet to stabilize diopside at elevated pressure.

Whereas the biotite-absent assemblages are defined by whole-rock chemistry (e.g. Fig. 12a), retrograde phengite-garnet-biotite assemblages form under static conditions to variable degrees by reaction and equilibration in local rock domains (Fig 12b). The retrograde reactions involve simultaneous dehydration and hydration reactions. For instance, the breakdown of the Al-celadonite component in phengite by decompression will release water (Fig. 10):

$$3Ace_{WM} = Phl_{Bt} + 2Or_{Kfs} + 3Qtz + 2H_2O \quad (8)$$

whereas growth of biotite by pressure or temperature release generally involves addition of water (Figs. 10 and 11):

$$\mathrm{Grs_{Grt}} + 2\mathrm{Py_{Grt}} + 2\mathrm{Or_{Kfs}} + 2\mathrm{H_2O} = \\ 2\mathrm{Phl_{Bt}} + 3\mathrm{An_{Plag}} + 3\mathrm{Qtz} \tag{9}$$

$$\mathrm{Ms_{WM}} + \mathrm{Phl_{Bt}} + 3\mathrm{Qtz} = \\ 2\mathrm{Or_{Kfs}} + \mathrm{Py_{Grt}} + 2\mathrm{H_2O}. \tag{10}$$

Notable is the retrograde shift of bulk composition in reacting domains towards higher Mg contents than the original bulk-rock composition, because Fe-rich garnet cores become isolated. Garnet rim composition becomes more magnesian in contact with Fe-Mg phases such as phengite and newly formed biotite, indicating the direction of crossing of Fe-Mg exchange reactions after the *PT* maximum.

GEOTHERMOBAROMETRY AND FLUID EVOLUTION

Methods

The *PT* conditions of metamorphism were estimated primarily by calculation of multivariant equilibria between mineral components using the internally consistent thermodynamic data set of Berman (1988; updated in 1992) and applying the Ge0-Calc software of Brown *et al.* (1988) to yield consistent data. The data set was in particular augmented by the compatible data for the mineral component Al-celadonite as derived by Massonne (1991, 1995). The following nonideal activity models were used for solid solutions: garnet (Berman, 1990), biotite (McMullin *et al.*, 1991), plagioclase and K-feldspar (Fuhrman & Lindsley, 1988), and phengite (Massonne, 1995). The thermodynamic data for Al-celadonite and the activity models for phengite (Massonne, 1995) are based on experimental data summarized by Massonne (1991). Ideal activity models were adopted for sphene and ilmenite, whereas the activities of kyanite, rutile and quartz were set to unity.

Biotite was not used in calculation of the maximum *PT* conditions, because it can be shown that this phase was, in most cases, not stable at these conditions. Generally the cores of garnet, plagioclase and white mica were correlated for the earliest recognizable equilibration stage. The water-independent equilibria (3), (4) and (5) were used as barometers. GASP barometry [reaction (3)] was found to be in good agreement with the calibration of Koziol & Newton (1988). The only reliable thermometer for the early equilibration stage was the two-feldspar thermometer. Exsolution lamellae in relic cores of large plagioclase and potassic feldspar grains in assemblages 1 and 2 were reintegrated using back-scattered electron (BSE) images as well as analyses with a wide electron beam. The reconstituted feldspar compositions (Fig. 2; Table 3) were used to derive temperatures by the two-feldspar thermometer of Fuhrman & Lindsley (1988), adapting both compositions to a ternary solvus (Kroll *et al.*, 1993). Finally, water-dependent reactions involving K-feldspar were calculated to estimate water activities at independently derived peak *PT* conditions. Together with reactions (1) and (2), these are, for stage I,

$$\begin{aligned} &\mathrm{Grs_{Grt}} + 2\mathrm{Ms_{WM}} + 3\mathrm{Qtz} = \\ &3\mathrm{An_{Pl}} + 2\mathrm{Or_{Kfs}} + 2\mathrm{H_2O} \end{aligned} \tag{11}$$

$$\begin{aligned} &3\mathrm{An_{Pl}} + 6\mathrm{Ace_{WM}} = \\ &\mathrm{Grs_{Grt}} + 6\mathrm{Or_{Kfs}} + 2\mathrm{Py_{Grt}} + 3\mathrm{Qtz} + 6\mathrm{H_2O}. \end{aligned} \tag{12}$$

For retrograde re-equilibration in local rock domains (stage II), biotite compositions were used in conjunction with

$$\begin{aligned} &\mathrm{Alm_{Grt}} + 3\mathrm{Ace_{WM}} = \\ &\mathrm{Ann_{Bt}} + \mathrm{Ms_{WM}} + \mathrm{Phl_{Bt}} + \mathrm{Qtz} \end{aligned} \tag{13}$$

$$\mathrm{Py_{Grt}} + 3\mathrm{Ace_{WM}} = \mathrm{Ms_{WM}} + 2\mathrm{Phl_{Bt}} + 6\mathrm{Qtz} \tag{14}$$

$$\mathrm{Alm_{Grt}} + \mathrm{Phl_{Bt}} = \mathrm{Py_{Grt}} + \mathrm{Ann_{Bt}} \tag{15}$$

$$\begin{aligned} &2\mathrm{Alm_{Grt}} + 3\mathrm{Ace_{WM}} = \\ &6\,\mathrm{Qtz} + \mathrm{Py_{Grt}} + \mathrm{Ms_{WM}} + 2\mathrm{Ann_{Bt}}\,. \end{aligned} \tag{16}$$

The Fe-Mg exchange reaction between garnet and biotite, as calculated in this study, yielded temperatures up to 50°C higher than conventional thermometers including good Ca corrections for garnet (Hodges & Spear, 1982; Ganguly & Saxena, 1984).

In assemblage 3, the following reactions may also be used for *PT*calculations using Ti-bearing phases:

$$\mathrm{Alm_{Grt} + 3Rt = 2Qtz + Ky + 3Ilm} \tag{17}$$

$$\mathrm{2Alm_{Grt} + Phl_{Bt} + 6Rt = 3Qtz + Py_{Grt} + Ms_{WM} + 6Ilm} \tag{18}$$

$$\mathrm{Alm_{Grt} + Ann_{Bt} + 6Rt = 3Qtz + Ms_{WM} + 6Ilm.} \tag{19}$$

Water-dependent reactions involving K-feldspar were again used for hygrometric estimates of the retrograde stage II:

$$\mathrm{Ann_{Bt} + Ms_{WM} + 3Qtz = 3Alm_{Grt} + 2Or_{Kfs} + 2H_2O} \tag{20}$$

$$\mathrm{Ms_{WM} + 3Ace_{WM} = 4Or_{Kfs} + Py_{Grt} + 4H_2O} \tag{21}$$

$$\mathrm{Alm_{Grt} + 3Ace_{WM} = Ann_{Bt} + 2Or_{Kfs} + Py_{Grt} + 3Qtz + 2H_2O} \tag{22}$$

$$\mathrm{3Ace_{WM} = Phl_{Bt} + 2Or_{Kfs} + 3Qtz + 2H_2O.} \tag{23}$$

For *PT* estimation of early retrograde equilibration in the single pyroxene-garnet symplectite found in assemblage 1, the conventional Fe-Mg exchange thermometer as calibrated by Ellis & Green (1979) was applied, and for pressure estimation the clinopyroxene-plagioclase-quartz barometer of Gasparik & Lindsley (1980) was used.

White mica may coexist with rutile and quartz during the prograde *PT*path, at peak conditions and during the retrogade *PT* path in all assemblages, thus providing an excellent opportunity to apply the Ti-muscovite-rutile-quartz barometry of Massonne *et al.* (1993). This method requires only the coexistence of rutile, white mica and quartz. This empirically calibrated barometer is based on the following multivariant reaction and uses a nonideal activity model for white mica:

$$\begin{gathered} Ms_{WM} + 2Rt = Ti\text{-}Ms_{WM} + 2Qtz \\ [Ti\text{-}Ms = KAl_3Ti_2SiO_{10}(OH)_2]. \end{gathered} \quad (24)$$

RESULTS AND DISCUSSION

Assemblage 1

Consistent temperatures of 830°C were derived by two-feldspar thermometry for four samples from different localities (A in Fig. 13) using reintegrated core compositions of coexisting plagioclase and K-feldspar with preserved exsolution lamellae. The assumed pressure of 22–23 kbar was derived by GASP barometry [reaction (3); equilibrium B in Fig. 13] with core compositions of garnet and plagioclase. Inclusions of phengite and paragonite in garnet, kyanite and zircon point to a lower-temperature equilibration event before trapping. Plagioclase inclusions are generally albite as in the matrix, but enclosed oligoclase was also found. This also points to lower pressures before trapping in an assumed prograde assemblage with garnet, quartz and kyanite. Independent proof is provided by Ti-muscovite barometry of phengite enclosed together with rutile in zircon (C1 in Fig. 13). It was this particular zircon which was dated by Kröner & Willner (1995) with the Pb-Pb evaporation method, yielding 340 Ma as the age of metamorphic zircon growth during the prograde path approaching the thermal peak at elevated pressures. No later modification of the inclusion composition appears to have occurred, as is also supported by the lack of cracks propagating towards these inclusions, or any sign of exchange with the host mineral. Hence garnet as well as kyanite and zircon grew on the late prograde path close to the pressure maximum. However, garnet core composition can be expected to have homogenized at the very high peak temperatures of >800°C by volume diffusion and thus should not be used to calculate prograde *PT* conditions.

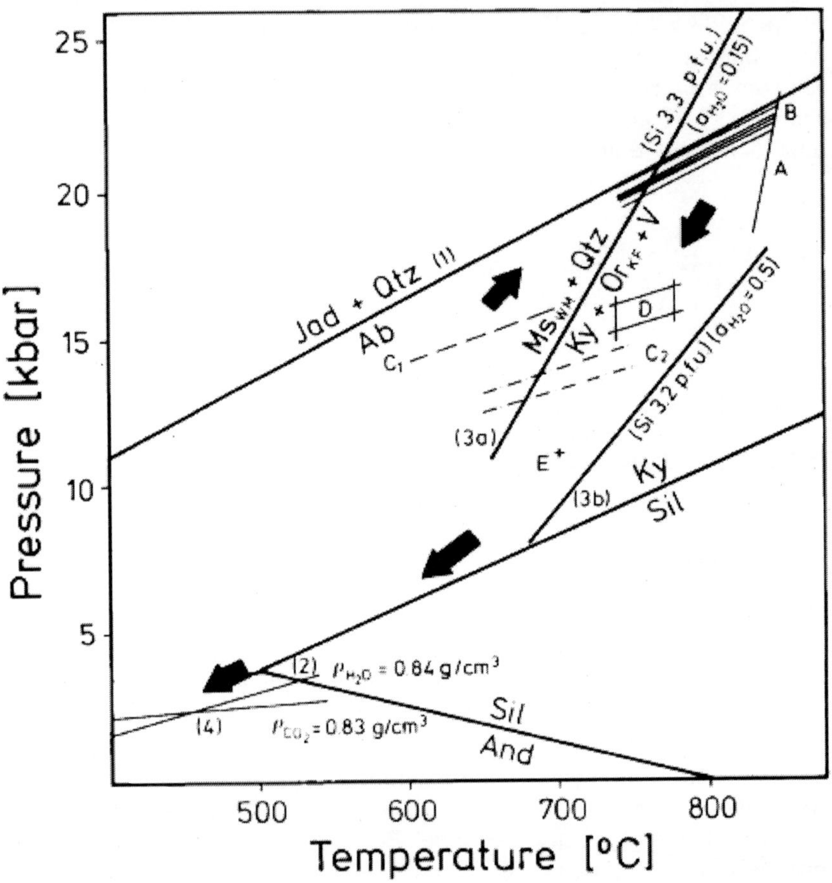

Figure 13. *PT* history for rocks of assemblage 1 (kyanite granulites). (For a discussion of equilibria A-E, see text.) 1, Holland (1980); 2, aluminosilicate stability field after Holdaway (1971); 3, reactions calculated with the modified thermodynamic data set of Berman (1988); 4, isochores of mean compositions of secondary fluid inclusions [Krentz *et al.*, 1990; Thomas, personal communication to Rötzler (1995)].

Very small quartz-K-feldspar-plagioclase and quartz-K-feldspar aggregates enclosed in garnets, as well as irregular granitic veins lacking biotite or white mica may be interpreted as relics of dehydration melts. This suggests that the breakdown of biotite and white mica may in part have involved melt-producing reactions such as

$$\mathrm{Bt + WM + Plag + Qtz = Grt + Kfs + L.} \tag{25}$$

At the maximum *PT* conditions considered here dehydration melts were produced experimentally by LeBreton & Thompson (1988) by the breakdown of biotite with intermediate Fe/Mg ratios. Such melts could have absorbed additional water, one of the possible reasons for the strong reduction of water activity in these rocks (see below).

Because maximum *PT* conditions are close to the upper stability field of albite, the question arises as to whether the jadeite stability field was ever crossed. No jadeite was trapped in any phase, although even earlier, lower-pressure prograde phases are present. If the rocks were ever subjected to the conditions of the jadeite stability field, the formation of jadeite must have been prevented for kinetic reasons.

The early retrograde formation of omphacite-plagioclase symplectites observed in one sample occurred at 745–780°C and 15–17 kbar, according to clinopyroxene-garnet thermometry (using garnet core compositions) and clinopyroxene-plagioclase-quartz barometry (field D in Fig. 13). The spread of omphacite composition is reflected by the range of *PT* data. The conditions of formation of retrograde white mica were estimated for two samples by Ti-muscovite barometry as 12–13 kbar at assumed temperatures of 700°C (curves C_2 in Fig. 13). A further retrograde *PT*coordinate at 11 kbar and 695°C was estimated in one sample from reactions (3) and (14), using rim compositions of garnet 1 and plagioclase in contact (E in Fig. 13). Barometry using rim compositions of plagioclase is difficult and hazardous owing to widely variable rim compositions. On the basis of the available data, the retrograde *PT* path must have involved an initial strong decompression of >10 kbar, accompanied by cooling of <100°C (see Fig. 13).

The calculated positions of the white mica + quartz breakdown curves [reaction (1)] suggest that water activities must have been below 0.15 at peak *PT* conditions (white mica unstable), increasing to at least 0.5 during retrograde formation of white mica. This early retrograde increase in water activity must reflect external hydrous fluid influx responsible for the growth of mica.

No alteration of kyanite to sillimanite or growth of sillimanite by any other reaction has ever been observed in the Central Erzgebirge. Hence it is suggested that the late retrograde *PT* path might generally have been restricted to the kyanite stability field. As in all other rocks of the Central

Erzgebirge, late secondary fluid inclusions were trapped at low pressures (Krentz *et al.*, 1990; Thomas, personal communication to Rötzler, 1995). Typical isochores for mean secondary hydrous and CO_2-rich fluid inclusions are shown in Fig. 13, suggesting that the rocks were below 3 kbar after crossing the 500°C isotherm, where greenschist-facies minerals such as chlorite, epidote and fine-grained white mica were formed, particularly in rocks deformed by the D_3 phase.

Assemblage 2 and metablastic gneisses

Here also the only reliable estimates of maximum temperature conditions are provided by two-feldspar thermometry, using relic integrated core compositions of coexisting plagioclase and K-feldspar. However, such relics are rare in assemblage 2; only in sample E29e could a temperature of 780°C be calculated for 20 kbar (A in Fig. 14). Such maximum pressures of 18–21 kbar may be derived by calculating the multivariant reaction (4), using core compositions of garnet, plagioclase and phengite (B in Fig. 14). These show little variation in each sample; moreover, the core compositions of matrix plagioclase and phengite are similar to those enclosed in garnet. Somewhat lower pressures of 17–18.5 kbar were calculated independently for the same phengites by Ti-muscovite barometry (D in Fig. 14). Calculated reactions involving sphene and rutile in sample E29e intersect reaction (4) at 760°C and 21 kbar (C in Fig. 14). These maximum *PT* conditions near the upper stability limit of albite are comparable with those found for assemblage 1. Again, all geothermobarometry depends on the existence and stability of plagioclase, which is also an abundant inclusion phase in garnet. These rocks show no sign that they were ever subjected to conditions of the jadeite stability field. Owing to the stronger retrograde overprint, it is hazardous to attempt derivation of maximum *PT* conditions for the widespread metablastic gneisses in a similar way.

For three samples of assemblage 2, water activities were calculated with the water-dependent reactions (2), (11) and (12). The estimated values of 0.15–0.4 well exceed those suggested for assemblage 1. Hence the existence of phengite in the presence of quartz at peak metamorphic conditions definitely depends on a higher water activity in these rocks.

Calculating early retrograde equilibria is difficult owing to the existence of two garnet generations and strongly variable compositions of plagioclase rims and biotite. However, in the overprinted metablastic gneisses a more restricted compositional variation in biotite indicates better re-equilibration. Multivariant reactions (13)-(16) were calculated for garnet and phengite rim compositions in contact with biotite, including pairs from

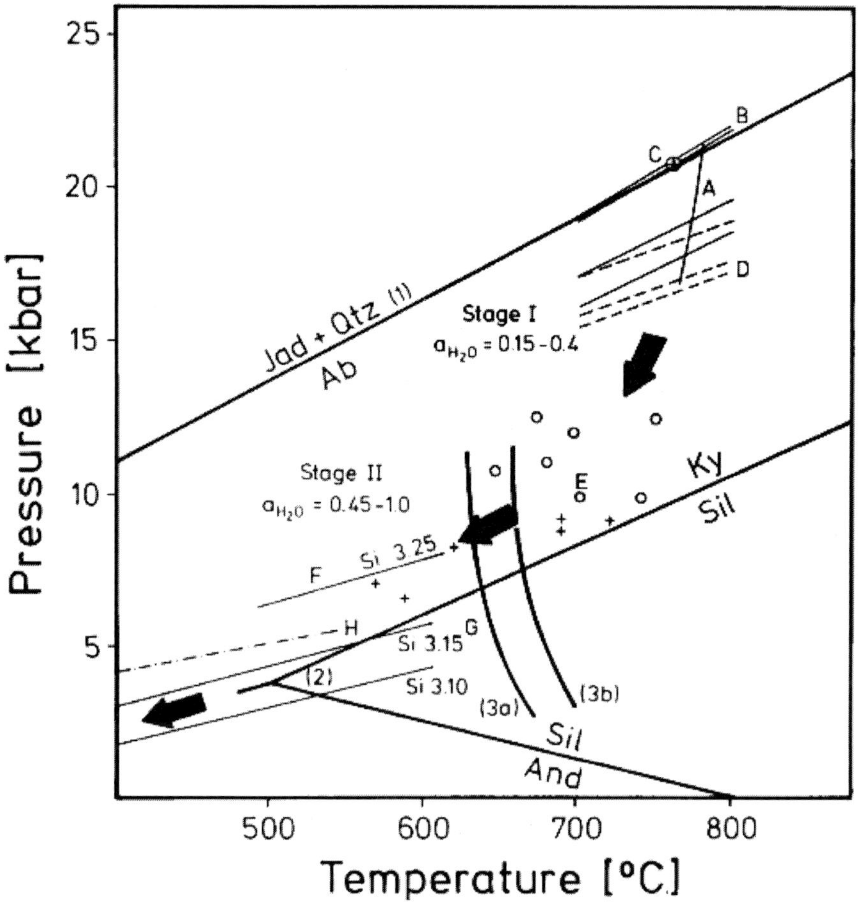

Figure 14. *PT* history for rocks of assemblage 2. (For discussion of equilibria A-H, see text.) 1, Holland (1980); 2, aluminosilicate stability field after Holdaway (1971); 3a, wet granite solidus, 3b, wet tonalite solidus according to Johannes (1984).

overprinted gneiss samples as well. *PT* coordinates scatter strongly at 580–750°C and 6–12 kbar (see symbols labelled E in Fig. 14). There is a tendency toward higher *PT* values for garnet 1 rims (circles for E in Fig. 14) as compared with those calculated from garnet 2 rims (crosses for E ofFig. 14). The water-dependent reactions (19)-(22) involving K-feldspar were calculated for hygrometric estimates at the *PT* values obtained. As in assemblage 1, a strong relative increase in water activities to values ranging between 0.45 and 1.0 is an obvious indication of water addition during retrograde biotite growth.

An independent corroboration of the second equilibration stage is given by migmatitic gneisses. Average Si contents of white micas in partial melts of tonalitic and granitic composition are 3.25 p.f.u. [F in Fig. 14; isopleth after Massonne (1991)], indicating a pressure of 8–9 kbar at the intersection with the wet granite and tonalite solidi. Conditions of 616–688°C and 8–8.5 kbar with aH2O = 0.8–1.0 may be calculated for the migmatitic gneiss E43e with granitoid composition, using multivariant equilibria. Hence widespread anatexis in metablastic gneisses is a retrograde phenomenon strongly related to abundant hydration.

Barometry with reactions involving the anorthite component of plagioclase gives only pressures in the detected range for the formation of garnet 2, when the rare rim compositions of plagioclase with the highest An contents are considered. This indicates that grossular-rich garnet 2 was in equilibrium with local late oligoclase.

The data again indicate early decompression of up to 10 kbar and subsequent cooling within the kyanite stability field. Later retrograde re-equilibration occurred, especially where discrete, ductile s_3 shear zones deform the rocks. Some recrystallized feldspars indicate that deformation already occurred above the 500°C isotherm. Ti-muscovite barometry of the recrystallized low-Si white micas (H in Fig. 14), as well as the respective Si isopleth for assemblages with biotite and K-feldspar (G inFig. 14), indicate that the rocks had attained pressures of 2.5–4 kbar after cooling to ~400–500°C.

Assemblages 3a and 3b

In these rocks, the number and chemical variation of phases in both metapelitic assemblages without K-feldspar is fairly restricted, in particular for a treatment with conventional geothermobarometry. On the other hand, Ti-muscovite barometry may be applied extensively. Using core compositions of matrix phengites and phengite inclusions in garnet, a wide pressure range between 12 and 24 kbar at an assumed maximum 700–800°C results (dashed curves B in Fig. 15), whereby the inclusion phengites consistently yield higher pressures in all samples (curves A inFig. 15). In Fig. 15 somewhat lower temperatures are assumed during equilibration of the inclusion phengites relative to the first equilibration of the matrix phengites in accordance to the observations made with the other assemblages. The following major conclusions may be extracted from these relationships and should also be applicable to all other rocks of the Gneiss-Eclogite Unit.

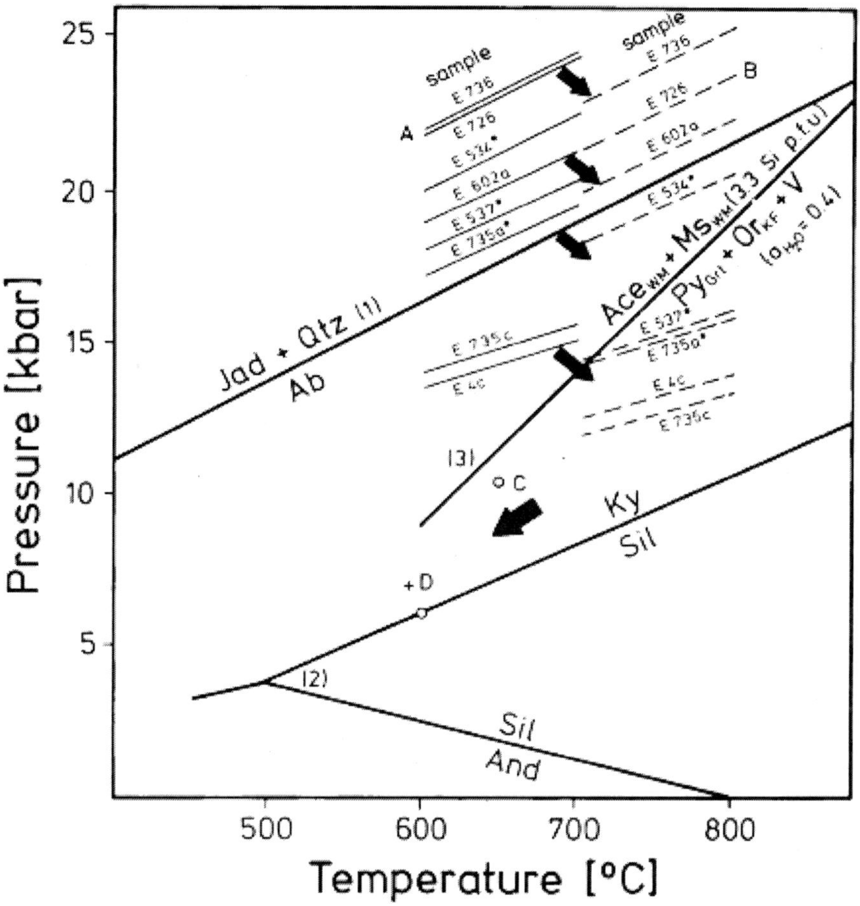

Figure 15. *PT* history for rocks of assemblages 3a and 3b. (For a discussion of equilibria A-D, see text.) 1, Holland (1980); 2, aluminosilicate stability field after Holdaway (1971); 3, reaction calculated with the modified thermodynamic data set ofBerman (1988), using average garnet and white mica compositions.

A large difference in the maximum pressures attained seems to exist between different rocks of the Gneiss-Eclogite Unit. Some acid rocks of the Gneiss-Eclogite Unit may in fact even have equilibrated in the jadeite stability field. Hence the quartzo-feldspathic rocks of the unit delineate a metamorphic array rather than a unique *PT* path at maximum pressures. The remarkable pressure differences consistently indicated by core compositions of the matrix phengites and of the inclusion phengites for different samples are independently corroborated by the evident differences in the Si contents of the white micas. These presumably record pressure differences, because they mostly occur in a limiting assemblage with garnet, kyanite and quartz. The resulting multivariant barometer reaction (see Fig. 10)

$$\mathrm{Ace}_{\ \mathrm{WM}} + \mathrm{Ky} = \mathrm{Ms}_{\ \mathrm{WM}} + \mathrm{Py}_{\ \mathrm{Grt}} + \mathrm{Qtz} \tag{26}$$

is also water independent (however, owing to the very low activities of pyrope in garnet this reaction was not used for barometry). There is no recognizable regional trend in the distribution of these calculated maximum pressures, because the tectonic position of the samples is rather difficult to correlate from outcrop to outcrop. However, if the different white mica compositions were actually an effect of differential preservation during uplift and variable diffusion kinetics at high temperatures, then the consistently similar trend between matrix and inclusion white micas should not be seen. White micas are rather large in rocks with assemblage 3, and diffusion of Al, Si or Ti would involve complex substitutions and, hence, would also be rather slow. Furthermore, retrograde reaction to biotite is very limited in these rocks.

Retrograde *PT* data, including the growth of biotite in assemblages 3a and 3b, fall into the same range as for all other gneisses: *PT* coordinate C inFig. 15 was calculated in the same way as for assemblage 2 in Fig. 14 , whereas point D was derived by calculation of equilibria (17), (18) and (19) with rutile and ilmenite.

The calculation of water-dependent reaction (2) with average mineral compositions of an assemblage including phengite with Si=3.3 p.f.u. and a theoretical K-feldspar (curve 3 in Fig. 15) indicates that water activities have generally exceeded 0.4 in all rocks of assemblage 3 at maximum *PT*conditions and during decompression.

Hence those assemblages lacking K-feldspar were in fact the most hydrous assemblages during maximum *PT* conditions among the rocks studied. Kyanite was never found altered to sillimanite in rocks of paragenesis 3a within the entire unit, again indicating that most probably the entire retrograde *PT* path lay within the kyanite stability field.

CONCLUSIONS

Most quartzo-feldspathic and metapelitic rocks of the Gneiss-Eclogite Unit in the Central Erzgebirge were subjected to HP-HT metamorphism leading to ubiquitous prograde biotite breakdown and the formation of grossular-rich garnet in granitoid rock compositions in the uppermost stability field of albite, with phengite in part stable and in part absent in the presence of quartz. Thus metamorphism under maximum *PT* conditions occurred within the extremely wide transitional field of the granulite, eclogite and amphibolite facies. The lower limit of stability of such Ca-rich garnets in albite-bearing rocks of granitic composition is probably ~10–12 kbar at 700°C, as such maximum conditions prevailed in the underlying Red and Grey Gneiss Unit where granitoid orthogneisses mostly lack garnet (Klemm, 1995; Rötzler, 1995; Kröner *et al.*, 1995). Separation of the pressure and temperature maximum–where at all possible–was minor, indicating no major thermal relaxation after the pressure peak, i.e. exhumation and cooling were sufficiently rapid to prevent the temperature increase that would have occurred during any delay near maximum metamorphic conditions.

Variable water activity governed the stability of three different, predominantly biotite-free acid assemblages at peak metamorphic conditions. Kyanite granulites, which may be completely devoid of any hydrous phase (particularly phengite), are the rocks with the lowest water activity (<0.15), followed by rocks with phengite stable together with quartz, garnet, plagioclase and K-feldspar. These indicate intermediate water activities (0.15–0.4). Metapelitic rocks without K-feldspar but with stable phengite must have had water activities well exceeding 0.4. This means that water activity was strongly reduced only in some rocks, and variable in most others at peak metamorphic conditions. Preservation of such short-range differences in a penetratively extremely deformed rock unit with subsequent fluid influx is good evidence that later retrograde fluid flow was strongly channelized.

Variable availability of water must have had a strong influence on the preservation of relic deformation features such as the HT mylonite fabric. Preservation of this fabric was only possible owing to (1) strong reduction of prevailing water activity after formation and (2) short duration of high temperatures. The deformation responsible for the penetrative s_2 foliation of the Gneiss-Eclogite Unit may be directly observed in this preserved relic HT mylonitic fabric and occurred after formation of the peak assemblage, because all maximum *PT* phases can show signs of this deformation. By contrast, retrograde biotite in the HT mylonites is always undeformed, indicating that the D_2 deformation, characterized by extreme flattening, is associated with the strong, near-isothermal decompression that followed the equilibration stage at maximum *PT*. This is interpreted as large-scale, deep-seated stretching within overthickened crust during an early retrograde extensional event. Tectonic transport was in an east-west direction, as indicated by the consistent direction of the stretching lineation in the entire unit.

During this event, rocks representing different depths of burial must have been brought together. The maximum pressures of garnet peridotites and most eclogites well exceed those of most acid rocks. Even the maximum pressures within the metapelitic gneisses were shown to differ by up to 8–10 kbar.

Retrograde biotite growth is observed in all rocks to a highly variable degree at intermediate pressures and temperatures between 600 and 750°C. Estimated water activities increased to 0.4–1.0 during this stage. Retrograde formation of tonalitic and granitic partial melts is observed in widespread migmatitic gneisses at the wet solidus. Such features imply pronounced external fluid influx at this stage during uplift. However, fluid flow was strongly channelized, as indicated by the extreme variability of overprint within these rocks. Willner (1994) has shown that the local presence of Cl-rich biotites within rocks of the Gneiss-Eclogite Unit indicates the preservation of locally reduced water activities. This must be a sign of localized fluid flow, because any low fluid influx would be sufficient to homogenize Cl activity gradients.

The strong zoning observed in all phases is normally an unexpected feature in HT rocks. The strong decrease in Fe/(Fe + Mg) of garnet in contact with Fe-Mg phases such as phengite and biotite, along with the marked decrease in garnet grossular component in all rocks studied, is consistent with a distinctive decompression path that is steeper in d*P*/d*T*slope than the curves of constant Fe-Mg fractionation for Fe-Mg exchange reactions. The same effect is prominent in biotites from rocks still preserving evidence of low water activity (Fig. 9), in which the corresponding increase in Fe/(Fe + Mg) is combined with decreasing Si on the order of 0.2 p.f.u. Massonne (1991)has shown that such a decrease of Si in pure phlogopite in the presence of phengite, K-feldspar and quartz could correspond to an ~10 kbar in isothermal pressure release.

Late reversal rims in garnet, in which Fe/(Fe + Mg) once again increases, reflect predominant cooling until a critical closing temperature for volume diffusion was reached. These rims are very narrow and indicate that cooling after the initial near-isothermal decompression was very rapid. Thus the observed zoning in garnet is a feature largely due to diffusion at very high temperatures that was frozen-in because of later rapid cooling rates. The restriction of late retrograde re-equilibration to local rock domains is a further corroboration of this conclusion.

A particularly puzzling phenomenon, intimately connected with retrograde fluid infiltration, is the growth of a second grossular-enriched garnet generation after grossular decrease and garnet resorption of the first generation. The possibility of a further strong increase in pressure after an initial decompression must be rejected owing to (1) the lack of analogous zonation features in all other phases, (2) the lack of evidence for a second decompression and resulting resorption phenomena in second-generation garnet, and (3) the proven local equilibration between the rims of coexisting phases and the very Ca-rich plagioclase rims after the major first decompression of the rocks at transient medium-pressure conditions. External influx of Ca-rich hydrous fluids on a rather local scale may be proposed to explain Ca enrichment in the garnet rims of plagioclase-free and hence Ca-poor assemblage 3a, where Ca-richer fluids would come from surrounding plagioclase-bearing rocks. However, it would be an improbable explanation for garnet 2 growth in the most abundant plagioclase-bearing rocks, because there is no evidence for a possible source of Ca-richer fluids on a regional scale.

As a third possibility, formation of the second grossular-rich garnet might be due to (1) kinetic control of the differential rate of material availability for the growth of new minerals, while the first garnet was resorbed during isothermal decompression, and (2) garnet growth during subsequent rapid cooling in a markedly changed local domain bulk composition. It is obvious that only a small volume of the existing albite reacted at the outermost rims, and volume diffusion within the albite grains was extremely slow. Thus only a very restricted amount of albite in the rock was available for the retrograde reaction to oligoclase. In contrast, breakdown of the Fe-Mg components in garnet to form new biotite would have taken place readily at appropriate conditions after pressure release and increase in water availability. This would also have favoured very rapid dissolution of the unstable grossular component. The combination with the concomitant retarded formation of more An-rich plagioclase should have led to a transient enrichment of Ca in the fluid phase and hence to a transient Ca-rich environment in local domains of the rock. During rapid cooling after decompression grossular-forming reactions would be cut with a flat angle. Locally therefore, a new stable grossular-rich garnet grew again in local equilibrium with a high Ca-plagioclase showing also local equilibration with biotite and phengite in the range 720–570°C and 6–8 kbar (Fig. 13).

In summary, the early stage of uplift and retrograde *PT* evolution can be characterized as follows:

1. A HP-HT equilibration event in a region of low geothermal gradient was followed by a near-isothermal decompression of as much as 10 kbar with a minor temperature decrease of <100°C.
2. The formation of the prominent penetrative s_2 foliation is bracketed by the above HP-HT event and medium-pressure retrograde equilibration with external fluid influx. A strong stretching of the crust occurred as well as a juxtaposition of rocks subjected to different maximum, but invariably high pressures. Hence the entire rock pile represents a major extensional shear zone in the lower part of continental crust with excess crustal thickness. Very different levels of this lower crust were amagalmated during the D_2deformation. The process ended with a major influx of hydrous fluids at medium-pressure conditions, leading to pronounced, but variable, re-equilibration.
3. The uplift at high temperatures must have been very rapid, because (a) strong mineral zoning during HT decompression is mostly still preserved, (b) HT deformation fabrics are locally preserved in dry rocks and (c) no major thermal relaxation, i.e. temperature increase, occurred after the pressure maximum.
4. The second part of the *PT* path monitoring the final exhumation of the Gneiss-Eclogite Unit has a different geometry and is characterized by less pronounced pressure release and extensive cooling. Discrete ductile shear zones with normal fault kinematics radial to the pronounced dome structures (Fig. 1; Krohe, 1997) developed during cooling to below the 500°C isotherm at 2.5–4 kbar pressure, as indicated by recrystallized low-Si white micas and secondary fluid inclusions. There are no signs that this late *PT*evolution ever took place outside the kyanite stability field, in contrast to the retrograde evolution of HP granulites in the nearby Granulite Massif (Rötzler, 1992) or Lower Austria (Carswell & O'Brien, 1993). Very narrow retrograde re-equilibration rims in garnet are due to cooling and indicate a rapid transition from high temperatures near 700°C to closing temperatures of 550–600°C for diffusion processes in garnet. Although both markedly different branches of the *PT* path indicate different confining conditions, it is unlikely that there was a long time gap between these two episodes.

CONSTRAINTS FOR VARISCAN GEODYNAMICS

This publication contributes to the understanding of a rather specific group of granulites that are widespread along the Variscan chain of Europe from Spain to Poland, the so-called 'group I granulites' of Vielzeuf & Pin (1989). These restricted and scattered occurrences are strikingly similar, although from widely separated areas of the continent. GASP barometry in kyanite granulites yields maximum pressures ranging from 13 to 20 kbar. Associated with these felsic HP granulites are mafic granulites, garnet peridotites and/or eclogites. A retrograde amphibolite facies overprint producing dominant migmatitic gneisses is observed, but a later HT-LP overprint is generally lacking. In particular, many group I granulite units are known surrounding the Bohemian Massif [see review by O'Brien & Carswell (1993)], where *PT* evolutions comparable with those in the Central Erzgebirge were derived, e.g. in the Polish Sudetes (Sowie Gory:O'Brien *et al.*, 1996; Snieznik: Steltenpohl *et al.*, 1993; Kryza *et al.*, 1996), the Saxonian Granulite Massif (Rötzler, 1992)or in Lower Austria (Carswell & O'Brien, 1993). Although this type of metamorphism was regarded as mainly Early Variscan in Western Europe by Vielzeuf & Pin (1989), the age of the metamorphic peak in the Central European occurrences seems to range from early Devonian to early Carboniferous, e.g. 400 Ma in Sowie Gory (O'Brien *et al.*, 1996) and 340 Ma in the Saxonian Granulite Massif (Von Quadt, 1993). Because high-pressure granulites seem to be rare world-wide, according to the review by Harley (1989), HP granulitic rocks must be of special importance in the evolution of the Hercynian belt, indicating a widespread specific geotectonic environment (O'Brien & Carswell, 1993).

As in the above occurrences, maximum peak *PT* conditions in the Gneiss-Eclogite Unit were dated at 340 Ma (Kröner & Willner, 1995; Kotková *et al.*, 1996; Willner *et al.*, 1996; $^{207}Pb/^{206}Pb$ ages of metamorphic zircons). Similar ages of 333–360 Ma (Sm-Nd mineral-whole-rock isochrons) and 348–355 Ma (Ar-Ar plateau ages for phengites) were provided bySchmädicke *et al.* (1995), dating eclogites of the Gneiss-Eclogite Unit. Abundant Ar-Ar plateau ages of white micas and hornblende consistently range around 338 Ma in the Gneiss-Eclogite Unit, as in the units below and above, similar to Rb/Sr ages of white micas (Werner *et al.*, 1996). This overlap derived from isotopic systems with different closure temperatures requires extremely fast tectonic exhumation. Exhumation must have been terminated by 325 Ma at the latest, as shown by the first regional unconformity (Kurze, 1966). Hence all processes related to exhumation occurred during the peak of the Variscan orogeny and were very rapid, favouring tectonic thinning at rates on the order of those proposed by England & Thompson (1986). The high maximum pressures recorded indicate derivation of the quartzo-feldspathic rocks from the lower part of a crust thickened to >60 km along a geothermal gradient slightly exceeding 10°C/km, thus somewhat higher than commonly inferred for typical subduction metamorphism in accretionary prisms. From the age constraints, exhumation rates of at least 4 mm/yr result.

The preserved pressure-temperature-deformation evolution is a history of continuous retrograde extension responsible for exhumation. The same *PT*path, with its characteristic geometry, was also recognized by Kotková (1993) for granulites on the Czech side of the Erzgebirge, also indicating that the typical conditions prevailing in the Gneiss-Eclogite Unit were similar over a regional extent. A comparable *PT* path was also detected in the HP granulites of Lower Austria, indicating widespread similar geotectonic conditions (Carswell & O'Brien, 1993).

HP metamorphism is commonly masked by an amphibolite-facies overprint, but fortunately, except for the easternmost Erzgebirge, not by the HT-LP metamorphism typical for the Moldanubian region further south. Hydrous fluid influx leading to a general increase of water activity and finally to abundant retrograde anatexis at the wet solidus in the most overprinted gneisses cannot solely be produced by dehydration reactions during decompression (mostly involving breakdown of the Al-celadonite component), as proposed, for instance, by Heinrich (1982) for HP metapelites in the vicinity of eclogites. Although some reactions of this type certainly occur, as shown above, there should at least be a retrograde reduction of water activity in many acid rocks. However, changes in water activity could not be recorded in the K-feldspar-free assemblages, and the 'driest' conditions invariably occurred at peak metamorphic conditions in all K-feldspar-bearing rocks. On the other hand, if the HP unit were thrust onto another crustal unit undergoing prograde dehydration during uplift, another source for hydrous fluid influx would be provided.

Continued crustal stacking must, in fact, be expected, to exhume rocks from the lower part of an overthickened crust to the surface in a single orogenic cycle. By pure extension of the entire crust only, such rocks would remain in the lower part of a normal isostatically equilibrated crust (Ellis, 1987). In the case of the Central Erzgebirge, it was shown by Rötzler (1995) and Klemm (1995) that the underlying Red and Grey Gneiss Unit was subjected to intermediate maximum pressures only, and also to lower peak temperatures. Such tectonic insertion of the HP-HT unit into intermediate crustal levels would also provide the possibility of more rapid cooling during further uplift, as indicated by the second branch of the retrograde *PT* path of the Gneiss-Eclogite Unit. Towards the overlying Mica-Schist-Eclogite Unit there is also a marked '*PT* discontinuity' to much lower peak conditions (Rötzler, 1995). Hence, during decompression from medium-pressure to low-pressure conditions (second branch of the *PT*path), the entire HP-HT unit was probably cooled from below and above.

Therefore the overall position of the entire HP-HT Gneiss-Eclogite Unit is one of a dismembered subhorizontal nappe unit sandwiched between other crustal units of contrasting *PT* evolution by ductile deformation processes occurring during uplift into mid-crustal levels, but definitely not at a coherent regional suture zone. Also, the Erzgebirge crystalline complex cannot be the basement for the neighbouring Saxothuringian basin with unmetamorphosed to low-grade sediments, because sedimentation was still continuing in the basin when HP metamorphism occurred in the crystalline complex. Juxtaposition of these contrasting terranes can be explained by large-scale movements with lateral components during uplift of the crystalline complex, in particular when rocks were still within the ductile deformational regime (Krohe, 1997).

The processes deduced from the Gneiss-Eclogite Unit, which obviously occurred in similar widespread areas within Central Europe, point to a prolonged period of crustal stacking and crustal thickening of vast volumes of continental crust, followed by very rapid exhumation of HP slices from the lower part of the thickened crust. Such features are of continental extent and must be due to continuous continent-continent collision during the Variscan orogeny, rather than subduction at a convergent plate margin before continent-continent collision.

Table 5. Mineral abbreviations

Ab	albite	Ky	kyanite
Ace	Mg-aluminium celadonite	L	melt
Alm	almandine	Ms	muscovite
An	anorthite	Or	orthoclase
And	andalusite	Phe	phengite
Ann	annite	Phl	phlogopite
Bt	biotite	Plag	plagioclase
Cpx	clinopyroxene	Py	pyrope
Di	diopside	Qtz	quartz
Grs	grossular	Rt	rutile
Grt	garnet	Sil	sillimanite
Ilm	ilmenite	V	vapour
Jad	jadeite	WM	white mica
Kfs	potassic feldspar		

ACKNOWLEDGEMENTS

This research project was supported by grants of the Deutsche Forschungsgemeinschaft (Ma 689/6 and Ma 689/10), which are gratefully acknowledged. The paper is also a contribution to the DFG priority programme 'Orogenic processes with particular reference to the Variscides'. J. Kopp is thanked for first introduction to the field, initiation of the project and support in numerous other ways. Careful reviews by G. T. R. Droop and P. J. O'Brien substantially improved the manuscript.

REFERENCES

1. Behr H. J., Fritsch E., Mansfeld L. Die Granulite von Zöblitz im Erzgebirge als Beispiel für Granulitbildung in tiefreichenden Scherzonen. *Krystallinikum* 1965;3:7-29.
2. Bergner R. Vergleichende Untersuchungen an temperaturbetont metamorphisierten Gesteinen ausgewählter Einheiten des Saxothuringikums.Freiberger Forschungshefte 1990;C434:6-66.
3. Berman R. G. Internally-consistent thermodynamic data for minerals in the system Na_2O-K_2O-CaO-MgO-FeO-Fe_2O_3- Al_2O_3-SiO_2-TiO_2-H_2O-CO_2. Journal of etrology 1988;29:445-522.
4. Berman R. G. Mixing properties of Ca-Mg-Fe-Mn garnets. American Mineralogist 1990;75:328-344.
5. Bernhardt H. J., Massonne H.-J., Reinecke T., Reinhardt J., Willner A. P. Digital element distribution maps, an aid for petrological investigations. Berichte der Deutschen Mineralogischen Gesellschaft, Beiheft 1 zum European Journal of Mineralogy 1995;7:28.
6. Brown T. H., Berman R. G., Perkins E. H. Ge0-Calc: Software package for calculation and display of pressure-temperature-composition phase diagrams using an IBM or compatible personal computer. Computers and Geoscience 1988;14:279-289.
7. Carswell D. A., O'Brien P. J. Thermobarometry and geotectonic significance of high-pressure granulites: examples from the Moldanubian zone of the Bohemian Massif in Lower Austria. Journal of Petrology 1993;34:427-459.
8. Ellis D. J. Origin and evolution of granulites in normal and thickened crust. Geology 1987;15:167-170.

9. Ellis D. J., Green D. H. An experimental study of the effect of Ca upon garnet-clinopyroxene Fe-Mg-exchange equilibria. Contributions to Mineralogy and Petrology 1979;71:13-21.

10. England P. C., Thompson A. B. Some thermal and tectonic models for crustal melting in continental collision zones. Collision Tectonics. In: Coward M. P., Ries A. C., editors. Geological Society of London, Special Publication. Vol. 19. 1986. p. 83-94.

11. Freyer G., Hoth K., Sehnert M. Kenntnisstand bis 1990/91 über die Fossilführung im Bereich des Erzgebirges. Zeitschrift für geologische Wissenschaften 1994;22:555-565.

12. Fuhrman M. L., Lindsley D. H. Ternary feldspar modeling and thermometry. American Mineralogist 1988;73:201-215.

13. Ganguly J., Saxena S. K. Mixing properties of aluminosilicate garnets: constraints from natural and experimental data, and applications to geothermobarometry. American Mineralogist 1984;69:88-97.

14. Gasparik T., Lindsley D. H. Phase equilibria at high pressure of pyroxenes containing monovalent and trivalent ions. Pyroxenes. In: Prewitt C. T., editor. Reviews in Mineralogy, Mineralogical Society of America. Vol. 7. 1980. p. 309-339.

15. Gerstenberger H. Autometasomatic Rb enrichments in highly evolved granites causing lowered Rb-Sr isochron intercepts. Earth and Planetary Science Letters 1989;93:65-75.

16. Harley S. L. The origin of granulites: a metamorphic prospective. Geological Magazine 1989;126:215-247.

17. Heinrich C. A. Kyanite eclogite to amphibolite facies evolution of hydrous mafic and pelitic rocks, Adula Nappe, Central Alps. Contributions to Mineralogy and Petrology 1982;81:30-38.

18. Hodges K. V., Spear F. S. Geothermometry, geobarometry and the Al2SiO5 triple point at Mt. Moosilauke, New Hampshire. American Mineralogist; 1982;67:1118-1134.

19. Hofmann J., Mathe G., Wienholz R. Metamorphose und zeitliche Stellung tektonometamorpher Prozesse im östlichen Teil des Saxothuringikums. Zeitschrift für geologische Wissenschaften 1981;9:1291-1308.

20. Holdaway M. J. Stability of andalusite and the aluminium silicate phase diagram. American Journal of Science 1971;271:97-131.

21. Holland T. J. B. The reaction albite = jadeite + quartz determined experimentally in the range 600–1200°C. American Mineralogist 1980;65:129-134.

22. Hoth K., Lorenz W., Berger H. J. Die Lithostratigraphie des Proterozoikums im Erzgebirge. Zeitschrift für angewandte Geologie 1983;29:413-418.

23. Johannes W. The significance of experimental studies for the formation of migmatites. In: Ashworth J. R., editor. Migmatites. Glasgow: Blackie; 1984. p. 36-85.

24. Klápová H. Eclogites of the Bohemian part of the Saxothuringicum. Rozpravy Ceskoslovenske Akademie Ved, Rada Matematickych a Prirodnich Ved 1990;100:1-86.

25. Klemm I. 1995. Druck-Temperaturentwicklung von Metamorphiten im östlichen Erzgebirge (Sachsen); p. 121. Unpublished Diploma Thesis, Ruhr-Universität Bochum.

26. Kossmat F. Über die Tektonik des Gneisgebietes im westlichen Erzgebirge. Zentralblatt für Mineralogie, Geologie und Paläontologie 1916;1916:135-144.

27. Kossmat F. Erscheinungen und Probleme des Überschiebungsbaus im variskischen Gebirge Sachsens und der Sudetenländer. Zentralblatt für Mineralogie, Geologie und Paläontologie 1925;1925:348-359.

28. Kotková J. Tectonometamorphic history of lower crust in the Bohemian Massif example of north Bohemian granulites. Czech Geological Survey Special Papers 1993;2:42.

29. Kotková J., Kröner A., Todt W., Fiala J. Lower Carboniferous HP event in the Bohemian Massif: evidence from North Bohemian granulites. Geologische Rundschau 1996;85:154-161.

30. Koziol A. M., Newton R. C. Redetermination of the anorthite breakdown reaction and improvement of the plagioclase-garnet-Al2SiO5-quartz geobarometer. American Mineralogist 1988;73:216-223.

31. Krentz O., Thomas R., Wiedemann R. Aussagen zur Regionalmetamorphose des mittleren und westlichen Erzgebirges mit Hilfe thermobarometrischer Einschlußuntersuchungen. Zeitschrift für geologische Wissenschaften 1990;18:315-326.

32. Krohe A. Krustendehnung im Erzgebirge? Ansätze zu einem neuen Konzept. Zentralblatt für Geologie und Paläontologie 1992;1992:180. (abstract).

33. Krohe A. Variscan tectonics of Central Europe: postaccretionary intraplate deformation of weak continental lithosphere. Tectonics 1997;15:1364-1388.

34. Kroll H., Evangelakis C., Voll G. Two-feldspar geothermometry: a review and revision for slowly cooled rocks. Contributions to Mineralogy and Petrology 1993;114:510-518.

35. Kröner A., Willner A. P. Magmatische und metamorphe Zirkonalter für Quarz-Feldspat-Gesteine der Gneis-Eklogit-Einheit des Erzgebirges. Terra Nostra 1995;95((8)):112.

36. Kröner A., Willner A. P., Hegner E., Frischbutter A., Hofmann J., Bergner R. Latest Precambrian (Cadomian) zircon ages, Nd isotopic systematics and P-T evolution of granitoid orthogneisses of the Erzgebirge, Saxony and Czech Republic. Geologische Rundschau 1995;84:437-456.

37. Kryza R., Pin C., Vielzeuf D. High-pressure granulites from the Sudetes (south-west Poland): evidence of crustal subduction and collisional thickening in the Variscan Belt. Journal of Metamorphic Geology 1996;14:531-546.

38. Kurze M. Die tektonisch-fazielle Entwicklung im Nordostteil des sächsischen Lineamentes. Freiberger Forschungshefte 1966;C201:1-58.

39. Le Goff E., Ballèvre M. Geothermobarometry in albite-garnet orthogneisses: a case study from the Gran Paradiso nappe (Western Alps). Lithos 1990;25:261-280.

40. LeBreton N., Thompson A. B. Fluid-absent (dehydration) melting of biotite in metapelites in the early stages of crustal anatexis. Contribution to Mineralogy and Petrology 1988;99:226-237.

41. Lorenz W. (compiler). Metallogenetisches Kartenwerk 1:100 000. Geologische Karte Erzgebirge/Vogtland. Berlin: Zentralinstitut für Geologie; 1982.

42. Lorenz W., Hoth K. Lithostratigraphie im Erzgebirge–Konzeption, Entwicklung, Probleme und Perspektiven. Abhandlungen des Staatlichen Museum für Mineralogie und Geologie zu Dresden 1990;37:7-35.

43. Massonne H.-J. 1991. High-pressure, low-temperature metamorphism of pelitic and other protoliths based on experiments in the system K2O-MgO-Al2O3-SiO2-H2O; p. 172. Unpublished Thesis of Habilitation, Ruhr-Universität Bochum.

44. Massonne H.-J. Thermochemical determination of water activities relevant to eclogitic rocks. Water-Rock Interaction. In: Kharaka Y. K., Maest A. S., editors. Proceedings of the 7th International Symposium on Water-Rock Interaction, Park City, Utah. Rotterdam: Balkema; 1992. p. 1523-1526.

45. Massonne H.-J. P-T evolution of eclogitic lenses in the crystalline complex of the Erzgebirge, Middle Europe: an example for high-pressure to ultrahigh-pressure metabasites incorporated into continental crust. Extended abstract 1. Workshop on Ultra-high-pressure Metamorphism and Tectonics, Stanford, CA, USA. Stanford: School of Earth Sciences; 1994:A29-A32.

46. Massonne H. J. Experimental and petrogenetic study of UHPM. In: Coleman R. G., Wang X., editors. Ultrahigh Pressure Metamorphism. Cambridge: Cambridge University Press; 1995. p. 33-95.

47. Massonne H.-J., Grosch U. P-T evolution of Paleozoic garnet peridotites from the Saxonian Erzgebirge and the Aheim region, W. Norway. In: Extended abstract, Sixth International Kimberlite Conference, Novosibirsk. Stuttgart: Schweizerbart; 1995. p. 353-355.

48. Massonne H.-J., Grosch U., Willner A. P. Geothermobarometrie mittels Ti-Gehalten in Kalihellglimmern. Berichte der Deutschen Mineralogischen Gesellschaft, Beiheft 1 zum European Journal of Mineralogy 1993;5:85.

49. McMullin D. W., Berman R. G., Greenwood H. J. Calibration of the SGAM thermometer for pelitic rocks using data from equilibrium experiments and natural assemblages. Canadian Mineralogist 1991;29:889-908.

50. Mehnert K. R. Migmatites and the Origin of Granitic Rocks Amsterdam: Elsevier; 1968.

51. O'Brien P. J., Carswell D. A. Tectonometamorphic evolution of the Bohemian Massif: evidence from high pressure metamorphic rocks. Geologische Rundschau 1993;82:531-555.

52. O'Brien P. J., Kröner A., Jaeckel P., Hegner E., Zelazniewicz A., Kryza R. Petrological and isotopic studies on Palaeozoic high pressure granulites with medium pressure overprint, Góry Sowie (Owl) Mts., Polish Sudetes. Journal of Petrology 1997. (in press).

53. Reinisch R. Erläuterungen zur Geologischen Karte von Sachsen, Nr. Leipzig: Geol. Landesamt Sachsen; 1929. 129, Blatt Zöblitz, 2nd edition [1st edition by Hazard (1884)].

54. Rötzler J. Zur Petrogenese im sächsischen Granulitgebirge. Die pyroxenfreien Granulite und die Metapelite. Geotektonische Forschungen 1992;77:1-100.

55. Rötzler K. Die PT-Entwicklung der Metamorphite des Mittel- und Westerzgebirges. 1995:220. Scientific Technical Report STR95/14, Geoforschungszentrum Potsdam.

56. Rötzler K., Schumacher R., Maresch W. V. PT-Entwicklung in Glimmerschiefern und Orthogneisen des Westerzgebirges. Berichte der Deutschen Mineralogischen Gesellschaft, Beiheft 1 zum European Journal of Mineralogy 1993;5:257.

57. Scheumann K. H. Die Rotgneise der Glimmerschieferdecke des sächsischen Granulitgebirges. Abhandlungen der sächsischen Akademie der Wissenschaften, mathematisch-naturwissenschaftliche Klasse 1935;87:251-286.

58. Schmädicke E., Okrusch M., Schmidt W. Eclogite-facies rocks in the Saxonian Erzgebirge, Germany: high pressure metamorphism under contrasting P-T conditions. Contributions to Mineralogy and Petrology 1992;110:226-241.

59. Schmädicke E., Mezger K., Cosca M. A., Okrusch M. Variscan Sm-Nd and Ar-Ar ages of eclogite facies rocks from the Erzgebirge, Bohemian Massif. Journal of Metamorphic Geology 1995;13:537-552.

60. Schreurs J. Prograde metamorphism of metapelites, garnet-biotite-thermometry and prograde changes of biotite chemistry in high grade rocks of West Uusimaa, Southwest Finland. Lithos 1985;18:69-80.

61. Sebastian U. 1995. Die Strukturentwicklung des spätorogenen Erzgebirgsaufstiegs in der Flöhazone–Ein weiterer Beitrag zur postkollisionalen Extension am Nordrand der Böhmischen Masse; p. 100. Doctoral Thesis, TU Bergakademie Freiberg.

62. Seltmann R., Bankwitz P., Hösel G. Structural environment of tin granites in the Erzgebirge. In: Pagel M., Leroy J. L., editors. Source, Transport and Deposition of Metals. Rotterdam: Balkema; 1991. p. 493-496.

63. Steltenpohl M. G., Cymerman Z., Krogh E. J., Kunk M. J. Exhumation of eclogitized continental basement during Variscan lithospheric delamination and gravitational collapse, Sudety Mountains, Poland. Geology 1993;21:1111-1114.

64. Vielzeuf D., Pin C. Geodynamic implications of granulitic rocks in the Hercynian belt. Evolution of the Metamorphic Belts. In: Daly J. S., Cliff R. A., Yardley B. W. D., editors. Geological Society Special Publications. Vol. 43. 1989. p. 343-348.

65. Von Quadt A. The Saxonian Granulite Massif: new aspects from geochronological studies. Geologische Rundschau 1993;82:516-530.

66. Werner O., Hess J. C., Lippolt H. J. Das variscische Abkühlungsmuster des Westerzgebirges–Ergebnisse isotopischer Mineraldatierungen. Terra Nostra 1996;96((2)):192-196.

67. Wienholz R., Hofmann J., Matte G. Über Metamorphose, Tiefenbau und regionale Position des Erzgebirgskristallins. Zeitschrift für geologische Wissenschaften 1979;7:385-395.

68. Willner A. P. Chlor in metamorphen Biotiten–Ein Überblick. Berichte der Deutschen Mineralogischen Gesellschaft, Beiheft 1 zum European Journal of Mineralogy 1994;6:312.

69. Willner A. P., Rötzler K., Maresch W. V., Krohe A., Kopp J. Hochdruckrelikte und Druck-Temperatur-Entwicklung in Orthogneisen des mittleren Erzgebirges. Berichte der Deutschen Mineralogischen Gesellschaft, Beiheft 1 zum European Journal of Mineralogy 1993;5:256.

70. Willner A. P., Rötzler K., Krohe A., Maresch W. V., Schumacher R. Druck-Temperatur-Deformations-Entwicklung verschiedener Krusteneinheiten im Erzgebirge. Eine Modellregion für die Exhumierung von Hochdruck-Gesteinen. Terra Nostra 1994;94((3)):104-106.

71. Willner A. P., Klemm I., Rötzler K., Maresch W. V. Petrologische Belege aus dem mittelkrustalen Niveau für eine variskische Krustenstapelung im Erzgebirge. Terra Nostra 1995;95((8)):138.

72. Willner A. P., Kröner A., Teufel S. Time of formation, peak of HP-metamorphism and cooling history of quartz-feldspar rocks from the Central Erzgebirge (Saxony/Germany). International Goldschmidt Conference, Heidelberg. Journal of Conference Abstracts 1996;1:675.

3

CHAPTER

Geological Exploration of Marble Deposits in Toto Area, Nasarawa State, Nigeria

Ali Moumouni[1], Nathaniel Goeter Goki[2], Mohammed Suleiman Chaanda[3]

[1]*Université Dandicko Dankoulodo de Maradi, Maradi, Niger*

[2]*Nasarawa State University, Keffi, Nigeria*

[3]*School of Geography, Earth and Environmental Sciences, University of Plymouth, Plymouth, UK*

ABSTRACT

Geological mapping as well as chemical analysis for marble deposit in the Muro-Obugu-Panda area of Toto Local Government, Nasarawa State, Nigeria, was carried out. The objectives of the study are to first map out the extent of the deposit and then evaluate its chemical characteristics. The geology of the area forms part of the Upper Proterozoic low-grade (green schist facies) schist belts of Central Nigeria, and consists of schists, banded iron, quartzite, and marble. Geological mapping revealed that the marble deposit stretches in a NE-SW direction, from Muro in the north to Kolo in the south, covering about 28 km^2. This deposit consists mainly of silicified gray and white varieties of marble. The mapping also identified three (3) exploration targets A—Panda, B—Obugu, and C— Soka. Chemical analysis of eight (8) selected samples shows that CaO content varies from 3.72 to 53.12 wt%. Other impurities are Fe_2O_3, Al_2O_3, MgO and SiO_2. Plot of Na_2O + K_2O vs. SiO_2 shows a positive correlation while that of CaO and MgO exhibits a well defined negative correlation with SiO_2 and this reflects the introduction of chert during the deposition of the parent limestone of the marble body. The variation plot of CaO vs. MgO shows an inverse relationship. This is reflected in the inverse relationship in the proportion of calcite to dolomite in the marble. All these results correlate with those found in the adjacent area by previous workers and this indicates the same extension of marble in these areas. The Muro marble can be used economically as raw materials for a variety of products such as fillers, glass, papers, lime, pesticide and sewage treatment, decorative construction, monuments, paint making and most of all very suitable for cement production.

Keywords: Marble, Muro, Obugu, Panda-Toto, Geochemistry

1. INTRODUCTION

Toto area of Nasarawa State, Nigeria, is endowed with an extensive marble deposit associated with schist, quartzite and banded iron formation. The area covers Katakwa sheet 228 NE, SW, NW, and SE 1:50,000. It falls within longitude 7°11' - 7°18' and latitude 8°7' - 8°23' (Figure 1).

It is accessible through Gadabuke-Muro-Soka and Toto-Ugya-Kolo-Panda-Ayiwawa-Ladaya feeder roads, which have their major linkages from Keffi-Nasarawa-Gadabuke-Toto and from Abuja-Abaji-Toto highway roads. The vegetation is the savannah type with thick forest and tall grasses. There exist streams where some form the major tributaries of River Benue. Geologically, the study area belongs to the schist belt of Nigeria. The major rock types include schists, banded iron, quartzites, and marble. Of particular interest is the extensive occurrence of marble in this area where mining at Soka, Obugu, Ayiwawa, Panda and Kolo near Ugya dates many years back. The objective of this study is to produce a geological map so as to delineate the extent and disposition of the marble deposit and then also to characterize its chemical composition being the first of its kind, hence this would subsequently determine the primary and secondary end user that would require it as a form of raw material or otherwise.

2. MATERIEL AND METHOD

In order to carry out the study, materiel such as a topographic map: Katakwa Sheet 228, NW, NE, SW and SE at 1/50,000 scale; a Global Positioning System (GARMIN 12 Model); a compass clinometer (SILVA); a geological hammer; some sample bags; a digital camera; a field note book; some writing materials were acquired for a successful conduct of the fieldwork.

The first stage of the field mapping on a scale of 1:50,000 involved traversing both across and along the strikes of the lithologic units. This was to have a full picture of the extent of the marble occurrence in the area. Coordinates, strikes and dips were recorded at each location. Samples were collected and properly labeled. The second stage involved a rigorous and systematic synthesis of all geological data which include lithologies, measured dip and strikes of foliation planes and all structural dispositions of the marble deposit and its relationship with the other rocks. The collected data consequently were plotted unto a GIS-based digitization packages, using Idrisi 32 and ArcView GIS 3.2a. The sheets covering the area were first scanned, cropped and exported in acceptable formats, and latter used to capture all the layers to build the geological map.

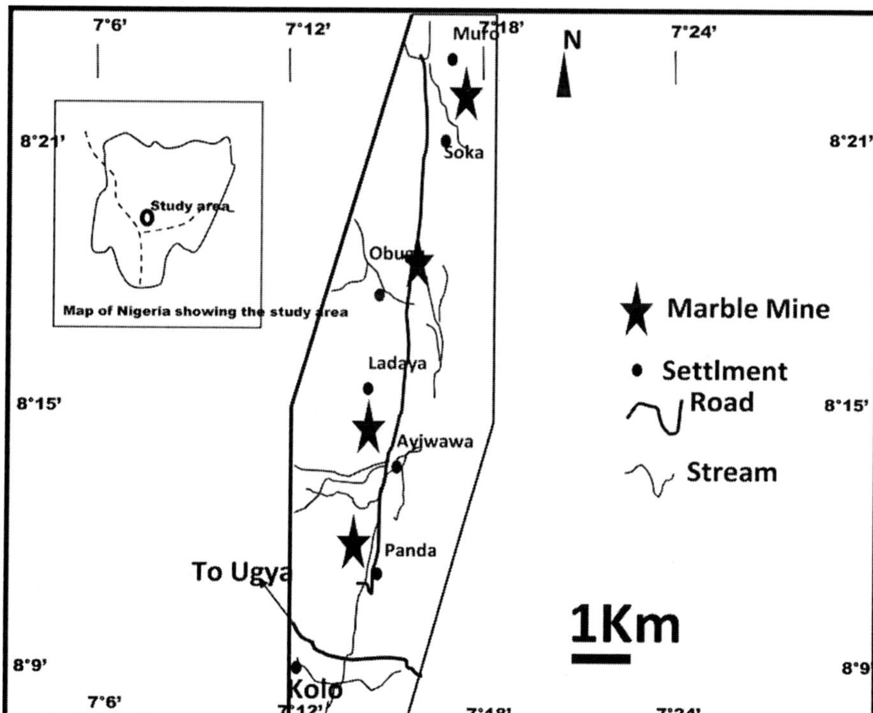

Figure 1. Location map.

The chemical analysis of eight (8) representative marble samples was carried out at the Obajana Cement Company Laboratory, using X-Ray fluorescence Spectrometry where oxides of Calcium, magnesium, iron, sodium and. The Loss of Ignition (LOI) was also determined for the same samples. The last stage involved data interpretation and report writing.

3. RESULTS AND DISCUSSION

3.1. Geology

Geologically, the study area belongs to the Schist Belts province, which is well developed in the western Nigeria, stretching from the northwestern, central to southwestern Nigeria [1] . The lithological variations of the schist belts include coarse to fine grained clastics, pelitics schists, phyllites, banded iron formation, carbonate rocks (marbles/dolomitic marbles) and mafic metavolcanics (amphibolites). References [2] and [3] , based on structural and lithological associations; suggest that there are different ages of sediments. However, [4] disagree with this conclusion and show that both series contained identical deformational histories.

Toto area falls specifically on the Central Schist Belt of Nigeria and the geology forms part of the Upper Proterozoic low-grade (green schist facies), which consists of phyllitic schist, quartzite, marble, banded iron formation and sandstone covering the solid geology in the some parts of the area [1] [5] . The northwestern and southwestern belts are known to be associated with gold, believed to be sourced from deep-seated faults and shear zones, while the central belts, though lacking in substantial gold, are known for supracrustal rocks associated with quartzitic banded irons interlayed with metamorphic marble stretching from Jakura Marble in Kogi State currently mined for cement production, to the Muro area marble in Nasarawa State. The marble occurs associated with steeply dipping quartzites, schists, banded iron and gneisses with prominent quasi-N-S trending strike directions and steep dips.

The area investigated consists of almost north-south to NE trending metasedimentary series including phyllitic schist, quartzite, banded iron, marble and undeformed sandstones (Figure 2). The quartzites, banded irons form elongate ridges while the marble and schists appear low-lying, except in areas where the deformation and folding have upthrusted them producing near vertical to steeply dipping foliation surfaces. Most foliation surfaces measured on all the lithologies dip to the west and north-west and in some case vertical, signifying probably that the whole complex structure may be a part of an arm of an anticlinal structure as evidenced by a corresponding eastward dip in the far western flank around Yerwa on quartzite and phyllitic schists. The following lithologies were identified.

3.1.1. Schists

The schists occupy the low-lying area mostly, but could be found uplifted and associated with quartzites on the ridges. They consist mineralogically of quartz-mica and phyllitic varieties and in some cases carbonaceous. The mica is mostly biotite and at contact with quartzite, may be rich in muscovite. Similar schists zones in Nigeria that are studied extensively are those of Igarra area [6] - [9] . Systematic mapping could not map visible contacts between the varieties due to deformation and extensive weathering. Foliation surfaces are very visible on these rocks with a general N-S strike and westward dips.

3.1.2. Quartzite

Granoblastic quartzite outcrops as elongate ridges consistent with the general structural grain alignment of the area with a near N-S and N-W strike, the whitish and light colored types predominate except in cases where the iron and dark colored mica are present. Texturally, the clearly banded quartzites consist of equigranular quartz and muscovite. The main ridges outcrop north of Muro Hill traceable along strike right down to Panda area. In places they have been disjointed by post emplacement NW-SE brittle deformations possibly explaining the general dismembered outcrops of iron rich quartzites. A sedimentary origin is corroborated by these texture and fabric but in places specially at stream near Obugu, highly jointed outcrops appear to be sedimentary in origin, signifying possibly that the metamorphism could have been related to hydrothermal activity during which large quartz vein have been emplaced and subjected to post metamorphic deformative phase.

3.1.3. Banded Iron

The iron formation occurs as bands of iron oxides and granular quartzite mostly in the Muro Hill and traceable along strike to Ayiwawa area. They have been faulted and distorted as shown on the geological map (Figure 4). Banding include a few millimeters to centimeters of iron oxides (hematite) intercalated with gray to slightly yellowish quartz (including muscovite) band as a result of weathering.

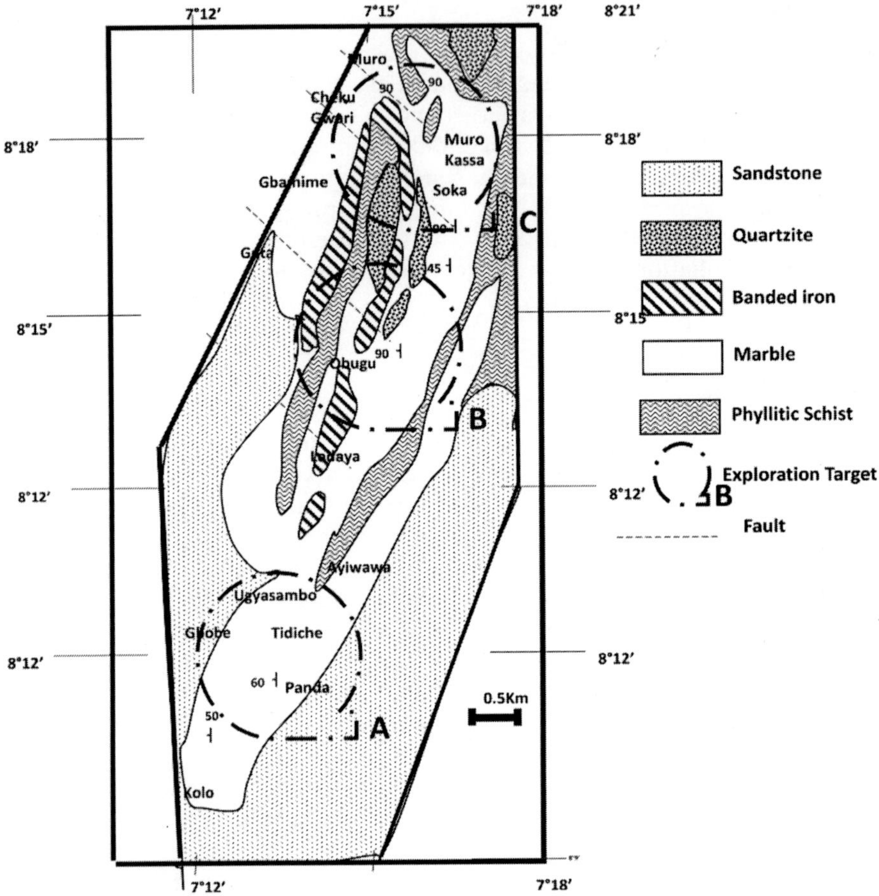

Figure 2. Geological map of the study area with exploration targets identified.

3.1.4. Sandstone

Continental sandy sediments are overlain on the metasediments towards the south (Figure 2). The sandstones in hand specimen appear highly ferruginized in some places but further away from the ridges, they are normal brown to light coloured.

3.1.5. Marble Deposits

Marble deposits cover a very wide expanse with limits starting from north of Muro and Soka with an average lateral width of about 4 kilometers stretching southwards for a distance of about 7 kilometers terminating around Kolo in the south of the project area (Figure 2). Mapable varieties of the marble include the gray and white which evidently are the most predominant. Silicified gray types appear impure and at contact with the phyllitic rocks, visible laminae of biotite and quartz can be seen. Two main types of marble can be distinguished in the study area. These include:

1) Silicified Gray Marble

The silicified gray variety in the project area is typically banded and complexly folded. Banding ranges from a few millimeters to 10 centimeters. It is by far the widest spread and outcrop with a consistent near N-S strike. The bands alternate with the white variety, which in places could be as thick as 3 meters. Near Soka, a N-S trending gray bands marble vertically dipping 8 meters thick outcrops.

Structures include several pinch and swell structures boudinage and various types of folds including open and isoclinal folds, chevron and drag folds (Figure 3). Compositionally, the gray marbles of this area are silicified and in some places they grade into the schist.

2) White Marble

White marble outcrops as thick layers interbanded with silicified gray varieties. It is predominant around Obugu through Ayiwawa, and Panda-Ugya areas where deep mined out pits reveal sections of over 10 meters with shallow overburden thicknesses of lower than 2 meters (Figure 4). At Obugu, a very sharp contact exists between the white type eastward of about 200 meters of exposed deposit apparently less deformed and the gray type to the west about 500 meters thick complexly deformed gray with layers vertically dipping. At Panda, large deposits of the white type are exposed with a more crystalline texture. Structures are rare on the white type but the thick bands are consistently north south trending and with vertical dips.

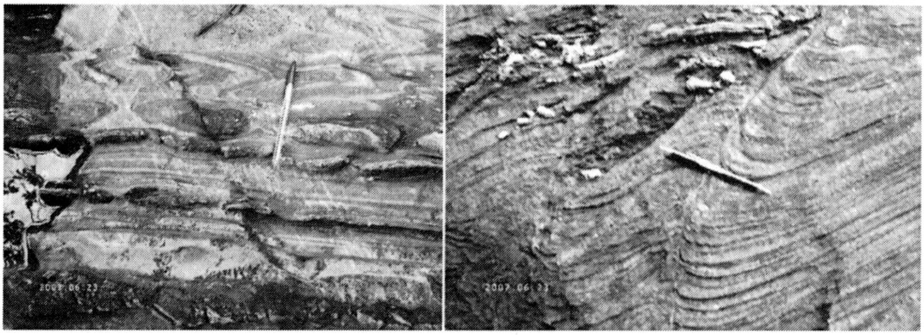

Figure 3. Pinch/swell and chevron/drag folds structures on the silicified marble in Muro.

Figure 4. Mine of white marble in the study area.

3.2. Geochemistry

Analytical Results

Eight samples (6 of white and 2 of dark marble) were collected and analyzed using X Ray Fluorescence Spectrometry. The results are presented in Table 1.

3.3. Discussion

In order to assess the quality of the Muro marble, there is need to look at the amount of CaO, MgO Fe_2O_3, MgO, and SiO_2. Iron and magnesium are admixed in the structural lattice of calcite as well as calc-silicate minerals. The theoretical composition of calcium carbonate is 56 wt% CaO and 44 wt% CO_2 [10] . On the other hand [11] indicated that pure carbonates are believed to have a total carbonate content ($CaCO_3$ or $CaMg(CO_3)_2$ of 70% and above while impure carbonates have between 40% - 70%. Other oxides like SiO_2, TiO_2, MnO, and Na_2O are usually less than 1%, and are regarded as constituting impurities %) and our present data set has some elements of similarity with the results obtained in the Gneiss and Marble from Igarra, Southwest of Nigeria [12] .

In most carbonates of metamorphic terrain, CaO and MgO are the most important elemental oxides. In calcitic type marble, CaO is usually in the order 50% - 54% while MgO is <15%. Dolomitic or Magnesian type marble on the other hand have CaO values generally in the range of 28% - 31% and MgO values in the range of 15% - 21% [13] .

Chemical analysis of the Muro marble shows CaO content ranging from 33.00 wt% to 53.12 wt% for the white marble and 3.72 to 5.12 wt% for the dark marble, meaning all CaO values fall below the theoretical value.

Evaluating chemical purity is important when assessing the marble's suitability for some specific uses. For example, very high chemical purity is required for filler applications. On the other hand, marble is widely used as a dimensional stone despite containing large amounts of fine grains of some sulphides (e.g. finely disseminated pyrite or marcasite). According to data of Table 1, the chemical composition of most Muro marble is not commercially suitable for uses other than a dimensional stone. The calcite content that has been calculated from the XRF analysis shows no sample has more than 97% calcite. Therefore, the Muro marble is not suitable at its natural state for use in the paint and plastic industry. Rather it can be used as dimension stones as shown on nearby house in the study (Figure 5, Panda). Also, Muro marble when polished can used for making tiles which is subsequently used in flooring homes, offices, industries and corporate bodies to give an aesthetic look. The polished are more costly than the unpolished.

Theoretically, the LOI of pure calcium carbonate is equal to 44 wt% of carbon dioxide. The L.O.I. represents mainly carbon dioxide, water or other volatile components, according to the purity of the calcium carbonate. Hence, it is a very useful applicable way that is used in confirming the CaO content and calculating its $CaCO_3$ content. The loss on ignition values range from 1.07 to 43.63 wt%, which means the proportion of the impurities, is variable with the marble type. The white type marble is hence purer than the gray one. And among the white ones samples S3, S2 and S1 are the purest respectively. The internationally recommended loss on ignition (LOI) value is around 40 wt% [14]. The white marble is the only type that fulfils the specifications for use as filler in paper and other industries.

An attempt to subject the data to a regression analysis on the geochemical data has revealed the following results: the plot of Na_2O + K_2O vs. SiO_2 (Figure 6) shows a positive correlation while that of CaO and MgO (Figure 7, Figure 8) exhibit a well defined negative correlation with SiO_2 and this reflects the introduction of chert during the deposition of the parent limestone of the marble body. The variation plot of CaO vs. MgO (Figure 9) shows an inverse relationship. This is reflected in the inverse relationship in the proportion of calcite to dolomite in the marble. All these results correlate with those found on the Itobe Marble by [15] in the adjacent area. This indicates the same extension of marble in the two areas.

Table 1. XRF results of the Muro marble.

S/No.	Sample	Chemical Parameters								
		SiO_2	Al_2O_3	Fe_2O_3	CaO	MgO	K_2O	Na_2O	S	LOI
S—Samples or white Marble										
1	S1	2.17	0.56	0.34	39.91	16.57	0.14	0.22	0.23	40.56
2	S2	0.31	0.14	0.11	52.16	4.12	0.05	0.03	0.09	43.20
3	S3	0.20	0.02	0.08	53.12	3.06	0.02	0.01	0.03	43.63
4	S4	22.69	4.52	2.23	33.00	11.09	0.61	0.98	0.02	25.80
5	S5	20.95	5.06	2.15	35.62	5.87	0.86	0.76	0.04	29.12
6	S6	12.57	2.91	0.93	46.13	4.86	0.49	0.15	0.20	31.80
H—Sample or silicified dark										
7	H1	80.65	5.96	0.62	5.12	0.28	1.08	0.02	0.01	6.40
8	H2	88.52	4.65	1.00	3.72	0.09	0.74	0.02	0.19	1.07

The protolith of the Muro marble like the Itobe marble may be a relatively pure limestone in which dolomite is present in varying proportions as found by [15]. It was most probably been deposited within shallow narrow basins which were initiated in structurally controlled depressions. The associated MgO in the marble must have been co-precipitated with the Ca from hyper-saline waters under anoxic conditions with micro-fauna and flora playing a vital role. The Muro marble occurs in association with Precambrian basement rocks comprising biotite schist, mica schist, quartz-schist, biotite hornblende schist, feldspathic quartzite, quartzitic schist, gneissic granite and minor intrusive rocks including pegmatite and quartz veins and based on the previous studies in adjacent areas (e.g. [15] - [18] etc.) in adjacent areas we can conclude that the Muro marble of metasedimentary origin. Similarly like the marble deposits in adjacent areas, the Muro marble can be used economically as raw materials for a variety of products such as fillers, glass, papers, lime, pesticide and sewage treatment, decorative construction, monuments, paint making and most of all very suitable for cement production.

Further work on the physical properties example porosity, specific gravity, shear strength, tensile and compressive strength, hardness and bulk density of the marble should be carried out which would be used to constrain its absolute resilience and durability before selection of its industrial usage.

Figure 5. Use of marble for decoration at a house in the study area.

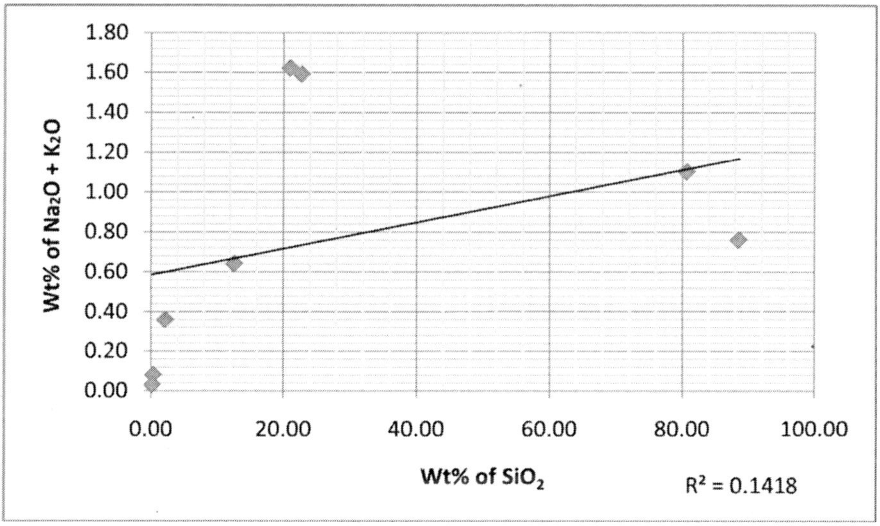

Figure 6. Plot of wt% $Na_2O + K_2O$ vs. SiO_2.

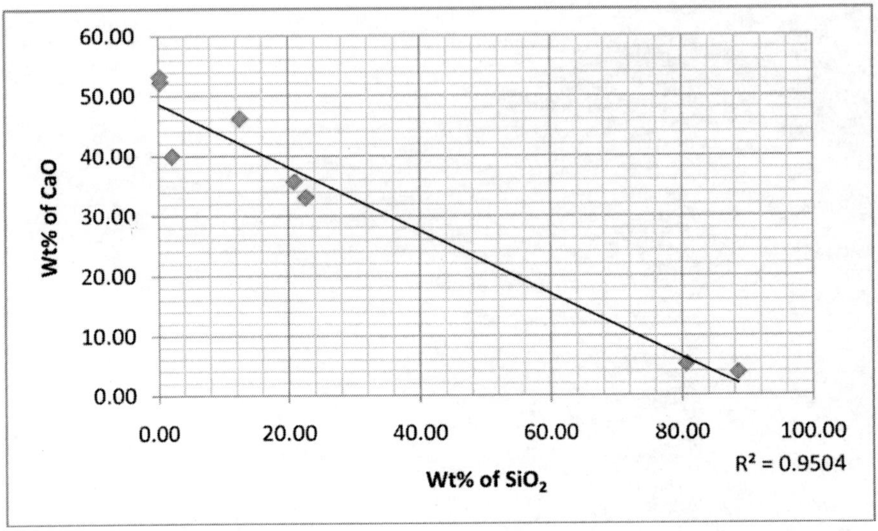

Figure 7. Plot of wt% CaO vs. SiO_2.

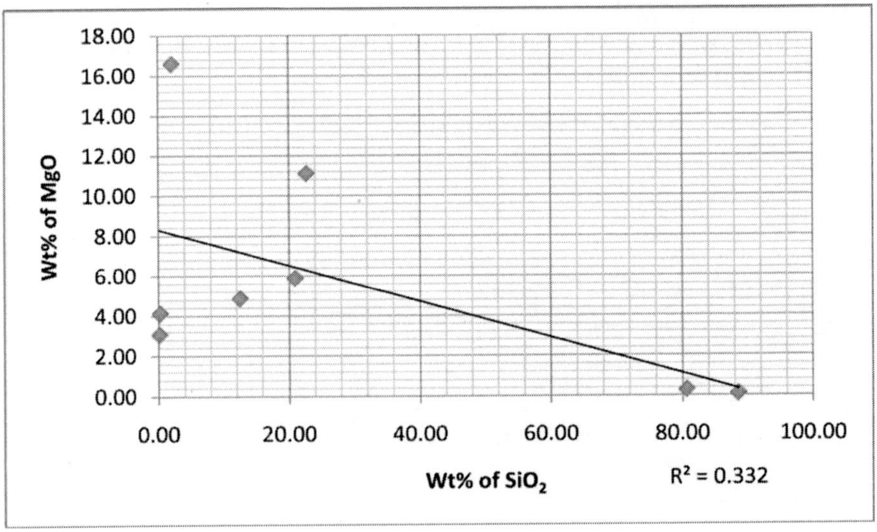

Figure 8. Plot of wt% MgO vs. SiO_2.

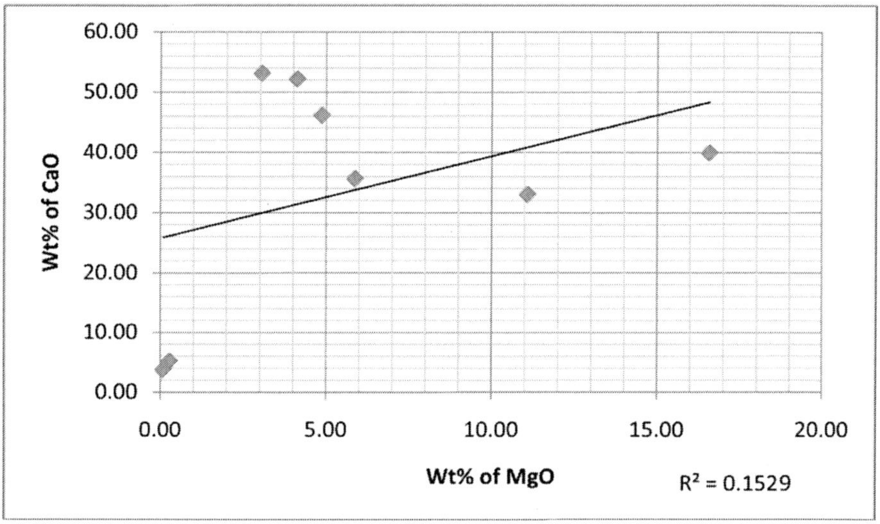

Figure 9. Plot of CaO vs. MgO.

4. CONCLUSION

Massive marble deposits of Toto area has been investigated through mapping and chemical characterization. The mapping shows that there are three (3) exploration targets successively around Panda, Obugu, and Soka. Chemical analysis shows that the marble deposits can be used economically as raw materials for a variety of products such as fillers, glass, papers, lime, pesticide and sewage treatment, decorative construction, monuments, paint making and most of all very suitable for cement production.

REFERENCES

1. Ajibade, A.C. (1976) Provisional Classification and Correlation of Schists Belts in Northwestern Nigeria. In: Kogbe, C.A., Ed., Geology of Nigeria, Elizabethan Pub. Co., Lagos, 88-90.

2. Grant, N.K. (1970) Geochronology of Precambrian Basement Rocks from Ibadan, South-Western Nigeria. Earth and Planetary Sciences Letter, 10, 19-38.

3. Turner, D.C. (1983) Upper Proterozoic Schist Belt in the Nigerian Sector of Pan African Province of West Africa. In: Kogbe, C.A., Ed., Geology of Nigeria, 2nd Revised Edition, Rock-view Limited, Jos, Nigeria.
4. Ajibade, A.C., Fitches, W.R. and Wright, J.B. (1979) The Zungeru Mylonites, Nigeria: Recognition of a Major Unit. Revue de Géologie et Géographie Physique, 21, 359-363.
5. Jones, H.A. and Hockney, R.D. (1964) The Geology of Parts of Southwestern Nigeria. Bulletin, Geological Survey of Nigeria, 31, 101.
6. Rahman, M.A. (1992) Precambrian Geology of Nigeria. Proceedings of High Grade Terrains Benin-Nigeria Traverse, Programme and Lecture Series, 150-200.
7. Odeyemi, I.B (1988) Lithostratigraphic and Structural Relationships of the Upper Precambrian Metasediments in Igarra Area, Western Nigeria. The Precambrian Geology of Nigeria, Geological Survey of Kaduna, 111-123.
8. Ekweme, B.N. (1990) Petrology of Southern Obodu Plateau Bamenda Massif, Southeastern Nigeria. In: Rocci, G. and Deschamps, M.., Eds., Recent Data in African Earth Sciences, CIFEG Occasional Publication, 22, 155-158.
9. Ekweme, B.N. (2000) Zircon Evaporation Ages from Obudu Plateau First Evidence of Archean Components in the Schist of Southeastern, Nigeria. 37th Nigerian Mining and Geosciences Conference, Jos, Book of Abstracts.
10. Talal, M.Q. (2000) Testing Jabal Farasan Marble Deposit for Multiple Industrial Applications. Arabian Journal for Science and Engineering, 33, 79-97.
11. Lipmann, F. (1973) Sedimentary Carbonate Minerals. Springer Verlag, Berlin Heidelberg, New York, 229 p.
12. Obasi, R.A. (2012) Geochemical and Appraisal of the Economic Potential of Calc-Gneiss and Marble from Igaara, Edo State, Southwest, Nigeria. Journal of Science and Technology, 2, 1018-1021.
13. Goldschmidth, J.R., Graff, L. and Joensu, O.I. (1955) The Occurrence of Magnesium Calcite in Nature. Geochimica et Cosmochimica Acta, 1, 212-230.
14. Boynton, R.S. (1979) Chemistry and Technology of Lime and Limestone. 2nd Edition, Wiley-Interscience, New York, 95-158.

15. Onimisi, M., Obaje, N.G. and Daniel, A. (2013) Geochemical and Petrogenetic Characteristics of the Marble Deposit in Itobe Area, Kogi State, Central Nigeria. Advances in Applied Science Research, 4, 44-57.

16. Okunlola, O.A. (2003) Geological and Compositional Investigation of Precambrian Marble Bodies and Associated Rocks in the Burum and Jakura Areas, Nigeria. Unpublished PhD Thesis, University of Ibadan, Ibadan, 250 p.

17. Emofurieta, W.O. and Ekuajemi, V.O. (1995) Lime Products and Economic Aspects of Igbetti, Ososo and Jakura Marble Deposits in Southwest Nigeria. Journal of Mining and Geology, 31, 89-97.

18. Elueze, A.A. and Okunlola, O.A. (2003) Compositional Features and Industrial Appraisal of the Metamorphosed Carbonate Rocks of Burum and Jakura Area, Central Nigeria. Mineral Wealth, 128, 41-54.

4

CHAPTER

The Lateral Boundary of a Metamorphic Core Complex: The Moutsounas Shear Zone on Naxos, Cyclades, Greece

Shuyun Cao[a], Franz Neubauer[a], Manfred Bernroider[a], Junlai Liu[b]

[a] Dept. Geography and Geology, University of Salzburg, Hellbrunnerstr. 34, A-5020 Salzburg, Austria

[b] State Key Laboratory of Geological Processes and Mineral Resources, China University of Geosciences, Xueyuan Rd. 29, Haidian, Beijing 100083, China

HIGHLIGHTS

- The Moutsounas shear zone bounds tilted Miocene sediments in eastern Naxos.
- Top-to-the-NNE shear associated with retrograde metamorphism (overprinting H-T deformation).
- Ductile and brittle deformation occurred mainly within non-coaxial regime.
- Chlorite thermometry reveals two temperature groups for late stages of shearing.
- Main deformation is related to the main stage of updoming between ca. 15 and 11 Ma.

ABSTRACT

We describe the structure, microstructures, texture and paleopiezometry of quartz-rich phyllites and marbles along N-trending Moutsounas shear zone at the eastern margin of the Naxos metamorphic core complex (MCC). Fabrics consistently indicate a top-to-the-NNE non-coaxial shear and formed during the main stage of updoming and exhumation between ca. 14 and 11 Ma of the Naxos MCC. The main stage of exhumation postdates the deposition of overlying Miocene sedimentary successions and predates the overlying Upper Miocene/Pliocene conglomerates. Detailed microstructural and textural analysis reveals that the movement along the Moutsounas shear zone is associated with a retrograde greenschist to subgreenschist facies overprint of the early higher-temperature rocks. Paleopiezometry on recrystallized quartz and calcite yields differential stresses of 20–77 MPa and a strain rate of 10^{-15}–10^{-13} s^{-1} at 350 °C for quartz and ca. 300 °C for calcite. Chlorite geothermometry of the shear zone yields two temperature regimes, 300–360 °C, and 200–250 °C. The lower temperature group is interpreted to result from late-stage hydrothermal overprint.

Keywords: Naxos metamorphic core complex; Moutsounas shear zone; Microstructure; EBSD texture; Paleopiezometry; Chlorite geothermometry

1. INTRODUCTION

Structural evolution of metamorphic core complexes (MCCs) is critical to understanding large-scale extension is accommodated exhumation of mid- to lower-crustal levels. A gneissic core is often overlain by a low-angle extensional detachment, that juxtaposes lower metamorphic grade to supracrustal rocks (e.g., Davis, 1983, Coney and Harms, 1984, Coney, 1987, Wernicke, 1985, Wernicke and Axen, 1988, Lister and Snoke, 1984,Lister and Davis, 1989 and Whitney et al., 2013). The development of an MCC is relative to the progression of different metamorphic stages during exhumation, producing a characteristic sequence of (overprinted) meso- and microstructures and textures within the low-angle extensional detachment along upper margins of the MCC. Structural observations, kinematic criteria, microstructural fabrics and metamorphic histories need to be examined in great detail within low-angle normal shear zones, as these features are often the key to constrain critical aspects of metamorphic core complex formation. Quantitative studies of microstructures and textures in MCCs often present important inferences on, e.g., deformation processes, deformation temperatures, strain state and deformation mechanisms and rheology that operated in exhumed shear zones on top of the MCC (e.g. Toy et al., 2008 and Whitney et al., 2013). Mylonitic rocks within a ductile shear zone are commonly characterized by a steady-state microfabric, e.g., with a stable grain size distribution, shape-preferred orientation and crystallographic preferred orientation (CPO). The balance between individual strain-accommodating processes, texture development and different deformation mechanisms are related to succession of physical conditions of deformation, e.g., temperature, strain rate, grain size, fluid activity (e.g. Hirth and Tullis, 1992, Rutter, 1998, Liu et al., 2002, Stipp et al., 2002 and Schenk et al., 2005) and second-phase particles (e.g. Herwegh and Kunze, 2002). The temperature is decreasing and the differential stresses and fluid flow typically increase within detachment zones during exhumation. It is also widely believed that a general rheological weakening associated with grain-size reduction by dynamic recrystallization occurs (e.g. de Bresser et al., 2001 and references therein). Asymmetric fabrics and textures relative to the meso-structural framework (e.g., foliation and lineation) may yield a criterion for the strain history and sense of shear (e.g. Wenk et al., 1987 and Bestmann et al., 2000).

MCCs typically occur within the south-central Aegean area (e.g. Lister et al., 1984, Urai et al., 1990, Wawrzenitz and Krohe, 1998, Brun and Sokoutis, 2007, Jolivet et al., 2010 and Grasemann et al., 2012). A classic metamorphic core complex is located on the island of Naxos in the Aegean Sea and numerous high-quality structural and microfabrics data have been published (e.g. Urai et al., 1990, Buick, 1991a, Buick, 1991b, Gautier et al., 1993, John and Howard, 1995, Vanderhaeghe et al., 2007, Lewis et al., 1998, Keay et al., 2001, Urai and Feenstra, 2001, Krabbendam et al., 2003, Schenk et al., 2007,Vanderhaeghe, 2004, Vanderhaeghe et al., 2007, Seward et al., 2009, Kruckenberg et al., 2010 and Kruckenberg et al., 2011). The Naxos-Paros ductile low-angle fault (e.g.Gautier et al., 1993), located along the northern margin of Naxos, is part of the North Cycladic Detachment System (Jolivet et al., 2010) (Figs. 1 and 2). A rolling hinge top-N displacement was proposed in recent models (Jolivet et al., 2009). Exhumation of the Naxos MCC was contemporaneous with migmatization and granitoid intrusion in the center of the dome. The Naxos migmatite contains subdomes, which formed by combination of upper crustal extension and deep crust contraction (Kruckenberg et al., 2010., 2011), The vast majority of the deformation in the migmatitic core is characterized by melt-present deformation conditions and microstructures show recrystallization at high-temperature. Melt-present deformation clearly was the key deformation process in the core of the dome throughout its history (e.g. Vanderhaeghe, 2004, Kruckenberg et al., 2010, Kruckenberg et al., 2011 and Siebenaller et al., 2013).

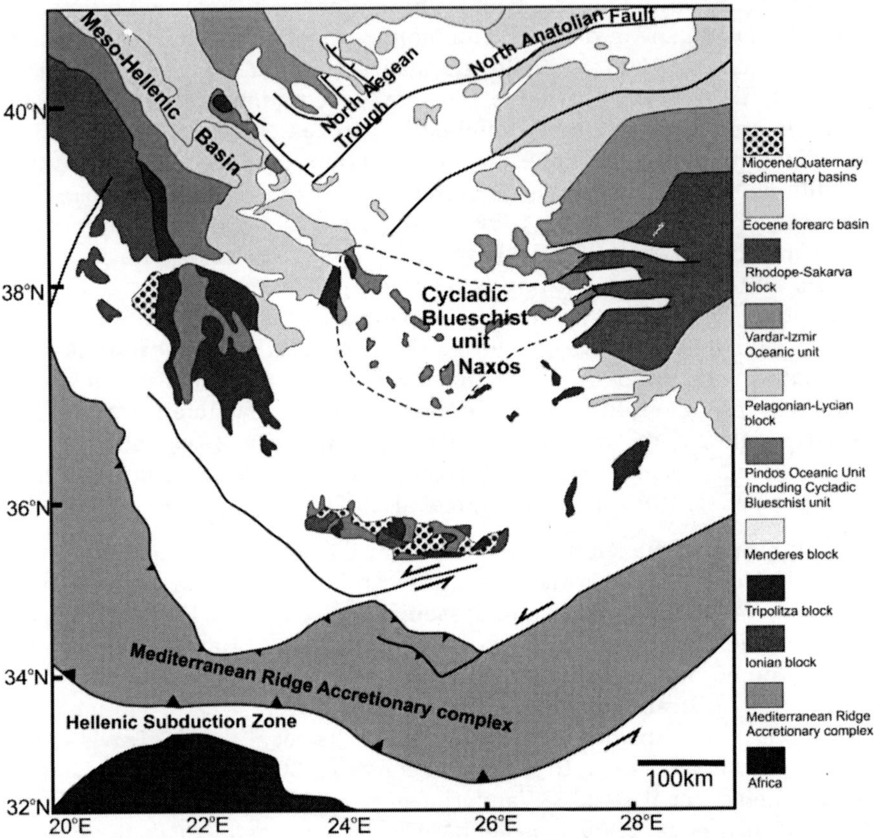

Figure 1. Simplified tectonic map of the Aegean region showing the main tectonic zones above the Hellenic subduction zone (modified after Jolivet et al., 2010 and Ring et al., 2010).

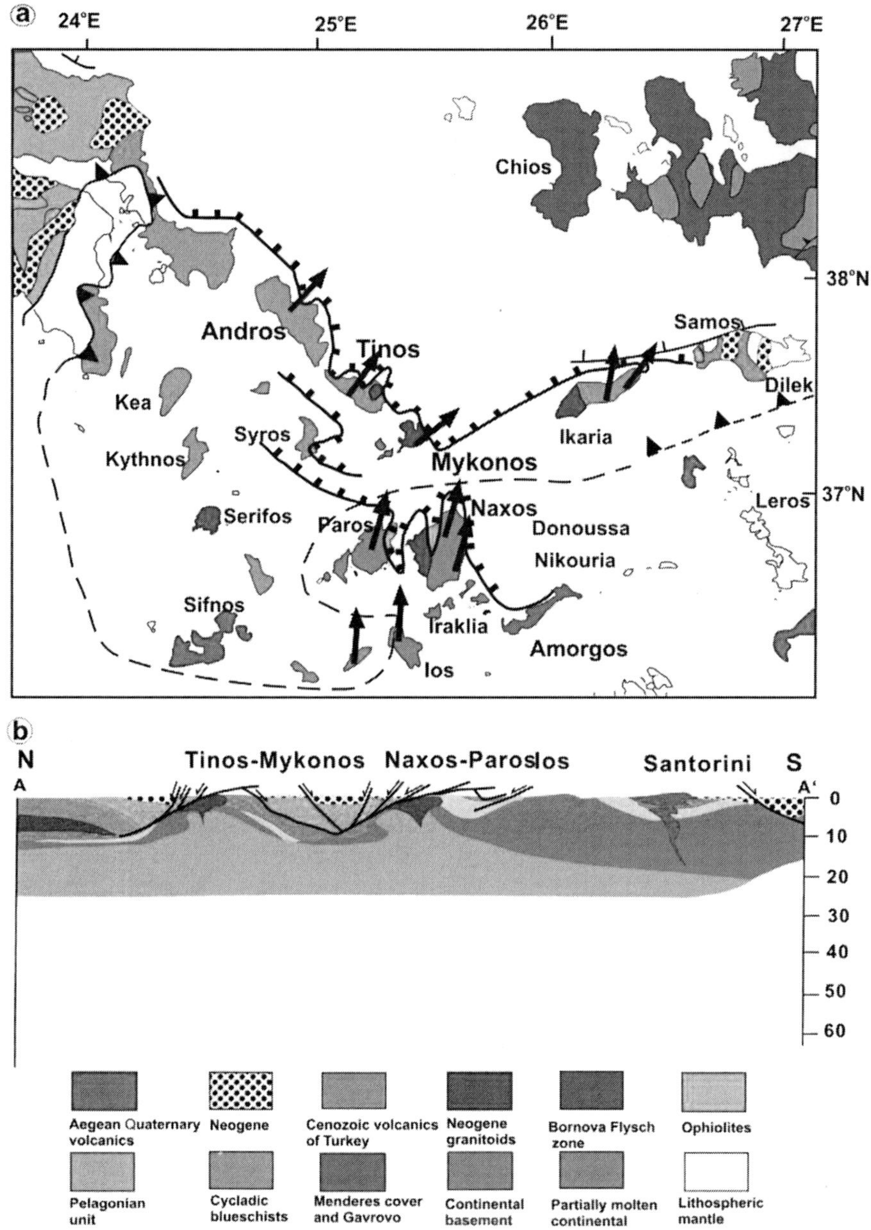

Figure 2. (a) Schematic geological map within the context of the North Cycladic Detachment System. The arrows indicate the sense of ductile shear in the lower plate (modified after Jolivet et al., 2010 and Huet et al., 2011). (b) Detail of N–S cross-section of the Aegean region centered on the Cyclades archipelago (modified after Jolivet et al., 2010).

Detailed information of the eastern boundary shear zone of the Naxos MCC is poorly known (e.g. Vanderhaeghe et al., 2007). The aim of this paper is to describe this low-angle ductile shear zone on the eastern boundary of the Naxos MCC and we denote this fault as the Moutsounas shear zone. Field structures combined with microstructural and textural analyses reveal the structural and microfabric evolution in terms of Naxos MCC exhumation, heating from an associated migmatite dome in the footwall and sedimentation on the upper plate. We also present preliminary paleopiezometric calculations and constraints of the temperature of deformation using chlorite thermometry and phengite barometry. These results, when combined with previous studies, provide new insight into the structural history of the Naxos MCC and into the mode and history of tectonic extension in the Aegean region during Neogene times.

2. REGIONAL GEOLOGICAL SETTING

The southern Aegean Sea is part of the Hellenic subduction system (Le Pichon and Angelier, 1979, Angelier et al., 1982, McClusky et al., 2000, Pe-Piper, 2000 and Wortel and Spakman, 2000) (Fig. 1). Earthquake focal mechanisms (e.g., Hatzfeld et al., 1993,Bohnhoff et al., 2004, Hollenstein et al., 2008 and Floyd et al., 2010) indicate retreat of the subduction zone to the SW, and associated extension in the Aegean region (Ring et al., 2010 and references therein). Pliocene to present-day arc-related volcanism developed to the south of the Attic Cycladic massif (Buick, 1991a). The Attic-Cycladic massif and its NW extension in the Pelagonian massif represent the submerged part of tectonic units striking NW–SE in mainland Greece, which turns to ca. E–W strike in the Cyclades. Miocene N–S extension in the Aegean Sea resulted in the sequential north to south formation of numerous MCCs in the Cyclades archipelago (Lister et al., 1984, Lee and Lister, 1992, Boronkay and Doutsos, 1994 and Doutsos and Kokkalas, 2001) along the North Cycladic, South and West Cycladic Detachments (Jolivet et al., 2010 and Grasemann et al., 2012). During Miocene extension, medium- to high-grade metamorphic rocks were deformed in a W–E to NW–SE compressive and NNE–SSW extensional ductile regime (Lister et al., 1984, Buick, 1991a, Buick, 1991b and Walcott and White, 1998). Extension in the region resulted in crustal thinning, tectonic unroofing, and emplacement of melts within the core complex (e.g., Kruckenberg et al., 2011). Tectonic unroofing continued in places until latest Miocene times, as recorded by apatite fission-track cooling ages (Hejl et al., 2002 and Brichau et al., 2010). Late Oligocene to Early Miocene crustal thinning was likely facilitated by the collapse of Eocene overthickened crust and aided by backarc extension related to subduction zone retreat, similar to the recent geodynamic setting (Lee and Lister, 1992, Jolivet and Patriat, 1999 and Jolivet et al., 2010).

The Alpine metamorphic history of metamorphic units of this region includes Meso-Hellenic high-pressure metamorphism (M1) at ca. 45 Ma and Neo-Hellenic Barrovian-type metamorphism (M2) between 23 and 16 Ma (Altherr et al., 1982, Andriessen et al., 1987, Wijbrans and McDougall, 1988, Okrusch and Bröcker, 1990, Avigad and Garfunkel, 1991 and Avigad, 1998). On the Greek mainland, the Alpine-metamorphic basement domain is juxtaposed against the external arcuate belt of unmetamorphic continental units (Sub-Pelagonian and Ionian units) and the Pindos ophiolite (Jacobshagen, 1986, Fig. 1).

3. GEOLOGY OF NAXOS

Naxos is the largest island of the Cyclades in the western Aegean Sea (Fig. 2a). The exhumation/uplift resulted in a north-south trending elongated dome (Lister et al., 1984,Jolivet et al., 2004, Jolivet et al., 2010 and Ring et al., 2010) (Fig. 3a). The main structure of the Naxos MCC is a migmatite-cored gneiss dome that is structurally overlain by a ductile low-angle normal fault in the east and a steep strike-slip-type fault zone in the west, the latter located in an N–S trending topographic depression. Within the MCC, continental basement, probably of Variscan age (Altherr et al., 1982, Strumpf, 1997 and Reischmann, 1998), is capped by a high-grade metamorphic cover succession of Mesozoic carbonates, fine-grained siliciclastic and tuffaceous rocks (Jansen, 1977 and Jacobshagen, 1986). The main rock types on Naxos are intercalated marbles and schists that became dismembered within the migmatitic central part (Fig. 3a; Jansen, 1973, Jansen and Schuiling, 1976 and Kruckenberg et al., 2011). The western part of Naxos exposes the Miocene Naxos granodiorite. The hanging wall unit of the MCC comprises remnants of unmetamorphic Mesozoic ophiolites and overlying Miocene and Pliocene sedimentary rocks.

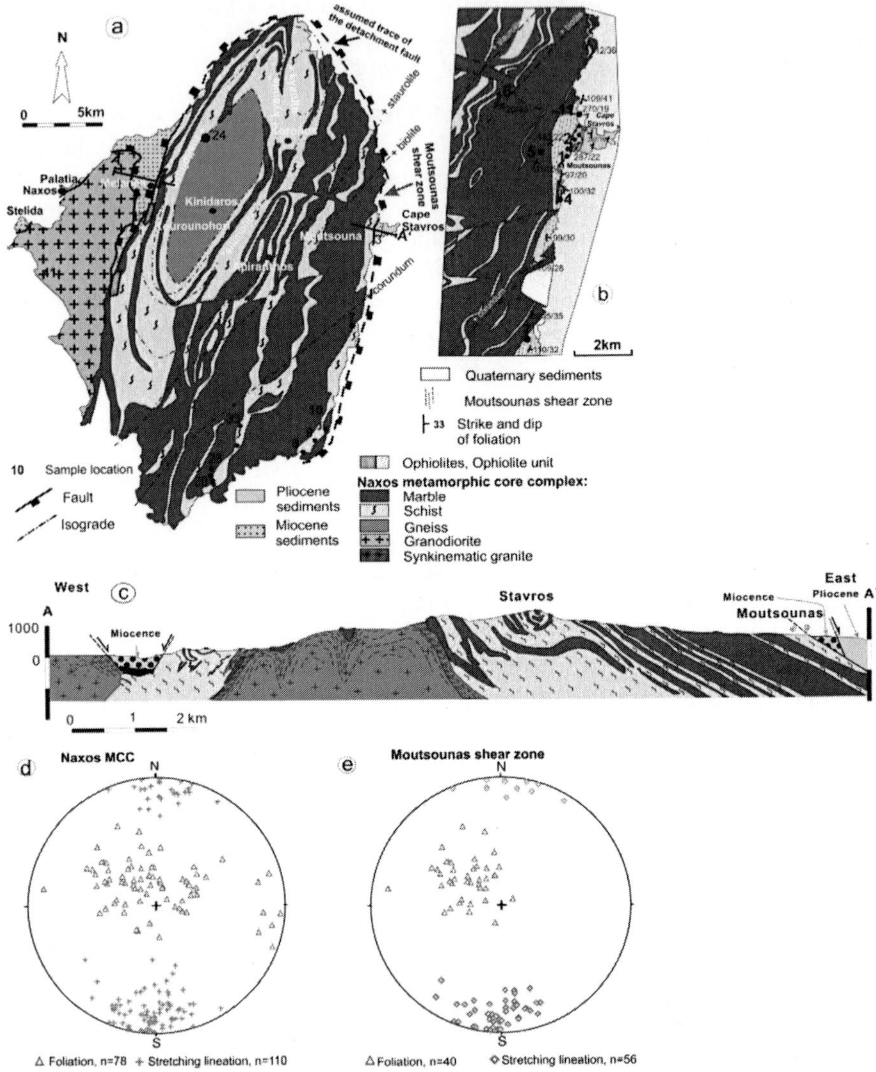

Figure 3. (a) Geological map of the Naxos MCC in the Cyclades Islands, Greece showing lithologies, metamorphic isogrades and faults with arrows for kinematics after John and Howard, 1995, Keay et al., 2001 and Koukouvelas and Kokkalas, 2003. (b) Detail of the eastern sectors of Naxos island with sample locations (modified after Jansen, 1977). (c) E–W cross section modified after Vanderhaeghe (2004) and Kruckenberg et al. (2010). Arrows indicate sense of shear retrieved from kinematic analysis of the ductile fabric. (d) Lower hemisphere, eual area stereoplot of foliation planes and stretching lineation (Naxos Island). (e) Stereoplot of foliation planes and stretching lineation (Moutsounas shear zone).

The I-type Naxos granodiorite batholith, now in tectonic contact with the Miocene clastic succession, was emplaced ca. 12–14 Ma ago according to U/Pb dating of zircons (e.g.Henjes-Kunst et al., 1988, Pe-Piper et al., 1997 and Keay et al., 2001). Rapid cooling of the batholiths is indicated by ca. 12.5 Ma $^{40}Ar/^{39}Ar$ ages from various minerals (Wijbrans and McDougall, 1988) and ca. 10 Ma K–Ar biotite cooling ages (Pe-Piper et al., 1997). Apatite fission track ages indicate that cooling to ca. 110 °C occurred at 8.2 Ma (Altherr et al., 1982), which was coeval with ongoing top-to-the-north low-angle normal faulting. Ultramylonite, pseudotachylyte and cataclasite formation close to the northern and western parts of the detachment fault record tectonic activity lasting until nearly surficial conditions were reached (John and Howard, 1995). Pseudotachylyte formation, which is locally intense in the Naxos granodiorite and along the tectonic contact below the ophiolite nappe, is estimated to have occurred at ca. 10 Ma (Andriessen et al., 1979). The reported ages constrain the final juxtaposition of the ophiolite nappe and the Miocene neo-autochthonous cover sequence, which apparently shows no thermal overprint (Kuhlemann et al., 2004 and Seward et al., 2009).

Two main successive Alpine metamorphic events affected the metamorphic zoning of the middle and lower units within the Naxos MCC (e.g. Andriessen et al., 1979, Urai et al., 1990 and Keay et al., 2001). The earlier high-pressure (HP) – low-temperature (LT) Eocene metamorphism (M1) phase is recorded by relics of blueschist facies minerals preserved in metamorphic rocks of southern sectors of the island (Wijbrans and McDougall, 1986, Avigad, 1998 and Avigad and Garfunkel, 1991). $^{40}Ar/^{39}Ar$ white mica ages of ca. 45 ± 5 Ma document an Eocene age for blueschist metamorphism (Andriessen et al., 1979, Wijbrans and McDougall, 1986 and Wijbrans and McDougall, 1988). In the higher grade metamorphic rocks of Naxos, no relics of an HP–LT metamorphism have been identified. Metamorphic rims of zircons yield U–Pb ages ranging from 57 to 41 Ma and are interpreted to reflect fluid-assisted zircon growth during blueschist metamorphism M1 (Martin et al., 2006 and Martin et al., 2008). The blueschist metamorphic phase (M1) was later overprinted by a widespread Barrovian type (MP–MT) metamorphism (M2) during the Miocene grading from greenschist facies to amphibolite facies conditions, and reaching partial melting conditions at the deepest structural levels, as attested by the abundant migmatization in the dome core (670 °C, 5–7 kbar) (Jansen and Schuiling, 1976, Buick and Holland, 1989, Wijbrans and McDougall, 1988, Avigad and Garfunkel, 1991, Duchêne et al., 2006, Martin et al., 2006 and Martin et al., 2008). Based on $^{40}Ar/^{39}Ar$ mineral ages, the timing of peak Barrovian-type M2 metamorphism was first estimated to have occurred between 20 and 15 Ma (Wijbrans and McDougall, 1988) and is consistent with new Rb/Sr ages for micas and garnets (Duchêne et al., 2006). External overgrowth rims of the zircons in the core migmatites have ages ranging from 19 to 14 Ma (Keay, 1998, Keay et al., 2001 and Martin et al., 2006). Finally, Martin et al., 2006 and Martin et al., 2008 found U–Pb zircon ages of ca. 16 Ma for the Barrovian type metamorphism. In the structurally higher metamorphic rocks, the age of MP–MT metamorphism M2 is constrained to between 27 and 19.9 Ma by $^{40}Ar/^{39}Ar$ phengite dating (Wijbrans and McDougall, 1988).

Previously buried lower crustal rocks were brought into contact with upper crustal units (hanging wall) along the gently N-dipping Naxos/Paros shear zone (Gautier et al., 1993), which is the ductile expression of the Naxos/Paros extensional fault system (Brichau et al., 2006). It has been convincingly argued that high-temperature metamorphism on Naxos developed as consequence of ductile extensional shearing (e.g. Buick, 1991a,Buick, 1991b and Gautier et al., 1993). Structural evidence indicates that the hanging wall of the core complex experienced large-scale top-to-the-north (ca. 010°) transport along a low-angle detachment fault (Gautier et al., 1990, Gautier et al., 1993 and Walcott and White, 1998) with evidence of deformation under ductile conditions (Urai et al., 1990,Buick, 1991b and Gautier et al., 1993) and subsequent brittle (Angelier et al., 1978 and John and Howard, 1995) deformation conditions. Extensional deformation started at or before the peak of high-temperature metamorphism, which has been dated at about 20–16 Ma on Naxos (Keay et al., 2001). Zircons from four granite intrusions clearly post-dating metamorphism were also dated by Keay et al. (2001) and range in age from 15.4 to 11.3 Ma, with the main period of magmatic activity at ca. 12 Ma. All zircon and apatite fission track ages on Naxos range between 11.8 ± 0.8 to 9.7 ± 0.8 Ma and 11.2 ± 1.6 to 8.2 ± 1.2 Ma and consistently decrease northwards in the direction of hanging wall transport (Hejl et al., 2002, Brichau et al., 2006 and Seward et al., 2009).Brichau et al. (2006) interpreted their results to support the scenario of an identical fault dip and a constant or slightly accelerating slip rate of ~6–8 km per Myr on the Naxos/Paros extensional fault system across the brittle/ductile transition, which separates the MCC from the overlying unit, during exhumation.

Only limited remnants of the hanging wall unit are exposed on the island of Naxos along the western and eastern margins of the Naxos MCC (Fig. 3a; see Jansen, 1977). Hanging wall units comprise an unmetamorphosed strongly disrupted ophiolite nappe (Jansen, 1977) and overlying Miocene and Pliocene sedimentary rocks. The ophiolitic members are serpentinite and basalt, with radiolarite in two spot-like outcrops. Overlying the ophiolite unit, Miocene and Pliocene sedimentary successions are a part of the hanging wall unit, which are exposed on several islands of the Cyclades (e.g., Gautier et al., 1993 and Jolivet and Patriat, 1999). Kuhlemann et al. (2004) considered the Miocene clastic succession as the neo-autochthonous cover of the imbricated ophiolite deposited after nappe formation, since both are unmetamorphosed and spatially associated. The Miocene deposits contain no components from the MCC of Naxos, which contrasts coarse-grained Pliocene deposits (Rösler, 1978). Therefore, the Miocene succession is considered to have been deposited prior to exhumation and uplift of the metamorphic complex. This is consistent with cooling ages of the presently exposed footwall rocks (Pe-Piper et al., 1997).

Based on fossil findings, Böger (1983) redefined Neogene formations of Naxos and two formations were distinguished, the Pesulia (23.5–15 Ma) and Eromonisia (11–7.0 Ma) Formations. The Pesulia Fm. comprises marine deposits and turbidites, containing Aquitanian to Early Burdigalian faunas (Angelier et al., 1978; ca. 23–19 Ma), which are well exposed in the center of the island in a fault-bounded block between the metamorphic core and the granodiorite, and in the Cape Stavros peninsula. By means of fission-track dating (30–500 Ma) of detrital zircons, Kuhlemann et al. (2004) report that the sediments were shed from an uplifting mountain region to the south of Naxos and found no high-grade metamorphic pebbles from the Naxos MCC. Seward et al. (2009)concluded from the lack of zircon fission-track ages younger than the stratigraphic age of the center of the island that this formation has not been exposed to temperatures greater than 170–200 °C (the lower bound of the zircon fission-track partial annealing zone) since its time of deposition. Apatite fission track ages from sandstones exposed on Cape Stavros peninsula are 19.8 ± 5.0 18.0 ± 9.6 Ma. From this region, Bargnesi et al. (2012)report (U–Th)/He zircon ages ranging from 270.0 ± 21.6 to 20.7 ± 1.7 Ma indicating a depositional age younger than 20.7 Ma.

Hejl et al. (2003) suggested that paleo-relief features at elevations of 650, 500 and 300–400 m have a maximum age of 8 Ma. Riedl, 1982a and Riedl, 1982b proposed an Early Miocene to Early Pliocene age for these paleo-erosional surfaces based on climatic and geomorphologic reasoning, which suggests that there has been limited erosion since that period.

4. MOUTSOUNAS SHEAR ZONE

The Mountains shear zone is an N-trending and E-dipping ductile to semi-ductile shear zone along the eastern boundary of the Naxos MCC (Figs. 3 and 4). The shear zone is variably thick, ca. 100–200 m in structural thickness, and similar fabrics can be traced from north of Moutsounas along coastal exposures to the southeastern edge of Naxos.

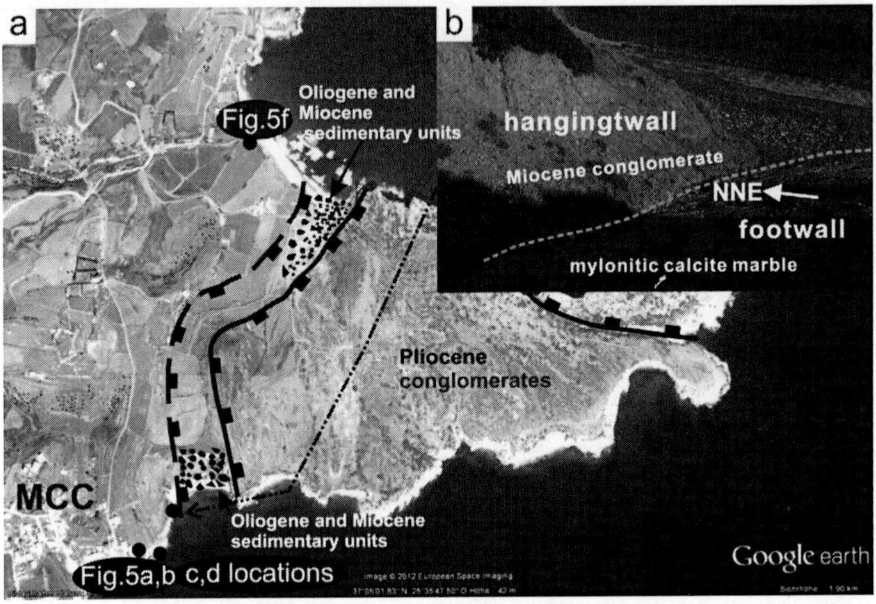

Figure 4. (a) Satellite image (Google Earth) of Moutsounas peninsula. View onto cape Stavros peninsula and geological structures. Islands in the background expose marble from the MCC and Oligocene-Pliocene sedimentary succession to the eastern margin. (b) Field coutcrop MCC/Moutsounas at location A. Upper boundary of the MCC with mylonitic calcite marble in the footwall and brittle-deformed Miocene conglomerate in the hangingwall.

In the field, the shear zone can be recognized by high-strain zones, fine grain sizes of rock forming minerals, and chloritization, which increases toward the top. Exposures are located between the biotite-in and corundum-in isogrades of the Barrovian type M2 metamorphism proposed by Jansen and Schuiling (1976) representing the chlorite-sericite zone, for which Andriessen et al. (1979) propose a temperature of 420–500 °C. The Moutsounas shear zone is located in the center of this zone and may represent, therefore, as a rough metamorphic temperature estimate, ca. 420–450 °C in terms of the M2 metamorphism. The southern edge of the Moutsounas shear zone is close to the location, for which Avigad (1998) found ca. 470 °C and 12 kbar for the M1 metamorphism.

The widespread development of mylonitic rocks derived from micaschist, marble and phyllite constitute the main part of the Moutsounas shear zones in the Naxos MCC (e.g.Figs. 4 and 5). Mylonitic quartzite and calc-schists are exceptionally well-developed and occur as discrete layers and lenses within the mylonitic micaschists and phyllites. Thin marble layers are prominent rock types along the Moutsounas shear zone. Marbles grade from basal coarse-grained mylonitic varieties into ultramylonites at the top of the shear zone. On an outcrop scale, S-L fabrics with the development of a mylonitic foliation and a pronounced NNW- to NNE-trending stretching lineation in the calcite-quartz-schist or micaschists are typical indicators of deformation (Fig. 5a and b). The orientations of stretching lineation and foliation (Fig. 3d) are similar to such reported in previous studies, e.g. Urai et al. (1990). Foliation is defined primarily by the preferred orientation of planar minerals (e.g. parallel alignment of mica, chlorite) and by flattened and elongated quartz ribbons, feldspar, chlorite and white mica grains or porphyroclasts. The lineation is defined by the preferred alignment of asymmetric feldspar porphyroclasts, quartz rods, stringers of quartz grains, elongated aggregates of recrystallized feldspars and smeared-out mica flakes. The foliation strikes NNE–SSW or N–S and dips moderately to the E or ESE at a low-angle in the schist, quartzite, marble and phyllite (Fig. 5d). These rocks show evidence of annealing during later heating and grade into retrogressed sheared rocks toward the Moutsounas Peninsula. Mineral stretching lineation in micaschists and phyllites plunges gently (<30°) to NNE or SSW.

The widespread occurrence of asymmetric fabrics, e.g. sheared veins, shear bands, sigmoid fabrics, sigma- and delta-porphyroclasts, asymmetric fringes around rigid clasts and oblique mineralized extensional veins, consistently indicate top-to-the-NNE shear. S–C fabrics are formed with S-planes defined by grain aggregates of elongated quartz and fractured plagioclase and C-planes by chlorite and biotite grains (Fig. 5a and c). In some cases, symmetrically strained mesoscopic structures such as pressure shadows around rigid objects and boudinage of quartz rods are observed. Additionally, late-stage, semi-ductile deformation is locally developed within the shear zone, generally with ~E–W trending steeply N-dipping quartz-filled extensional veins (Fig. 5d).

The shear zone is structurally overlain by hydrothermally altered Miocene conglomerates, which contain no pebbles from the Naxos MCC but exclusively from the ophiolitic hanging wall unit. Miocene conglomerates, mudstones and sandstones are exposed on the northern and southern edge of the Moutsounas Peninsula (Vanderhaeghe et al., 2007). Their bedding is variable in orientation but generally dips toward the NW and oblique to the E-dipping detachment fault. Outcrop-scale faults also dip to the east. The Miocene succession is overlain by subhorizontal or gently W-dipping conglomerates of suggested Pliocene age that form the main portion of the Moutsounas peninsula and which contain numerous clasts, mainly marble, derived from the MCC. The Pliocene conglomerates are gently folded and are cut by numerous NW-trending and steeply SW-dipping normal faults.

These sedimentary data indicate that exhumation of the Naxos MCC postdates the deposition of Miocene successions and predates Pliocene rocks.

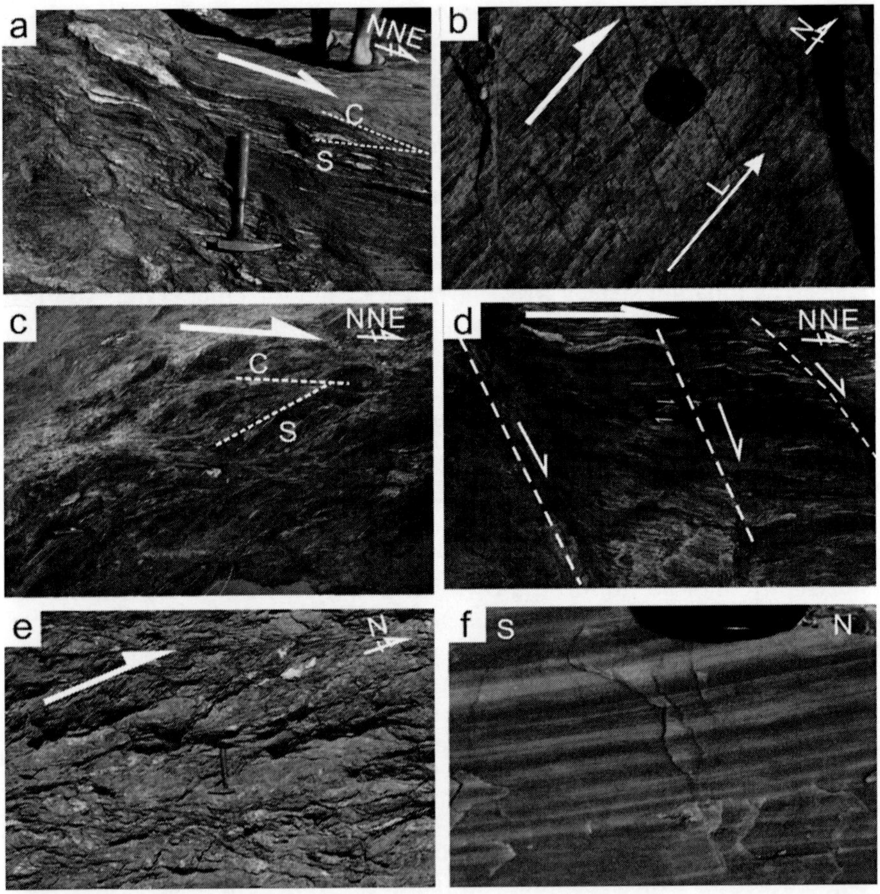

Figure 5. Macroscopic structural evidence of deformation, observed in the Moutsounas shear zone on top of the Naxos MCC. (a) Micaschist with mylonitic foliation and stretching lineation viewed on foliation surface. Shear bands (S–C fabrics) indicating top-to-the-north shear. (b) Quartzite mylonites with a well-developed stretching lineation. (c) S–C fabrics indicating top-to-the-north shear in the mylonitic phyllite. (d) Shear fabrics with shear bands in the schist and late-stage semi-ductile top-to-the-north shear-extensional normal fault filled with quartz. (e) Foliation boudinage and top-to-the-north shear bands in the greenschist facies mylonitic schists. (f) Mylonitic calcite marble with pronounced foliation.

5. MICROSTRUCTURES

5.1. Microstructures and textures: approach

The Early Miocene (M2) microfabrics are superimposed on earlier fabrics that formed during Eocene M1 metamorphism. For the comparison of microstructures and textures, we included also samples far outside of the retrogressed shear zone and we distinguish, therefore, three groups of rocks from the Naxos MCC (Fig. 3a and b):

Group 1: The main sample set is from the Moutsounas shear zone. The samples of the Moutsounas shear zone could represent variable overprint on Eocene M1 and Early Miocene M2 metamorphic fabrics according to metamorphic isogrades proposed by Jansen and Schuiling (1976).

Group 2: By comparison, a second sample set is represented rocks with a high-temperature migmatitic fabric within a thick aplitic dyke exposed at the northwestern margin of the migmatite dome. The migmatization has been dated at ca. 14–19 Ma (Martin et al., 2006). A description of migmatite structures is outside the scope of this paper.

Group 3: Rocks underwent high-pressure metamorphism and remained unaffected by the late-stage shearing of the Moutsounas shear zone. These rocks are mainly exposed in southern Naxos and exhibit high-pressure metamorphism (Avigad, 1998) of Eocene age (Wijbrans and McDougall, 1986 and Wijbrans and McDougall, 1988). Retrogression during a subsequent stage of top-N shearing is moderate.

All thin sections of samples were cut parallel to the kinematic X–Z section, which is parallel to the stretching lineation and normal to the foliation.

5.2. Moutsounas shear zone (Group 1)

Microstructural investigations were concentrated on the mylonitic quartzites, micaschists and calcite-marbles from the Moutsounas shear zone.

5.2.1. Mylonitic schists and quartzitic phyllites

Core-mantle structures of quartz present a bimodal grain-size distribution that change gradually to a completely dynamically recrystallized microstructure. Undulatory and inhomogeneous or patchy extinction, grain elongation and subgrain formation are common in porphyroclastic quartz grains (Fig. 6a). The quartz porphyroclasts are elongated and some are fully recrystallized into fine-grained elongated crystals that locally show well-developed individual grain shape alignment oblique to the mylonitic

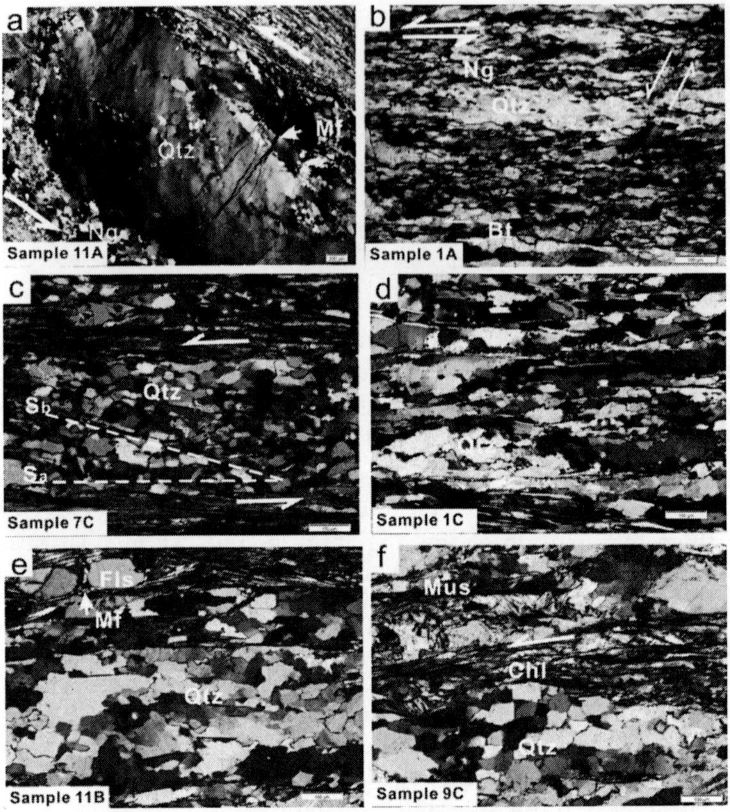

Figure 6. Microscopic deformation characteristics of mylonitic rocks within the Moutsounas shear zone. Crossed polarized light for all photomicrographs. (a) Small dynamically recrystallized grains, subgrains and large quartz porphyroclasts in the micaschist. Elongated quartz porphyroclast displays inhomogeneous extinction. The late-stage of microfractures developed in the porphyroclastic grain. (b) Elongated quartz grains with polygonized shape boundaries and recrystallized aggregates. Recrystallization results in the core-mantle structure in the mylonitic quartzite. (c) Mylonitic micaschist showing angular relationships between the main (Sa) and the oblique (Sb) foliation indicating the shear sense. The Sa is mainly defined by the alignment of micaceous and flattened quartz and feldspar grains and the Sb by the shape preferred orientation of dynamically recrystallized quartz grains and mica grains. d) and (e) Elongated quartz aggregates and ribbons in quartz-rich mylonites with irregular or serrated grain boundaries by grain boundary migration recrystallization and some quartz ribbons replaced by recrystallized grains. (f) Partly annealed quartz microstructures. Quartz aggregates with relatively irregular or straight boundaries and 120° triple junctions. Sheared chlorite and white mica aggregate band. Abbreviations: Bt–biotite, Chl–chlorite, Fls–feldspar–Mf, microfracture, Mus–white mica, Ng–recrystallized new grain, Qtz–quartz.

layering (Fig. 6b). In some samples, quartz grains are highly recrystallized, where fewer porphyroclasts are present (Fig. 6c–h). Polycrystalline and elongated quartz aggregates are common due to recrystallization. The preferred grain shape alignment (S_b) is oblique to the foliation (S_a) (Fig. 6c). The obliquity between foliation and S_b ranges between 5 and 35° in Type II S–C mylonites (Lister and Snoke, 1984). The most abundant quartz grains show irregular, lobate or serrated grain boundaries, which imply deformation at a relatively high temperature by grain boundary migration recrystallization (Fig. 6d and e). Locally, quartz grains also show polygonal shapes with relatively irregular or straight boundaries and 120° triple junctions (Fig. 6f), suggesting that quartz in the micaschist-quartzite underwent extensive recrystallization.

Early formed quartz microstructures were subsequently overprinted by low-temperature deformation, as evidenced by ribbon quartz grains that are partly recrystallized small-scale serration of grain boundaries, and rare occurrences of small grains in the triple junctions of larger quartz grains (Fig. 6f). Evidence for a late deformational overprint under decreasing temperature conditions is also observed in the intracrystalline microstructure, such as deformation lamellae and undulose and patchy extinction. Subgrains, as well as the deformation lamellae and deformation bands, are developed. Elongated (ribbon- or sigmoid-type) aggregates are common in the mylonites and are aligned parallel to the macroscopic foliation. Some polygonized grains occur in pressure shadows on both sides of feldspar grains (Fig. 7b and c). These have been interpreted as a major imprint by later lower temperature deformation (e.g. Brunel, 1980 and Law, 1986), that likely correlates with the formation of bulging recrystallization (BLG) at greenschist-facies metamorphic conditions. A similar overprint also occurs in the southeastern part of the shear zone, where the low-temperature recrystallization of quartz occurs along discrete shear surfaces crosscutting the high-temperature deformation fabric. Microstructures also show obvious micro-faults and intragranular microfractures overprinting plastically deformed and recrystallized quartz porphyroclasts in the quartzite samples (e.g. Fig. 6a) at high angle to foliation planes. These microfractures are mostly restricted to a single grain within quartz ribbon or porphyroclastic grains, where they are approximately normal to the basal plane.

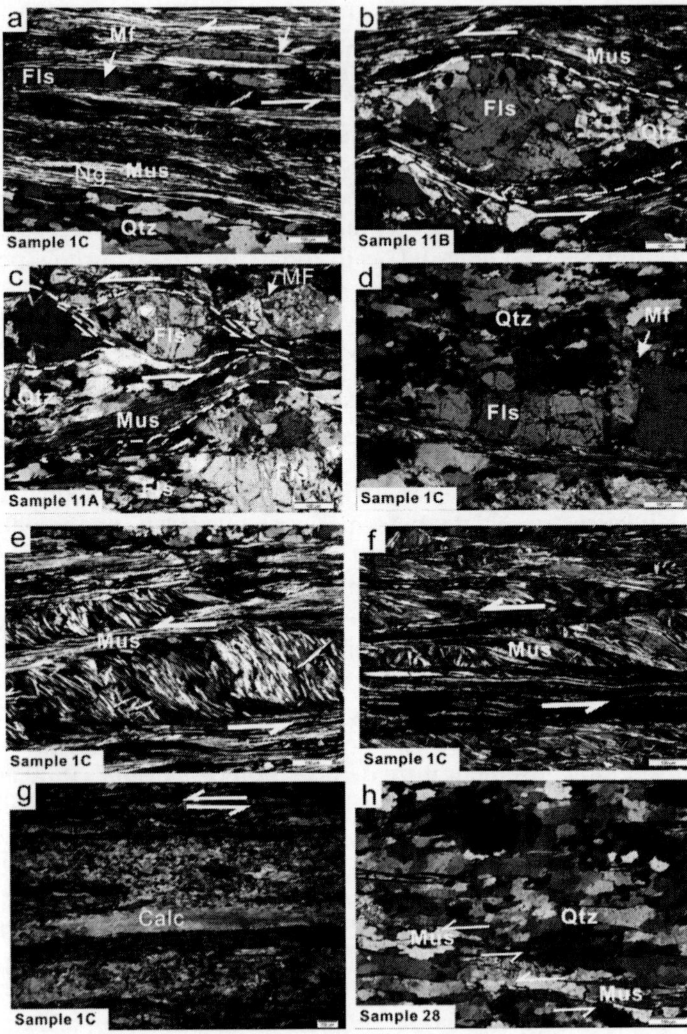

Figure 7. Microscopic deformation characteristics of mylonitic rocks. (a) – (g) are from the Moutsounas shear zone except (f), which is from the southern high-pressure belt. (a) Extremely elongated feldspar and recrystallized muscovite oriented parallel to the stretching lineation. (b) and (c) Feldspar porphyroclasts with asymmetric fabrics, indicating top-to-the-north shear sense. (d) Feldspar elongated with late stage extensional fractures oriented perpendicular or sub-perpendicular to the stretching lineation, which is filled by muscovite, quartz and calcite. (e) and (f) Recrystallized muscovite in the matrix oriented along the defined mylonitic foliation and lineation, S–C fabrics indicating the shear sense. (g) Extremely elongated calcite grain with straight to diffuse and generally irregular grain boundaries. (h) Quartz ribbons or aggregates with serrated or sutured grain boundaries. For abbreviations, see Fig. 6. Calc–calcite.

Feldspar grains in the mylonites of the Moutsounas shear zone are characterized by crystal plastic deformation (Fig. 7). In the shear direction, feldspar porphyroclasts are elongated and lenticular in shape with irregular and serrated grain boundaries. Some feldspar grains are extremely elongated and flattened into foliation, with aspect ratios of more than 5:1, and with long axes parallel to lineation (Fig. 7a). Undulatory and inhomogeneous extinction are common in porphyroclastic feldspar grains. In places, irregular grain boundaries are surrounded by a matrix of recrystallized feldspar grains. Recrystallization occurs around porphyroclasts with asymmetric fabrics, indicating a sinistral or top-to-the-north shear sense (Fig. 7b and c). Elongated porphyroclasts, as well as the cores of many mantled augens, show little evidence of internal deformation. In contrast, semi-brittle microstructures are characterized by the microfractures in some feldspar porphyroclasts. Microfractures often have infillings of quartz and mica grains (Fig. 7a–d).

White mica grains in the mylonitic schists are 50–200 μm long and a few tens of micrometers thick. The white mica matrix crystals are oriented the mylonitic foliation and lineation in the shear zone (Figs. 6e, 7a–c, e, f). Two types of white mica grains exist: coarse grains are randomly distributed in the rock and have a weak shape-preferred orientation, and fine grains are generally concentrated in small-scale (centimeters to millimeters wide) shear zones, whose original shape has been extensively modified by recrystallization. In some low-strain domains, mica-fishes are parallel to the foliation, forming a typical S–C fabric. The C-planes are defined by the fine-grained white mica aggregates and obliquely shaped white mica with opaque mineral aggregates forming the S-planes. Sometimes, opaque minerals are concentrated along the (001) planes of white mica. The primary host white mica has been progressively deformed by recrystallization to an aggregate of fine-grained mica neoblasts. Hence, some primary grains cannot longer observed optically.

Calcite grains in mylonitic calc-phyllites are variably elongated and show grain boundaries ranging from straight to diffuse and irregular (Fig. 7g) and later fine recrystallized grains occur around elongated relict grains producing a typical core-mantle structure. Twins and undulose extinction are conspicuously absent in these calcite grains. Some larger grains preserve several sets of twins or polyphase twinnings (twins in twins) at different angles, potentially indicating multiple deformations. Type III and Type IV twins (Ferrill et al., 2004) are common and only occur in large grains, which have not undergone complete recrystallization, and record, therefore, evidence of earlier deformation at higher temperature. However, most calcite twins are thin and straight, similar to type I and type II of Ferrill et al. (2004), and thus these are likely related to low-temperature deformation. In some samples, the elongation of calcite grains and rotation of initial twins into parallelism with the orientation of foliation. Their disc-shape appearance is consistent with pressure solution, strongly enhanced at the quartz/mica interface to calcite.

5.2.2. Mylonitic calcite marble

Fine-grained microstructures of calcite within the ultramylonites are characterized by complete recrystallization with only a few porphyroclasts remaining that have straight to slightly irregular grain boundaries (Fig. 8a–c). The recrystallized grains are slightly elongate or sub-equant. These ultramylonites are microstructurally more homogeneous than other types, and consist of very fine-grained calcite grains. The microstructure of mylonitic marble is typical of low-strained mylonite where porphyroclasts are abundant and appear completely surrounded by dynamically recrystallized grains producing a typical core-mantle structure (Fig. 8d–f). The coarse grains are characterized by undulatory extinction and kinking, both indicators of plastic deformation (Fig. 8e). Twins are widespread with regular as well as irregular shapes. Most twins are oriented anticlockwise to the long axis of the porphyroclasts and generally terminate at grain boundaries. Twinning systems intersect each other and some polyphase twins are common. Thick bent or curved twins inside the coarse-grained calcite likely formed during the first stages of deformation and were rotated toward parallelism with the shear zone boundary during further strain (Fig. 8e and f). Rotation and stretching of twinned porphyroclasts produced a very strong shape-preferred orientation with the orientation of the longer axes parallel to the lineation. Curved grain boundaries are indicative of grain boundary migration recrystallization. The presence of some subgrains also indicates that intracrystalline deformation mechanisms along with subgrain rotation were active. The microstructures suggest that the grain size reduction of calcite was predominantly the result of subgrain rotation recrystallization accompanied by twin boundary migration.

5.3. Aplitic dyke (Group 2) and high-pressure belt (Group 3)

Microstructures show strong plastic deformation of the main mineral phases in samples of the sheared aplite dyke located near internal migmatitic structure from the central portion of the Naxos (Group 2) and those from the southern part of the high-pressure (Group 3). The most common microstructure in the sheared aplitic dyke (Group 2) is the quartz ribbon. Elongated quartz grains compose quartz banding in a matrix of sheared feldspar grains. Such quartz ribbons form strong linear fabrics on

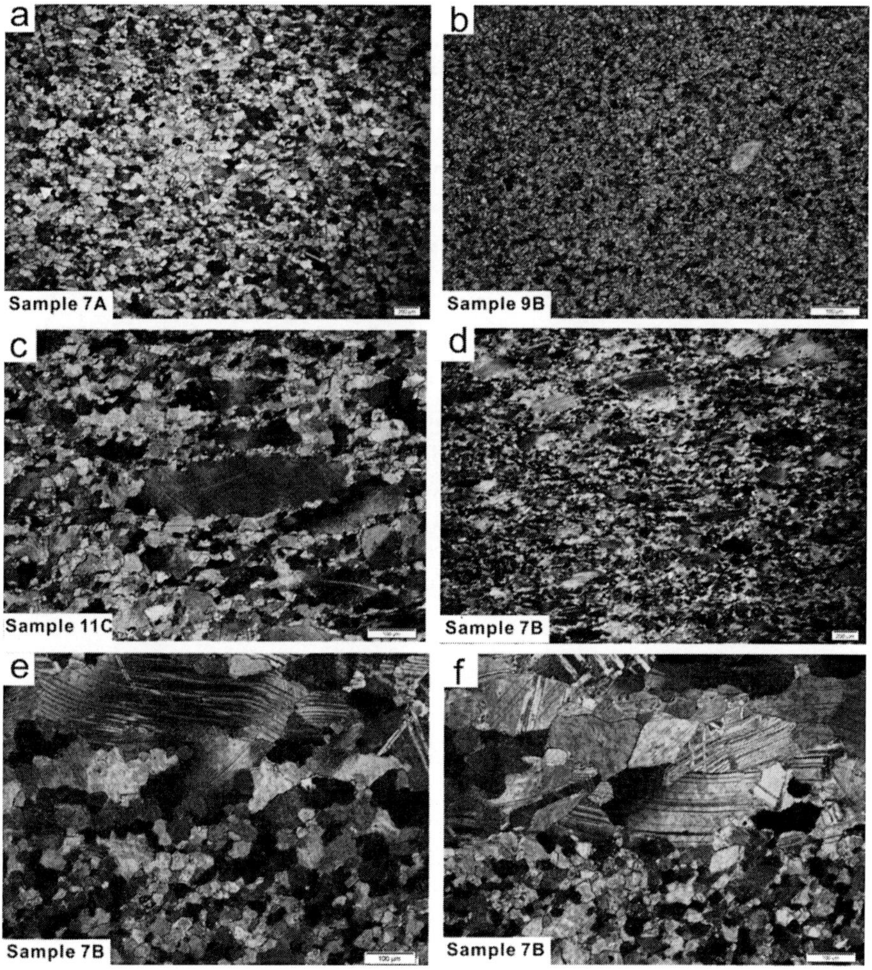

Figure 8. Microscopic deformation characteristics of marble mylonites within the Moutsounas shear zone. (a) and (b) show completely recrystallized calcite in the ultramylonites of marble. (c) Ultramylonites with less porphyroclasts and completely surrounded by dynamically recrystallized grains producing a typical core-mantle structure. (d) Recrystallization of calcite in the mylonitic marble by subgrain rotation results in the typical core-mantle structure. (e) Twins and undulose extinction are conspicuously absent in the bent calcite porphyroclasts. (f) Polyphase twins at different angles.

the hand specimen scale. Feldspar grains have irregular forms and subgranular shapes. They were elongated and have intragranular deformation. These evidences suggest that the rocks were deformed at high-or medium-temperature solid-state plastic deformation.

In the high-pressure belt (Group 3), most quartz grains have typically polygonal aggregate shapes and are parallel or subparallel to the major foliation and lineation of the rocks (Fig. 7h). Some quartz ribbons or aggregates with serrated grain boundaries occur in pressure shadows on both sides of the feldspar grains. Some irregular or sutured grain boundaries of the quartz grains imply the dominant role of grain boundary migration recrystallization, which suggests high-temperature grain boundary migration recrystallization (e.g., Stipp et al., 2002 and Passchier and Trouw, 2005). White mica is overgrown by dynamically recrystallized quartz grains. They also occur in ribbons of elongate rectangular grains or sigma-type aggregates aligned parallel to the macroscopic foliation in the mylonites. Localized low-temperature microstructures are observed, as undulatory extinction in some of the recrystallized quartz grains indicates a progressive deformation increment after their recrystallization.

6. CRYSTALLOGRAPHIC PREFERRED ORIENTATIONS (CPOS)

6.1. EBSD method

Crystallographic preferred orientations (CPOs) were measured using an electron-backscatter diffraction (EBSD) detector mounted on a tungsten filament Hitachi S-3400N-II scanning electron microscope (SEM) at the China University of Geosciences, Beijing (Liu et al., 2012). Detailed optical microstructural observations and CPO analysis were conducted on thin sections. All thin sections were cut parallel to the kinematic *XZ*plane (*X* — parallel to lineation, *Y* — parallel to foliation and normal to lineation, and *Z* — normal to foliation). Highly polished thin sections were prepared without cover and polished using a colloidal silica suspension with a particle size of 500 nm for 1–3 h. Thin sections were put in the scanning electron microscope chamber at a 70° tilt angle with the rock lineation (structural *X* reference direction) parallel to the SEM *X*-axis. The electron backscatter patterns were acquired at a low acceleration voltage of 15 kV and a beam working distance of ca. 18–20 mm. Conducting resin tapes attached to the sample surface surrounding the measurement area were used to reduce charging effects. The EBSD data acquisition was done using both point scan and mapping modes. Indexing was accepted when at least six detected kikuchi bands matched with those in the standard reflector file for the analyzed mineral phases and indexed points with a Mean Angular Deviation (MAD) larger than 1.2 (between detached and simulated patterns) were eliminated to avoid suspicious indexing. EBSD analysis was completed using the HKL Channel 5 software package. Pole figures were plotted in lower hemisphere equal area stereographic diagrams with the trace of the mylonitic foliation (S) and the stretching lineation (L) as reference directions. Systematic mis-indexing were noted in automated orientation maps and such data was replaced by zero solution pixels. A detailed description of the EBSD technique can be found in, e.g., Prior et al. (1996).

6.2. Quartz CPOs

CPOs of quartz were investigated mainly on mylonitic quartzite, micaschists and calcite-schists from the Moutsounas shear zone (Fig. 9, Group 1). For reference, we also analyzed textures of the quartz outside of the shear zone from the sheared aplite dyke with internal migmatitic structure (sample 24; mylonitic gneiss) (Group 2) and the southern part high-pressure belt (sample 33A, mylonitic micaschist) (Group 3) in order to compare them with textures within the Moutsounas shear zone. The volume fraction of quartz in these samples mostly varies from 40 to 90 percent. All textures of quartz were measured from the homogenously distributed areas. The quartz measurements from these samples result in different c-axis patterns that reflect different quartz microstructures. The primarily observed c-axis distributions can be broadly classified into four main types of fabric. These are: (1) Y maxima, (2) crossed girdles, (3) single girdles, and (4) weak girdles plus dispersion. The results of the textures are depicted in Fig. 9.

Group 1: Moutsounas shear zone (Miocene)

A well-developed quartz c-axis crossed girdle pattern occurs in quartzite mylonite (sample 1C), which is characterized by a maximum in the centre (*Y*) and the weaker maxima distributed on a great circle (*XZ*), between the maximum, with a large opening angle of 65° and a smaller opening angle of 20° and 45°. The fabric patterns are also commonly held indicative for high to middle temperature deformation with a dominant combined prismatic <a> and rhomb <a> slip (e.g. Schmid and Casey, 1986). The weak submaxima were attributed to a low-temperature superimposition during subsequent deformation. The fabrics are ascribed to a basal <a> slip during low-temperature non-coaxial deformation (Stipp et al., 2002), which is consistent with microscopic observations. The quartz c-axes crossed girdle patterns are also observed in quartzitic mylonites (samples 7C and 9C). A

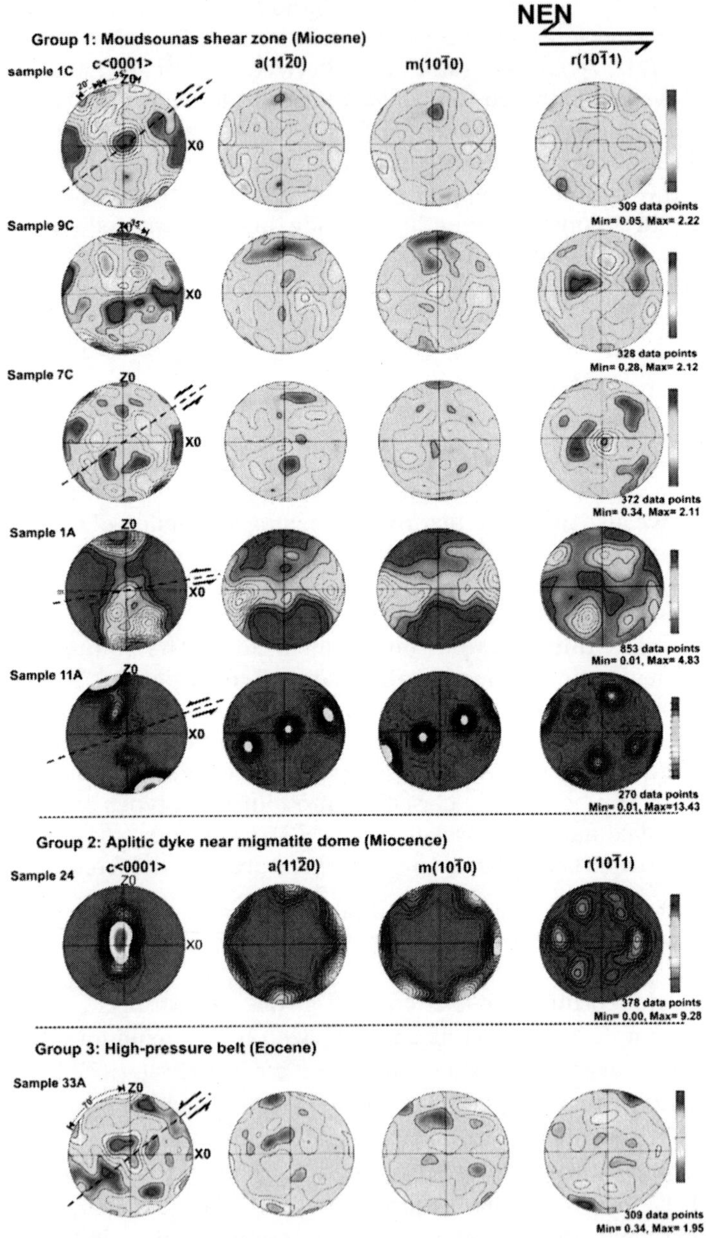

Figure 9. EBSD measured quartz textures from the mylonitic rocks. Equal area projections, lower hemisphere, contoured at interval uniform distribution. Foliation is horizontal and lineation is in the plane in the E–W direction. The number of quartz analysis (n) in each sample is shown.

weak maximum is close to Y and double maxima are distributed on a great circle with a 35° opening angle (sample 9C). The girdle fabric in sample 7C presents maxima in the *XZ*-plane at an intermediate orientation to *X* and *Z*. The patterns indicate activity of basal <a> slip and weak rhomb <a + c> slip systems with less frequent prism <a> slip, suggesting a strong low-temperature deformation overprint on a medium-temperature deformation. Asymmetric single girdle fabrics passing through the *Y*-direction are observed in the mylonitic quartzite (sample 1A) and micaschist (sample 11A). The concentrations of c-axes are close to the *Z*-axis. The girdle patterns show clockwise rotation with respect to the *XZ* plane. In contrast to the patterns of the other samples, these patterns here suggest dominant activation of the <a> basal slip of the deformation at lower temperatures, possibly under greenschist facies conditions. These patterns are slightly asymmetric, recording a significant component of noncoaxial deformation, consistent with top-to-the-NNE sense of shear.

Group 2: Aplitic dyke near migmatite dome (Miocene)

Quartz c-axes with *Y*-axis maximum fabrics are observed in the mylonitic aplitic gneiss (sample 24) from the north-western part of the MCC, with strong stretching lineation, as well as in fewer samples taken from the core of MCC. The a-axes present maxima in the*XZ*-plane. This configuration of c-axis point maxima to the *Y*-direction displays prismatic <a> slip toward a coaxial deformation. Such a situation can have resulted from high-temperature deformation under amphibolite-facies metamorphic conditions.

Group 3: High-pressure belt (Eocene)

Well-developed quartz c-axes crossed girdle pattern occurs in mylonitic micaschist (sample 33A), which presents a maximum in the centre (*Y*) and weaker maxima with large opening angles distributed on a great circle (*XZ*) (with an opening angle of 70°). The patterns are also commonly indicative of high- to medium-temperature deformation with a dominant prism<a> and rhomb <a> slip. The pattern is also asymmetric and records a significant component of non-coaxial overprint deformation, consistent with top-to-the-NNE sense of shear.

6.3. Calcite CPOs

C-axis textures and microstructures in calcite marble samples from the Moutsounas shear zone (Group 1), which were sheared and extensively recrystallized under greenschist facies conditions, were analyzed. All samples of calcite textures were chosen to represent within the Moutsounas shear zone (Group 1). Samples comprise: 1) ultramylonites completely dominated by dynamically recrystallized grains (Group 1a, samples 1D, 7A, 7C, 10C, 11C, 9B); and 2) mylonites characterized by core-mantle textures with oblique porphyroclasts (Group 1b, samples 5C, 6C, 7B, 7D); 3) mylonitic calcite-schists (Group 1c, samples 1C, 11A, 9C). CPOs analyses of calcite reveal variable pattern types in these samples and a different orientation of the fabric elements. Results of calcite CPO patterns are shown in Figs. 10 and 11.

Group 1a: Calcite CPOs of ultramylonites

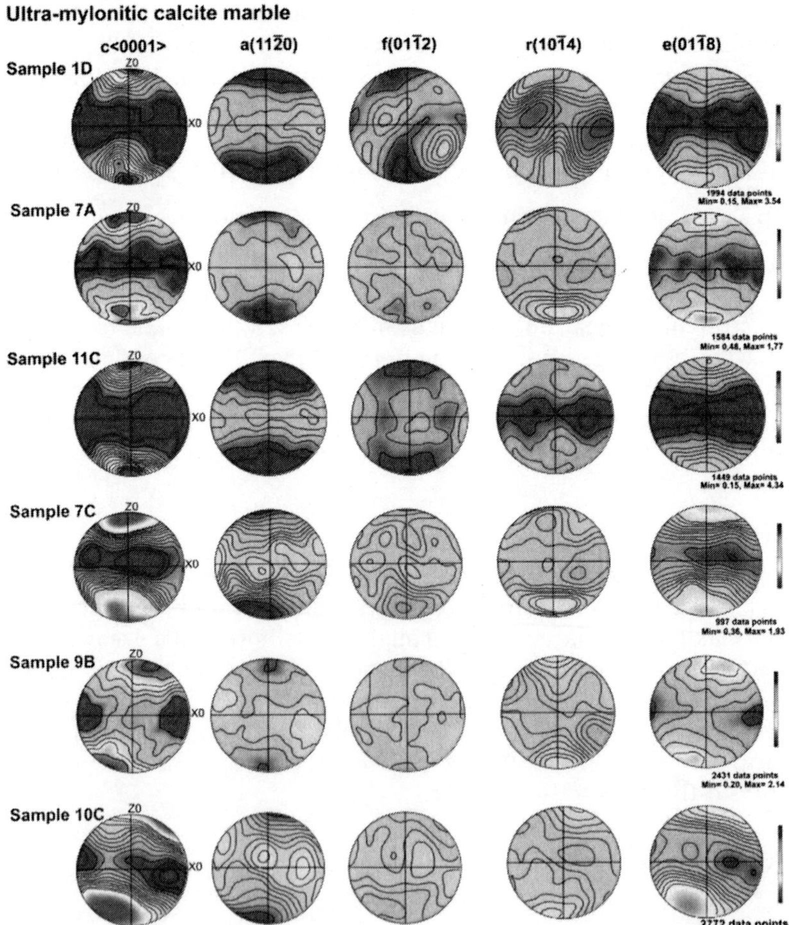

Figure 10. EBSD measured calcite textures from the mylonitic rocks. Equal area projections, lower hemisphere, contoured at interval uniform distribution. Foliation is horizontal and lineation is in the plane in the E–W direction. The number of calcite analysis (n) in each sample is shown.

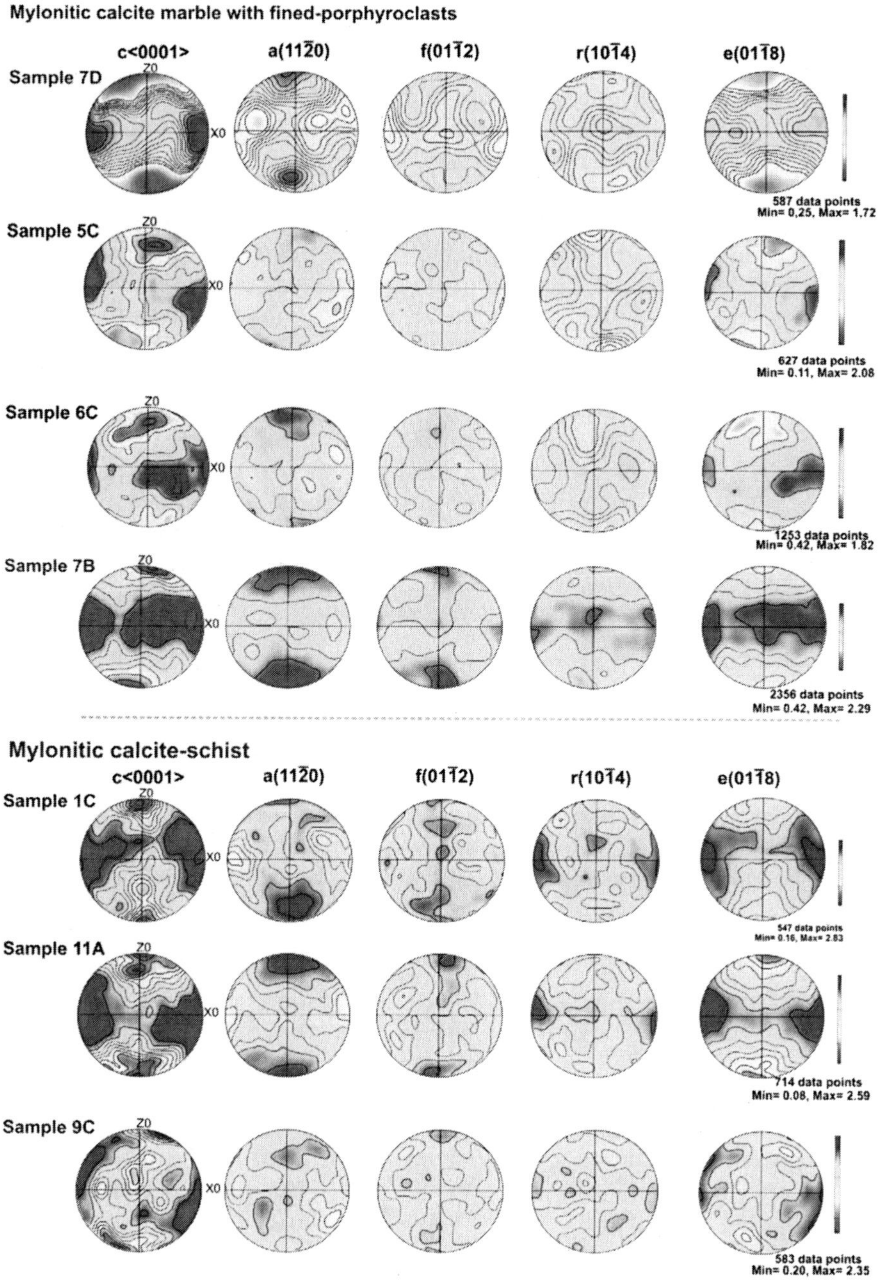

Figure 11. Contoured diagrams of calcite grain texture data of the marble mylonites and mylonitic micaschists.

The CPOs of ultramylonite samples (Fig. 10) show a fully orthorhombic symmetry with respect to the shear plane, with a classical broad c-axis – <0001> maximum oriented normal to the stretching lineation and shear plane (XY), an a-axis girdle parallel to the shear plane with a higher abundance in the stretching lineation. E-planes (f-twin) poles exhibit a weak single maximum similar to the c-axis perpendicular to the foliation plane and stretching lineation (Fig. 10). The {r} pole plot displays a weak maximum and a broad girdle parallel to the <c> poles. The CPO patterns of samples 9B and 10C display a monoclinic fabric with a c-axis distribution oblique to the shear plane. Asymmetric c-axis point maxima are rotated with respect to the shear plane, which gives a sinistral sense of shear for these samples in agreement with the general kinematics of the shear zone. The a-axes display a weak broad girdle perpendicular to the main <c> maximum and show three prominent clusters within the shear plane (*XY*). The {r} pole plot displays a weak maximum and a broad girdle parallel to the <c> poles. The e-planes (f-twin) poles exhibit a weaker single maximum similar to the c-axis pole plot. The most suitable slip system that can be deduced from this CPO is *c*<a> accompanied by *e*<a>.

Group 1b: Calcite CPOs of low-strained mylonites

CPOs are characterized by a *c*-axis- <0001> maximum normal or subnormal to the shear plane accompanied by a weaker girdle of *c*-axis in or close the *YZ* plane (Fig. 11). The <a> axes are distributed within a girdle in or close to the XY shear plane with a maximum in the *X*-direction. The *f* <e>-twin poles are weakly distributed on a similarly oriented axis. The CPO patterns with a c-axis distribution are oblique to the shear plane. The c-axis point maximum rotated opposite to the sense of shear, and consequently the asymmetry of the c-axis point maximum with respect to the shear plane is a shear sense indicator.

Group 1c: CPOs of calcites from mylonitic calc-schist

CPO exhibits a weak c-axis girdle distribution around the lineation (in the YZ plane) with a c-axis- <0001> maximum close to the *Z*-axis of finite strain (Fig. 11). The pole figures of the *a*-axis girdle is oriented parallel to the shear plane and the f-twin poles are weakly distributed on a similarly oriented axis. The calcite grains underwent continuous internal deformation and rotation during shearing, which produced a monoclinic CPO and, consequently, the asymmetry of the c-axis point maximum with respect to the shear plane is a shear sense indicator.

7. THERMOBAROMETRIC CALCULATIONS

Beside white mica, chlorite is one of the most common minerals formed during very low-grade and low-grade metamorphism (especially in greenschist facies) to moderate thermobarometric conditions (150–450 °C, and up to a few kilobar) (Cathelineau and Nieva, 1985 and Cathelineau, 1988). To get constraints on the temperature conditions of deformation during the high-pressure (M1) metamorphism that proceeded overprinting thermal metamorphism (M2), and especially during shearing and cooling, thin polished sections were prepared from the Moutsounas shear zone and high-pressure rocks for the analysis of chlorites and white mica.

7.1. EPMA methodology

Electron microprobe analysis (EMPA) of white mica and chlorite were carried out on a JEOL electron microprobe (JXA-8600) at the Department Geography and Geology, University of Salzburg, using a wavelength dispersive system. We used an acceleration voltage of 15 kV and a sample current of 40 nA. Natural and synthetic mineral standards were used to calibrate the microprobe and raw data was reduced using standard ZAF correction. The detection limits (2σ) for the elements Si are Al are 0.06 wt% and 0.04 wt%, respectively, for Na, K, Mg, Mn, and Fe are 0.025 wt%. Table 2 shows selected results of individual analyses of white mica and Table 3 show the results for chlorite including calculated temperatures (see below). Structural formulae have been calculated on the basis of 28 oxygens for chlorites, and of 22 oxygens for white mica. Three different types of composition chlorite geothermometers were applied in this work: (1) geothermometer of Cathelineau (1988) (T1) based on the number of tetrahedral Al; (2) geothermometers of Kranidiotis and MacLean (1987) (T2) and Jowett (1991) (T3) based on the number of tetrahedral Al and the value Fe/(Fe + Mg). The Cathelineau (1988) (T1) equation yields results very similar to the Kranidiotis and MacLean (1987) and Jowett (1991) methods (Table 3, results of geothermometers T1, T2 and T3, respectively).

We also applied phengite barometry on white mica based on the experimental calibration of Massonne and Schreyer, 1987 and Massonne and Szpurka, 1997 because of the potential preservation of phengite-rich white mica formed during the high-pressure metamorphism M1 predating the activity of the Moutsounas shear zone.

Table 1. List of investigated samples and methods applied to individual samples. Group 1, 2, 3 refers to structural units as defined in the main text. EBSD, electron back-scattered diffraction; EPMA, electron microprobe analysis.

Sample no.	Rock type	Group	Longitude	Latitude	Microstructures	EBSD texture	Paleopiezometry		EMPA	
							Quartz	Calcite	White mica	Chlorite
NAX-1A	Mylonitic quartzite	Group 1	37° 05′00.7″	25° 35′17.9″	x	Quartz	x		x	x
NAX-1B	Mylonitic marble	1			x					
NAX-1C	Mylonitic phyllite	1			x	Quartz, calcite	x		x	x
NAX-1D	Ultramylonitic calcite-marble	1			x	Calcite		x		
NAX-2A	Mylonitic marble	1	37° 04′46.8″	25° 35′20.9″	x					
NAX-5C	Mylonitic marble	1			x	Calcite				
NAX-6C	Mylonitic marble	1	37° 05′00.7″	25° 34′35.5″	x	Calcite				
NAX-7A	Ultramylonitic marble	1			x	Calcite		x		
NAX-7C	Ultramylonitic marble	1	36° 59′43.0″	25° 33′46.3″	x	Quartz	x		x	x
Nax-7D	Mylonitic marble	1			x	Calcite				
NAX-7B	Mylonitic marble	1			x	Calcite		x		
NAX-8	Mylonitic phyllite	1	36° 57′34.8″	25° 32′14.7″	x					
NAX-9A	Banded sericite quartzite	1	36° 57′48.3″	25° 33′01.7″	x			x		
NAX-9B	Ultramylonitic marble	1			x	Calcite		x		
NAX-9C	Mylonitic quartzphyllite	1			x	Quartz, calcite	x	x	x	x
NAX-10c	Ultramylonitic marble	1	36° 58′28.5″	25° 35′26.1″	x	Calcite		x		
NAX-11A	Micaschist	1	37° 05′24.5″	25° 35′30.0″	x	Quartz, calcite	x			
NAX-11B	Sheared quartzitic phyllite	1			x		x		x	x
NAX-11C	Ultramylonitic marble	1			x	Calcite		x		
NAX-17A	Ultramylonitic schist	1	37° 05′43.4″	25° 26′14.4″	x					
NAX-17B	Calcite marble	1			x					
NAX-24	Mylonitic aplite gneiss	Group 2	37° 28′31.0″	25° 27′59.6″	x	Quartz				
NAX-28	Quartz-bearing micaschist	Group 3	36° 56′06.1″	25° 28′30.3″	x		x			
NAX-29A	Micaschist	3	36° 56′16.43″	25° 28′26.9″	x				x	x
NAX-29C	Quartzitic micaschist	3			x		x			
NAX-33A	Phyllonitic micaschist	3	36° 58′13.6″	25° 28′53.4″	x	Quartz	x			
NAX-33B	Quartzitic micaschist	3			x		x			

Table 2. Representative compositions of white mica.

Sample no.	29A-1	29A-2	29A-3	29A-4	29A-5	29A-6	29A-7	29A-8	29A-9	29A-10	29A-11	29_12	1C-2	1C-3	1C-4	7C-2	7C-3	7C-4	7C-5	7C-6	7C-7	7C-10
Chemical composition (wt.%) (assuming 22 oxygen)																						
SiO_2	48.81	48.36	49.67	48.28	49.94	50.40	49.34	49.55	50.03	50.03	48.53	47.57	45.54	45.75	50.74	49.23	48.34	47.75	49.33	49.84	51.99	47.21
Al_2O_3	25.49	28.12	25.75	26.50	26.15	25.31	25.55	25.63	25.88	25.99	28.66	29.96	24.64	25.53	25.27	24.01	24.00	24.66	23.33	23.74	23.45	23.06
MgO	2.65	2.32	2.71	2.68	2.86	2.96	2.67	2.64	2.75	2.88	2.23	1.92	5.60	4.36	3.50	3.90	4.01	4.79	4.85	4.97	4.10	5.45
Na_2O	0.19	0.27	0.14	0.26	0.19	0.21	0.13	0.17	0.30	0.26	0.30	0.50	0.19	0.35	0.12	0.14	0.16	0.20	0.10	0.10	0.07	0.18
CaO	0.09	0.13	0.02	0.18	0.05	0.02	0.12	0.00	0.00	0.00	0.05	0.06	0.06	0.07	0.02	0.12	0.15	0.17	0.08	0.09	0.00	0.10
TiO_2	0.19	0.25	0.14	0.25	0.14	0.18	0.22	0.17	0.25	0.21	0.20	0.11	0.18	2.32	0.12	0.20	0.19	0.19	0.21	0.16	0.14	0.15
FeO	3.41	2.74	3.10	2.86	2.77	3.37	3.17	3.34	3.09	2.76	2.08	1.88	4.89	3.46	2.03	3.59	3.51	4.86	4.11	3.99	2.90	4.93
MnO	0.00	0.00	0.03	0.01	0.06	0.00	0.00	0.04	0.01	0.00	0.01	0.00	0.00	0.00	0.02	0.06	0.02	0.05	0.08	0.03	0.06	0.08
Cr_2O_3	0.00	0.02	0.03	0.01	0.02	0.02	0.03	0.01	0.01	0.00	0.00	0.01	0.16	0.17	0.11	0.08	0.06	0.10	0.05	0.11	0.23	0.12
K_2O	10.44	10.12	10.45	10.32	10.35	10.53	10.26	10.60	10.44	10.34	10.38	9.96	8.11	8.17	10.33	8.60	7.77	7.04	8.79	9.01	10.49	7.71
Total	91.28	92.33	92.04	91.35	92.53	92.99	91.49	92.15	92.76	92.47	92.44	91.98	89.36	90.19	92.25	89.92	88.19	89.82	90.93	92.04	93.43	89.00
Structural formulae on the basis of 22 negative charges																						
Si	6.85	6.67	6.89	6.76	6.87	6.93	6.89	6.88	6.89	6.89	6.67	6.56	6.55	6.48	6.97	6.95	6.92	6.75	6.92	6.91	7.09	6.78
Al	4.22	4.57	4.21	4.37	4.24	4.10	4.20	4.20	4.20	4.22	4.64	4.87	4.18	4.26	4.09	3.99	4.05	4.11	3.86	3.88	3.77	3.91
Mg	0.55	0.48	0.56	0.56	0.59	0.61	0.56	0.55	0.56	0.59	0.46	0.39	1.20	0.92	0.72	0.82	0.86	1.01	1.01	1.03	0.83	1.17
Na	0.05	0.07	0.04	0.07	0.05	0.06	0.03	0.04	0.08	0.07	0.08	0.13	0.05	0.10	0.03	0.04	0.04	0.06	0.03	0.03	0.02	0.05
Ca	0.01	0.02	0.00	0.03	0.01	0.00	0.02	0.00	0.00	0.00	0.01	0.01	0.01	0.01	0.00	0.02	0.02	0.03	0.01	0.01	0.00	0.02
Ti	0.02	0.03	0.01	0.03	0.01	0.02	0.02	0.02	0.03	0.02	0.02	0.01	0.02	0.25	0.01	0.02	0.02	0.02	0.02	0.02	0.01	0.02
Fe	0.40	0.32	0.36	0.33	0.32	0.39	0.37	0.39	0.36	0.32	0.24	0.22	0.59	0.41	0.23	0.42	0.42	0.57	0.48	0.46	0.33	0.59
Mn	0.00	0.00	0.00	0.00	0.01	0.00	0.00	0.00	0.00	0.00	0.00	0.00	0.00	0.00	0.00	0.01	0.00	0.01	0.01	0.00	0.01	0.01
Cr	0.00	0.00	0.00	0.00	0.00	0.00	0.00	0.00	0.00	0.00	0.00	0.00	0.02	0.02	0.01	0.01	0.01	0.01	0.01	0.01	0.03	0.01
K	1.87	1.78	1.85	1.84	1.82	1.85	1.83	1.88	1.83	1.82	1.82	1.75	1.49	1.48	1.81	1.55	1.42	1.27	1.57	1.59	1.83	1.41
Sample no.	7C-9	7C-11	7C-14	7C-15	7C-16	7C-17	9C-1	9C-2	9C-3	9C-4	9C-5	9C-6	9C-7	9C-8	9C-9	11B-1	11B-2	11B-3	11B-4	11B-7	11B-8	11B-9
Chemical composition (wt.%) (assuming 22 oxygen)																						
SiO_2	48.91	47.32	49.05	48.19	47.02	46.48	48.62	48.78	48.90	43.00	50.04	44.64	47.04	49.27	47.15	49.84	49.33	49.58	47.82	45.78	41.71	41.52
Al_2O_3	23.88	23.04	24.48	23.77	22.86	24.56	26.23	26.30	26.44	24.21	26.59	25.01	25.34	26.46	24.93	25.98	25.88	26.66	25.55	25.51	24.05	27.18
MgO	4.21	5.93	4.41	4.25	4.03	5.34	3.06	3.55	2.80	6.26	2.97	5.43	3.63	3.03	2.87	3.54	3.96	3.33	4.99	5.43	8.77	7.21
Na_2O	0.21	0.20	0.14	0.14	0.17	0.30	0.25	0.18	0.30	0.44	0.25	0.29	0.23	0.15	0.14	0.16	0.18	0.11	0.21	0.18	0.07	0.29
CaO	0.19	0.13	0.13	0.17	0.20	0.10	0.20	0.13	0.33	0.29	0.08	0.24	0.11	0.05	0.07	0.04	0.08	0.06	0.07	0.02	0.04	0.08
TiO_2	0.21	0.22	0.21	0.28	0.22	0.11	1.16	0.26	0.16	0.35	0.18	0.09	0.21	0.13	0.16	0.26	0.22	0.24	0.14	0.28	0.13	0.21
FeO	3.33	5.30	3.73	5.07	4.79	4.98	2.74	3.12	3.18	8.80	2.13	7.21	4.29	2.63	2.42	2.18	2.59	2.07	4.05	4.80	8.60	7.55
MnO	0.03	0.05	0.07	0.03	0.02	0.03	0.03	0.06	0.02	0.06	0.01	0.05	0.07	0.03	0.04	0.01	0.00	0.00	0.01	0.05	0.04	0.03
Cr_2O_3	0.17	0.14	0.11	0.15	0.05	0.03	0.19	0.16	0.22	0.22	0.19	0.28	0.18	0.08	0.04	0.08	0.10	0.18	0.21	0.36	0.24	0.10
K_2O	8.15	8.03	8.14	8.48	8.05	6.33	9.80	9.96	8.96	6.79	10.32	7.54	9.21	10.21	9.66	10.25	9.58	9.97	8.98	8.66	6.31	6.66
Total	89.29	90.37	90.48	90.53	87.40	88.26	92.29	92.49	91.30	90.41	92.76	90.77	90.32	92.04	87.49	92.34	91.94	92.20	92.03	91.06	89.96	90.83
Structural formulae on the basis of 22 negative charges																						
Si	6.93	6.73	6.87	6.83	6.88	6.68	6.72	6.74	6.80	6.25	6.85	6.40	6.69	6.81	6.85	6.86	6.81	6.81	6.65	6.48	6.08	5.96
Al	3.99	3.86	4.04	3.97	3.94	4.16	4.27	4.28	4.33	4.15	4.29	4.23	4.25	4.31	4.27	4.21	4.21	4.32	4.19	4.26	4.13	4.60
Mg	0.89	1.26	0.92	0.90	0.88	1.14	0.63	0.73	0.58	1.36	0.61	1.16	0.77	0.62	0.62	0.73	0.82	0.68	1.04	1.15	1.91	1.54
Na	0.06	0.06	0.04	0.04	0.05	0.08	0.07	0.05 -	0.08	0.13	0.07	0.08	0.06	0.04	0.04	0.04	0.05	0.03	0.06	0.05	0.02	0.08
Ca	0.03	0.02	0.02	0.03	0.03	0.02	0.03	0.02	0.05	0.05	0.01	0.04	0.02	0.01	0.01	0.01	0.01	0.01	0.01	0.00	0.01	0.01
Ti	0.02	0.02	0.02	0.03	0.02	0.01	0.12	0.03	0.02	0.04	0.02	0.01	0.02	0.01	0.02	0.03	0.02	0.02	0.01	0.03	0.01	0.02
Fe	0.39	0.63	0.44	0.60	0.59	0.60	0.32	0.36	0.37	1.07	0.24	0.86	0.51	0.30	0.29	0.25	0.30	0.24	0.47	0.57	1.05	0.91
Mn	0.00	0.01	0.01	0.00	0.00	0.00	0.00	0.01	0.00	0.01	0.00	0.01	0.01	0.00	0.00	0.00	0.00	0.00	0.00	0.01	0.00	0.00
Cr	0.02	0.02	0.01	0.02	0.01	0.00	0.02	0.02	0.02	0.02	0.02	0.03	0.02	0.01	0.00	0.01	0.01	0.02	0.02	0.04	0.03	0.01
K	1.47	1.46	1.45	1.53	1.50	1.16	1.73	1.76	1.59	1.26	1.80	1.38	1.67	1.80	1.79	1.80	1.69	1.75	1.59	1.56	1.17	1.22

Table 3. Selected data on chemistry of chlorites and calculated temperatures of formation (T1 to T3, see text).

Location	Group 3 (high-pressure belt)					Group 1 (Moutsounas shear zone)																	
Sample no.	29A-01	29A-02	29A-03	29A-04	29A-05	1C-01	1C-02	9C-01	9C-02	9C-03	9C-04	9C-05	11B-01	7C-01	7C-02	7C-03	7C-04	7C-05	7C-06	7C-07	7C-08	9C-05	9C-06
Microfabric type[a]						Type I	Type I	Type I	Type I	Type I	Type I	Type I	Type I	Type I	Type I	Type I	Type I	Type II	Type II	Type II	Type II	Type II	Type II
Chemical composition (wt.%) (assuming 28 oxygen)																							
SiO_2	25.06	25.78	25.38	25.16	25.16	26.88	26.50	26.53	27.35	26.36	25.62	26.72	26.58	28.87	28.79	27.76	29.38	28.29	29.18	28.20	27.94	30.17	29.60
Al_2O_3	20.91	20.64	20.74	21.07	21.37	19.68	19.70	19.15	19.23	19.67	19.20	19.68	20.85	18.10	17.97	17.96	17.48	17.22	16.71	17.96	15.46	18.02	19.43
MgO	14.50	14.14	14.09	14.70	14.24	17.73	17.85	15.22	14.94	16.01	15.95	15.03	18.52	17.27	16.69	17.53	17.40	16.53	15.37	17.54	13.57	13.28	13.45
Na_2O	0.00	0.03	0.05	0.04	0.01	0.02	0.04	0.10	0.12	0.08	0.07	0.11	0.07	0.22	0.41	0.11	0.49	0.39	0.67	0.15	0.71	1.07	0.63
CaO	0.03	0.06	0.13	0.04	0.20	0.07	0.07	0.35	0.14	0.15	0.29	0.17	0.21	0.16	0.37	0.39	0.45	0.49	0.56	0.14	0.50	0.53	0.54
TiO_2	0.05	0.02	0.04	0.06	0.07	0.06	0.03	0.01	0.05	0.02	0.03	0.03	0.05	0.03	0.05	0.01	0.03	0.00	0.09	0.05	0.09	0.00	0.02
FeO	24.68	25.06	25.22	25.01	24.86	19.72	20.48	21.81	21.79	22.29	22.09	22.95	19.15	19.81	19.17	19.29	18.81	19.31	21.41	19.59	19.54	20.97	20.23
MnO	0.15	0.16	0.19	0.15	0.18	0.27	0.20	0.33	0.35	0.39	0.42	0.35	0.14	0.37	0.29	0.29	0.28	0.30	0.25	0.33	0.23	0.29	0.33
Cr_2O_3	0.01	0.02	0.04	0.02	0.02	0.11	0.05	0.18	0.45	0.07	0.09	0.05	0.02	0.05	0.00	0.00	0.01	0.01	0.07	0.05	0.08	0.11	0.20
K_2O	0.00	0.00	0.00	0.00	0.00	0.00	0.00	0.00	0.06	0.00	0.00	0.00	0.02	0.00	0.00	0.00	0.00	0.00	0.07	0.00	0.30	0.21	0.00
Total	85.40	85.90	85.89	86.25	86.12	84.55	84.92	83.68	84.48	85.02	83.77	85.10	85.62	84.88	83.73	83.35	84.33	82.52	84.39	83.99	78.42	84.65	84.42
Structural formulae on the basis of 28 negative charges (Al^{IV} = 8-Si)																							
Al^{IV}	1.29	1.23	1.27	1.31	1.30	1.15	1.19	1.12	1.06	1.18	1.20	1.13	1.23	0.96	0.93	1.02	0.90	0.93	0.87	0.99	0.78	0.79	0.87
Si	5.41	5.54	5.47	5.39	5.39	5.70	5.62	5.77	5.87	5.65	5.59	5.73	5.55	6.08	6.14	5.96	6.20	6.14	6.26	6.01	6.43	6.42	6.27
Al	5.32	5.22	5.27	5.32	5.40	4.92	4.93	4.90	4.87	4.97	4.94	4.98	5.13	4.50	4.51	4.55	4.35	4.41	4.23	4.51	4.19	4.52	4.85
Mg	4.67	4.53	4.52	4.69	4.55	5.61	5.65	4.93	4.78	5.12	5.19	4.81	5.76	5.43	5.30	5.61	5.48	5.35	4.92	5.57	4.66	4.21	4.24
Na	0.00	0.01	0.02	0.02	0.01	0.01	0.02	0.04	0.05	0.03	0.03	0.05	0.03	0.09	0.17	0.05	0.20	0.16	0.28	0.06	0.32	0.44	0.26
Ca	0.01	0.01	0.03	0.01	0.05	0.02	0.02	0.08	0.03	0.03	0.07	0.04	0.05	0.04	0.08	0.09	0.10	0.11	0.13	0.03	0.12	0.12	0.12
Ti	0.01	0.00	0.01	0.01	0.01	0.01	0.01	0.00	0.01	0.00	0.01	0.01	0.01	0.00	0.01	0.00	0.01	0.00	0.01	0.01	0.01	0.00	0.00
Fe	4.46	4.50	4.54	4.48	4.46	3.50	3.64	3.96	3.91	3.99	4.03	4.12	3.34	3.49	3.42	3.47	3.32	3.51	3.84	3.49	3.76	3.73	3.58
Mn	0.03	0.03	0.03	0.03	0.03	0.05	0.04	0.06	0.06	0.07	0.08	0.06	0.02	0.07	0.05	0.05	0.05	0.05	0.05	0.06	0.05	0.05	0.06
Cr	0.00	0.00	0.01	0.00	0.00	0.02	0.01	0.03	0.08	0.01	0.01	0.01	0.00	0.01	0.00	0.00	0.00	0.00	0.01	0.01	0.02	0.02	0.03
Temperature (T°C)																							
T1	354	335	346	358	358	308	321	298	281	317	326	303	333	246	238	266	227	237	218	258	190	193	217
T2	328	316	324	331	331	290	299	288	277	300	306	293	305	250	245	262	236	244	235	258	217	221	236
T3	359	340	351	363	363	310	323	302	285	320	329	307	334	249	241	268	230	240	222	261	195	199	222

[a] Microfabric type in sheared samples from the Moutsounas shear zone: type I- coarse-grained microlithons; type II- sheared with very fine-grained mixtures of white mica and chlorite.

We selected five chlorite-bearing phyllitic samples for this analysis (Table 1). Sample 29A is from the high-pressure rocks (Group 3) in the southern part of Naxos, which was chosen to test the methodology. Samples 1C, 7C, 9C, 11C are from sheared low-grade metamorphic rocks (Group 1) from the surroundings of Moutsounas. Sample 11C is from the shear zone in southeastern Naxos (Group 1).

7.2. Thermometry and barometry results

We first describe results from Group 3, sample 29A (Fig. 12a). White mica and chlorite of this sample are coarse-grained and are seemingly in textural equilibrium according to the detailed backscatter electron (BSE) images (Fig. 12). White mica shows high Si contents (atoms per formula unit – a.p.f.u.), range 6.56–6.93, with the majority of data at ca. 6.90 Si apfu; Table 2), as is expected for high pressure rocks (Fig. 13a). Application of phengite barometry on white mica at a temperature of 360 °C yields a main range of pressure ca. 8–12 kbar (Fig. 13b) in accordance with peak P of 12 kbar reported byAvigad (1998). In the Moutsounas shear zone (Group 1 samples), we observed chlorite in two microfabric types (e.g., samples 1C, 7C, 9C, 11B) (Fig. 12b and c): (1) early formed coarse-grained chlorite within microlithons, sometimes associated with white mica (Type I), and (2) very fine-grained sheared band mixtures of white mica and chlorite, with apparent disequilibrium boundaries (Type II). Chlorite thermometry yields results ranging from 320 to 360 °C (Table 3) (Fig. 14). In most cases, the chlorite geothermometers yield temperatures ca. 290–320 °C for microfabric Type 1, and ca. 200–260 °C for microfabric Type II. In the case of sample 7C, chlorite from coarse microlithons yields temperatures ranging from 190 to 268 °C, and these are lower than in other samples suggesting sub-greenschist facies conditions. White mica yields a wide range of compositions with a dominance of Si at ca. 6.8–6.9 a.p.f.u. and has, therefore, phengitic compositions, and muscovite grains with a low Si a.p.f.u. at ca. 6.1 (Fig. 13a). Some phengitic compositions with high-Si white micas also occur in the sheared zones (microfabric Type 2) (Fig. 13a,Table 2). We note that phengite is, therefore, inherited from high-pressure metamorphism M1, and is only rarely reset to low-pressure white mica during shearing and exhumation.

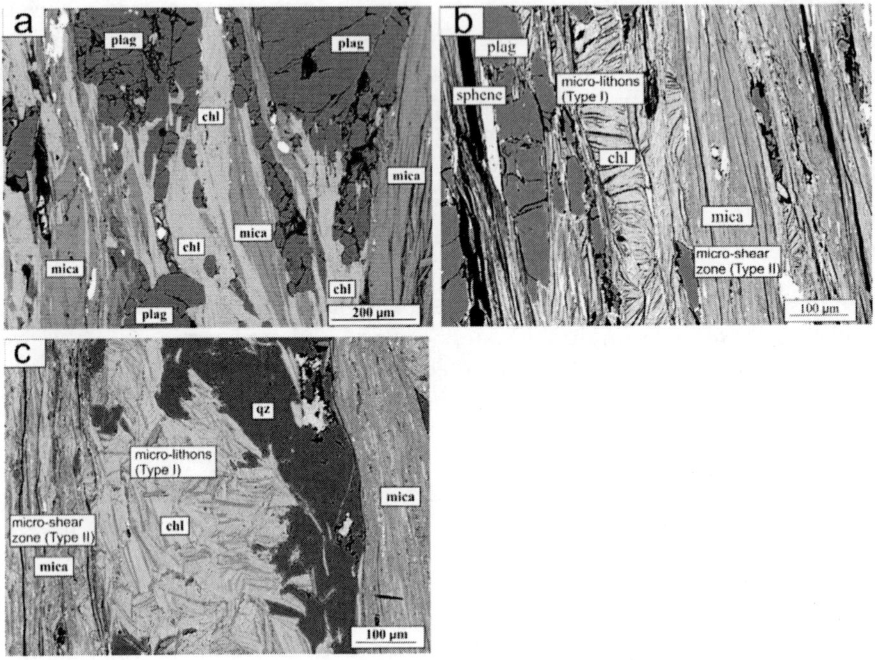

Figure 12. Back-scatter electron microscopy images of microfabrics containing white mica and chlorite: (a) Fabric dominated by high-pressure metamorphic overprint M1 (Group 3, sample 29A). (b) and (c) are representative sheared phyllitic rocks from the Moutsounas shear zone (Group 1, samples 7C and 1C).

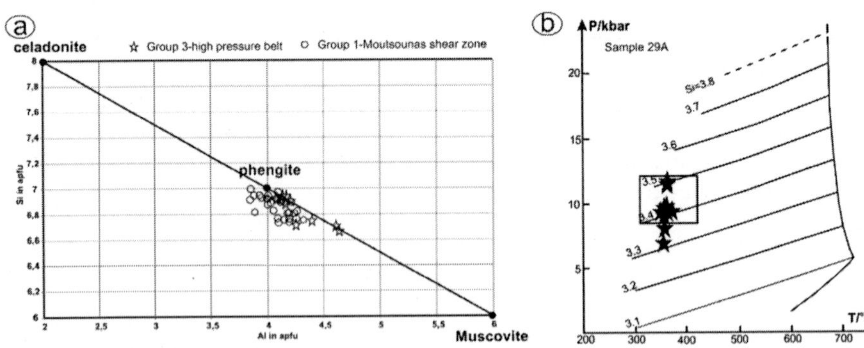

Figure 13. (a) Classification of the representative white mica phases. Chemical analyses are summarized inTable 2. (b) Phengite geobarometry (after Massonne and Schreyer, 1987).

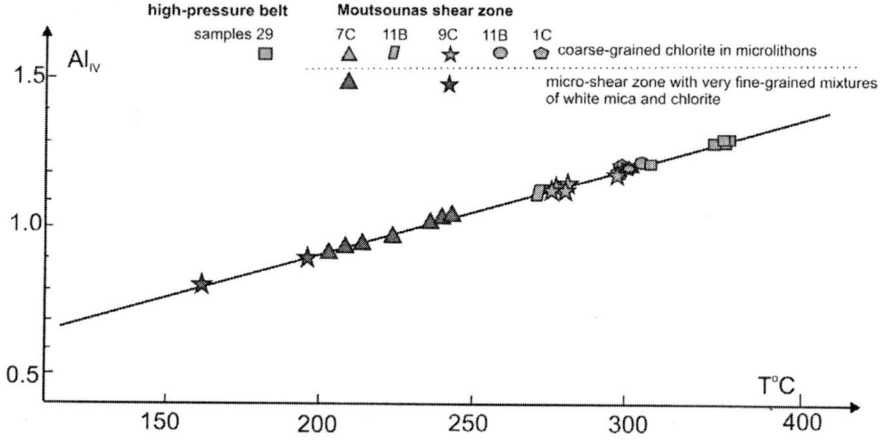

Figure 14. Plot of parameter Al^{IV} vs. temperature for chlorite from high-pressure belt (Group 3) and Moutsounas shear zone (Group 1).

8. PALAEOPIEZOMETER

8.1. Flow stress estimate from recrystallized grain size

Based on the grain size of dynamically recrystallized quartz and calcite, the piezometer can be used to calculate the differential stress from the grain size at a given temperature (e.g., Hacker et al., 1992, Gueydan et al., 2005, Warren and Hirth, 2006, Austin and Evans, 2007, Mehl and Hirth, 2008 and Okudaira and Shigematsu, 2012). Under the obtained differential stress conditions, the strain rate can be estimated. The piezometers used were established for quartz (Twiss, 1977, Mercier et al., 1977, Koch, 1983, Stipp and Tullis, 2003 and Stipp et al., 2010) and for calcite (Rutter, 1995). A suite of previously described quartzite, quartz-rich schists and calcite marbles of the Moutsounas shear zone were used for paleo-piezometric analysis. The samples are representative of syntectonic microstructures. For the present study, only samples with a uniform recrystallized grain size were considered for paleostress estimation. The diameters of equant recrystallized grains were measured optically on a standard petrographic microscope (for details, see also Behrmann and Seckel, 2007). Each grain was measured parallel and perpendicular to the macroscopic foliation. A minimum of 80 grains were measured in each sample. The mean grain size is calculated using the formula of Behrmann

(1985). The results of analysis and differential flow stresses from dynamically recrystallized grains are given in Table 4 for quartz and in Table 5 for calcite.

Table 4. Paleopiezometry data for quartz and deduced strain rates.

Sample no.	Grainsize (μm)	Differential stress (MPa)[a]						Strain rate $\dot{\varepsilon}$ (1/s)[b], 300 °C			Strain rate $\dot{\varepsilon}$ (1/s), 350 °C			Strain rate $\dot{\varepsilon}$ (1/s), 400 °C	
	Average D	$\sigma^{(a)}$	$\sigma^{(b)}$	$\sigma^{(c)}$	$\sigma^{(d)}$	$\sigma^{(e)}$	$\dot{\varepsilon}^{(1)}$	$\dot{\varepsilon}^{(2)}$	$\dot{\varepsilon}^{(3)}$	$\dot{\varepsilon}^{(1)}$	$\dot{\varepsilon}^{(2)}$	$\dot{\varepsilon}^{(3)}$	$\dot{\varepsilon}^{(1)}$	$\dot{\varepsilon}^{(2)}$	$\dot{\varepsilon}^{(3)}$
1A	27	65	72	37	107	54	3.03E-15	1.89E-14	7.12E-15	2.95E-14	1.84E-13	6.79E-14	2.04E-13	1.27E-12	6.79E-14
1C	24	69	77	40	118	57	3.91E-15	2.30E-14	8.44E-15	3.80E-14	2.23E-13	8.06E-14	2.63E-13	1.55E-12	8.06E-14
7C	30	60	68	34	95	50	2.29E-15	1.52E-14	5.89E-15	2.23E-14	1.48E-13	5.62E-14	1.54E-13	1.02E-12	5.62E-14
9C	39	50	56	28	70	41	1.07E-15	8.45E-15	3.53E-15	1.04E-14	8.21E-14	3.37E-14	7.22E-14	5.69E-13	3.37E-14
11A	25	67	75	38	113	55	3.46E-15	2.10E-14	7.78E-15	3.36E-14	2.03E-13	7.43E-14	2.33E-13	1.41E-12	7.43E-14
11B	48	44	49	25	56	35	6.29E-16	5.59E-15	2.46E-15	6.11E-15	5.43E-14	2.35E-14	4.23E-14	3.76E-13	2.35E-14
28	27	65	73	37	107	54	3.07E-15	1.91E-14	7.17E-15	2.98E-14	1.85E-13	6.85E-14	2.06E-13	1.28E-12	6.85E-14
29C	33	56	62	32	83	46	1.66E-15	1.19E-14	4.74E-15	1.61E-14	1.15E-13	4.52E-14	1.12E-13	7.97E-13	4.52E-14
33B	28	62	70	36	101	52	2.63E-15	1.69E-14	6.46E-15	2.55E-14	1.64E-13	6.17E-14	1.77E-13	1.14E-12	6.17E-14
33A	64	36	40	20	40	29	2.82E-16	3.00E-15	1.43E-15	2.74E-15	2.91E-14	1.37E-14	1.90E-14	2.02E-13	1.37E-14

[a]Stress estimated using differential piezometer, $\sigma^{(a)}$ – Twiss (1977), $\sigma^{(b)}$ – Twiss (1980), $\sigma^{(c)}$ – Mercier et al. (1977), $\sigma^{(d)}$ – Ord and Christie (1984), and $\sigma^{(e)}$ – Stipp and Tullis (2003).

[b]Differential strain rate calculated based on the stress estimated by theTwiss (1980), $\acute{\varepsilon}^{(1)}$ strain rates calculated reference to Hirth et al., 2001, $\acute{\varepsilon}^{(2)}$ strain rates calculated reference to Paterson and Luan (1990 $\acute{\varepsilon}^{(3)}$ strain rates calculated reference to Koch (1983).

Table 5. Paleopiezometry data for calcite and deduced strain rates.

Samples no.	Grainsize (μm)	[a]Differential stress (MPa)		[b]Strain rate $\dot{\varepsilon}$ (1/s), 300 °C		Strain rate $\dot{\varepsilon}$ (1/s), 250 °C	
	Average D	$\sigma^{(1)}$	$\sigma^{(2)}$	$\dot{\varepsilon}^{(1)}$	$\dot{\varepsilon}^{(2)}$	$\dot{\varepsilon}^{(1)}$	$\dot{\varepsilon}^{(2)}$
1D	55	45	28	1.21E-13	3.56E-15	2.68E-15	3.56E-15
7B	18	14	10	1.52E-13	4.09E-14	3.36E-15	4.09E-14
7A	87	70	41	1.11E-13	1.33E-15	2.45E-15	1.33E-15
9B	50	41	26	1.23E-13	4.26E-15	2.73E-15	4.26E-15
9C	35	28	19	1.33E-13	9.47E-15	2.94E-15	9.47E-15
7a	47	38	24	1.25E-13	4.93E-15	2.77E-15	4.93E-15
9A	50	40	25	1.24E-13	4.44E-15	2.74E-15	4.44E-15
10	52	42	26	1.23E-13	4.03E-15	2.72E-15	4.03E-15
7B	48	39	25	1.24E-13	4.65E-15	2.75E-15	4.65E-15
11C	54	44	27	1.22E-13	3.66E-15	2.69E-15	3.66E-15

[a] Stress estimated using differential piezometer, $\sigma^{(1)}$- Rutter (1995), $\sigma^{(2)}$- Schmid et al. (1980.

[b] Differential strain rate ($\acute{\varepsilon}$) calculated based on the stress estimated by the Rutter (1995), $\acute{\varepsilon}^{(1)}$ strain rates calculated reference to Walker et al. (1990), $\acute{\varepsilon}^{(2)}$ strain rates calculated reference to Herwegh et al. (2005).

In quartz, the differential flow stress ranges between between 36 and 69 MPa using the equations of Twiss (1977), between 40 and 72 MPa by the method of Twiss (1980), between 29 and 54 MPa according to Stipp and Tullis (2003), and between 20 and 40 MPa using the equation of Mercier et al. (1977). In the marbles, the stress similarly ranges between 28 and 70 MPa by the Rutter (1995) piezometer and between 19 and 41 MPa using the equation of Schmid et al. (1980). The results of Ord and Christie

(1984)yield a flow stress range of 40–118 MPa that seems to be too high. However, uncertainties still exist in the estimation of the flow stress from the grain size, which is influenced by the presence of other phases such as mica and chlorite, as well as the chemical environment during deformation.

8.2. Strain rate

The relationships between temperature, microstructures and CPO patterns corresponding to the dominant slip systems and thermobarometric calculations of chlorites indicate deformation under greenschist-facies conditions at temperatures of ca. 300–400 °C in the Moutsounas shear zone. We constructed deformation mechanisms of quartz aggregates at temperatures of 300, 350–400 °C. Using our flow stress and deformation temperature estimates, we calculated strain rates using the quartz flow law from the works of Hirth et al., 2001 and Paterson and Luan, 1990, and Koch (1983) to cover the range of available experimental results. When applied to the differential stress estimates from dynamically recrystallized grain sizes of quartz with the range covered by both piezometers (Twiss, 1980 and Stipp and Tullis, 2003), strain rate estimates from the Moutsounas shear zone range from 8.5×10^{-15}–2.3×10^{-13} S^{-1} at 300 °C, 6.1×10^{-15}–$1.8 \times 10^{--13}$ S^{-1} at 350 °C, and 8.1×10^{-14}–1.6×10^{-12} S^{-1} at 400 °C deformation conditions (Table 4).

The microstructures suggest that the grain size reduction of calcite was predominantly the result of subgrain rotation recrystallization accompanied by twin boundary migration suggests that mylonitization temperatures were higher than 250 °C (e.g., Burkhard, 1993, Ferrill et al., 2004 and Schenk et al., 2005). Using our flow stress and deformation temperature estimates, we calculated strain rates of calcite using the flow law from the works of Walker et al. (1990) and Herwegh et al. (2005) to cover the range of available experimental results. The strain rate of calcite lies in the range of 1.1×10^{-13}–1.52×10^{-13} S^{-1} at 300 °C and 2.5×10^{-15}–2.9×10^{-15} S^{-1} at 250 °C (Walker et al., 1990) and 4.1×10^{-14} S^{-1} to 1.3×10^{-15} S^{-1} and 4.1×10^{-14}–4.9×10^{-15} S^{-1} at 250 °C (Herwegh et al., 2005) with the Rutter (1995) piezometer (Table 5).

9. DISCUSSION

9.1. Evolution of microstructures and overprinting deformation

Microstructural analyses of the exhumed ductile Moutsounas shear zones provided insights into ductile deformation processes in the deep crust. The evolution of macro- and microstructures dominantly interpret the plastic deformation of major mineral phases (e.g. quartz, feldspar, white mica and calcite) in response to a progressive deformation history. The optical strain features including undulose extinction, planar deformation bands and subgrain formation, the development of quartz ribbons subdivided into subgrains or grains, as well as the presence of strain features in the newly recrystallized grains, indicate that dislocation creep was accompanied by dynamic recovery and dynamic recrystallization. Feldspar grains in the mylonites are occasionally characterized by irregular and serrated grain boundaries. Undulatory and inhomogeneous extinction are common in porphyroclastic feldspar grains. Dynamic recrystallization occurs in porphyroclast tails with asymmetric fabrics, which extend into layers defining the mylonitic foliation. Intense grain-size reduction of calcite generated ultramylonites. Twinning, twin migration and subgrain rotation are important processes of dynamic recrystallization of calcite in the marble mylonites within the Moutsounas shear zone. Certain rocks from the high-strain zone are completely transformed into extremely fine grains (e.g. white mica).

It is noteworthy that the microstructural and textural analysis has revealed the co-existence of microstructures ranging from middle-/high-temperatures to subsequent low-temperature within the Moutsounas shear zone. Quartz displays abundant evidence for early high-temperature plastic deformation (e.g. dynamic recrystallization, ribbon and polygonal grains with straight grain boundaries forming 120° triple junctions). These features suggest annealing during subsequent metamorphism and deformation at lower amphibolite facies M2 conditions (Buick, 1991a). The abundance of microstructural styles supports the later low-temperature deformational overprint during shearing. Elongated quartz ribbons, undulose and inhomogeneous extinction and deformation lamellae in quartz grains, as well as large amounts of extremely fine grains replacing ribbons, are direct evidence for low-temperature shearing. Low-temperature grain boundary migration processes or bulging recrystallization are therefore interpreted as responsible for the generation of such grains. Certain minerals such as white mica show extensive grain-size reduction in mylonites, which become ultramylonitic in some instances.

Dislocation creep, accompanied by dynamic recrystallization, was the principal deformation mechanism. Paleostress estimations of quartz and calcite suggest a flow stress within the range of 20 and 77 MPa during mylonitization. These estimates also are similar to the flow stress estimate of 28–62 MPa in quartzites, and 31–56 MPa in the marbles reported from

Iraklia, a small Aegean island located to the south of Naxos (Behrmann and Seckel, 2007). Schenk et al., 2005 and Schenk et al., 2007 report very low flow stresses of ca. 2 MPa at high temperatures in marble within marbles of the migmatite dome in the center of the Naxos MCC followed by high flow stresses of 48 MPa in low-temperature (271 ± 15 °C) shear zones. Although access of hydrous fluids into the Moutsounas shear zone is documented particularly in micaschists and phyllites, water had no weakening effect in associated marbles as experiments reveal (de Bresser et al., 2005).

Based on the dynamic recrystallization grain size of quartz and the deformation temperature, we estimate that the strain rate for the mylonitic rocks of Moutsounas shear zone were 10^{-13}–10^{-15} s^{-1} for 350 °C. At high temperatures, the estimated strain rate bySchenk et al. (2007) is similar at 4×10^{-15} s^{-1}. Our strain rate estimates are in good agreement with the expected tectonic strain rates and are also similar to that of 10^{-12}–10^{-14} s^{-1} reported by Behrmann and Seckel (2007) from Iraklia is land to the south of Naxos. However, the strain rate results indicate that they have no obviously different strain trend within this structural section, so more and wider testing of experimental flow laws to the well-constrained natural rocks in Naxos seems necessary.

9.2. Significance of quartz *c*-axes patterns

Prior to this study, the textural development of quartz in the Moutsounas shear zone was largely undocumented, although numerous high-quality data on structures and microfabrics have been published from the Naxos MCC (e.g., Buick, 1991b, Urai and Feenstra, 2001, Schenk et al., 2005, Schenk et al., 2007, Jolivet et al., 2010 and Kruckenberg et al., 2011). Quartz *c*-axes from the exhumed metamorphic rocks of Naxos have revealed that crystallographic preferred orientations preserve the record of polyphase deformation in these units. Asymmetric CPO patterns of quartz indicate a late top-to-the-N or NNE shear, which is consistent with field and microstructure observations. The activation of slip systems of quartz is temperature-sensitive because the *c*-axis orientation of quartz changes with temperatures (e.g., Stipp et al., 2002 and Passchier and Trouw, 2005). Prism <a> slip is commonly observed in high-grade metamorphic rocks (e.g., Egydio-Silva et al., 2002 and Otani and Wallis, 2006) and the basal <a> slip is observed in low-grade or overprinted metamorphic rocks (e.g. Toy et al., 2008, Cao et al., 2010, Cao et al., 2011a and Cao et al., 2011b). Quartz from the granitic gneisses and quartz-rich mylonites in the Naxos MCC exhibits numerous slip systems, including high-temperature slip systems: combined prisms <a> and rhomb<a> and low-temperature basal <a>slip. The development of strong quartz textures close to the migmatite dome and in the southern high-pressure belt of the Naxos MCC indicates high-middle temperature at amphibolite facies metamorphic conditions. High-temperature conditions were likely sufficient to influence deformation at higher structural levels, thereby facilitating localization of ductile strain. Medium-temperature shearing is responsible for the generation of the point maxima, which are characterized by ribbon quartz grains or grain aggregates. Low-temperature shearing is evidenced by muscovite grain size reduction and the formation of mica fish, and is also reflected in the quartz microfabrics.

Quartz *c*-axis fabric open-angles have been proposed as a potential thermometer byKruhl (1996) with opening angles increasing as the component of prism <c> and prism <a> slip becomes more important with increasing deformation temperatures. The Y point maxima (Group 2) suggests deformation-induced grain boundary migration recrystallization under high- or medium-temperature conditions from 300 °C to 600 °C (e.g. Stipp et al., 2002), which is consistent with optical observations. These texture patterns imply that the high-temperature deformation is mostly preserved in the migmatite dome and high-pressure belt and locally in the eastern Moutsounas shear zone. It is also noteworthy that these fabrics are formed by Y maxima superimposition onto crossed girdles or a single girdle. The superimposition is well supported by microstructural studies, which reveal that most sheared rocks from the high-pressure belt and the Moutsounas shear zone are relics of early high-temperature deformation. The high-temperature microstructures and textures are partly or entirely transposed by subsequent late low-temperature shearing. Therefore, the high-temperature conditions should be the results of an early event (M1 or peak conditions of M2). As suggested byToy et al. (2008), the strong Y-maxima of quartz from the mylonites and ultramylonites result from high ductile shear strains under amphibolite facies conditions deep within the crust. The development of these strong fabrics represents combined prism <a> and rhomb <a> slip. During exhumation, basal <a> slip is likely to become the more favored system, but the inherited strong high-temperature fabrics mean that prism <a> slip is likely to continue in these strongly oriented rocks. Basal <a> slip is dominant at lower temperature simple shear and at faster strain rates, causing a *c*-axis fabric maximum near the *Z*-axis of the finite strain ellipsoid.

9.3. Significance of calcite textures

Plastic deformation of calcite occurs in low- and high-temperature regimes (e.g. Wenk et al., 1987, Leiss and Barber, 1999, Pieri et al., 2001a, Pieri et al., 2001b, Liu et al., 2002,Leiss and Molli, 2003, Bestmann et al., 2000 and Romeo et al., 2007). The marble mylonites from the Moutsounas shear zone present different calcite texture types, which are related to meso- and microstructures and textures. Deformation accommodated twin and grain boundary migration, subgrain rotation, e-twinning slip and intracrystalline slip observed in the microstructures are important processes of dynamic recrystallization, and suggest that the texture patterns are probably influenced by dynamic recrystallization. These observations suggest that calcite crystals were deformed by intracrystalline plastic deformation (dislocation creep and twinning), which was accommodated by dissolution-precipitation. A progressive development of c-axes of calcites from the ultramylonites, mylonites and mylonitic calc-schist is well exhibited. The occurrence of the c-axis and e-pole great circle girdle textures normal to lineation appears to be produced in the initial stages of deformation. Subsequent progressive deformation rapidly gives way to other mechanisms, with strain increments giving rise to different LPOs (the c-maximum normal to lineation) and markedly changing the main shape fabric. At a low strain, the observed crystal flow behavior is best characterized by deformation through grain boundary sliding associated with dislocation creep processes, and dynamic recrystallization by subgrain rotation and grain bulging dominates (e.g. Schmid et al., 1977, Walker et al., 1990 and Romeo et al., 2007). On the other hand, in coarse-grained rocks, deformation is associated with dislocation creep processes and dynamic recrystallization by grain boundary migration (Schmid et al., 1980 and Rutter, 1995). Fabric softening (i.e. LPO) leads to crystal alignment in easy slip orientations (e.g., Poirier, 1980, Rutter et al., 1994,Rutter, 1998, Pieri et al., 2001a and Pieri et al., 2001b). These CPO patterns suggest that high-temperature deformation is controlled by slip on {0001} <11–20> (*c*<a>) and a low-temperature deformation is characterized by twinning on {−1018}<11–20>(*e*<a>). The (*c*<a>) and *e*<a> could have been active during deformation. However, *c*<a> is possibly the dominant slip system deformation at high shear strain dynamic recrystallization, because the {c} maximum is stronger than the {e} one in the *Z*-axes normal to the shear foliation. In this case, the texture of calcite was controlled by *c*<a>, combined with a low-temperature slip *e*<a>, which can be interpreted as the texture developed during exhumation. The different textures in the mylonite and ultramylonites also suggest a different amount of shear strain in which the concentration of deformation could have played a role. The amount of dynamic recrystallization increases with strain (e.g., Pieri et al., 2001b and Barnhoorn et al., 2004). The different rocks (e.g., mylonites and ultramylonites) experienced different strains under different strain regimes, which resulted from contrasting rheological behavior during deformation. The wider maxima of CPOs (e.g. samples 7A, 10C) could be related to a certain degree of thermal annealing after deformation as suggested by Barnhoorn et al. (2005). The asymmetry of the c-axis point maximum with respect to the shear plane has also been observed in porphyroclasts and recrystallized grains (e.g. sample 7A), indicating a sinistral sense of shear in agreement with the general kinematics of the shear zone.

Microfabrics and patterns of textures in calcite mylonites are sensitive to final stages of deformation (e.g. Lister and Williams, 1979). Therefore, the microfabrics and texture patterns of calcite can here document the kinematic character during the late, low-temperature stage of exhumation. It seems probable that the marbles have undergone a component of early coaxial deformation during the exhumation of the MCC. Simple shear is recorded in the mylonites and locally in ultramylonites. Compared to plastically deformed marble mylonites, non-coaxial deformation components on the calc-schists cannot be completely ruled out, but their effects are not obvious. Strain concentration during extension under retrograde conditions resulted in the localization of deformation in a narrow zone (Bestmann et al., 2000), where the imposed bulk strain is partitioned into simultaneously deforming coaxial and non-coaxial zones and pervasive dynamic recrystallization occurred in the shear zone.

9.4. Timing of the Moutsounas shear zone

No detailed petrological P-T data and only a few geochronological mineral ages are available from the Moutsounas shear zone ss mentioned earlier. Along the Moutsounas shear zone, the M2 temperature should diminish southward and the Eocene M1 metamorphism should dominate. Consequently, the temperature before retrogression after M1 and M2 metamorphism is similar along the strike of the Moutsounas shear zone. The following geochronological mineral ages are known from the Moutsounas shear zone. Andriessen et al., 1979 and Andriessen, 1991 reported four K–Ar white mica ages including 37.9 ± 0.9, 36.8 ± 1.1, 32.2 ± 0.9 and 41.8 ± 1.3 Ma from the southern portion of the Moutsounas shear zone $^{40}Ar/^{39}Ar$ phengite ages for the M1 metamorphism were dated at 45 ± 5 Ma by Wijbrans and McDougall, 1986 and Wijbrans and McDougall, 1988.Seward et al. (2009) report a zircon fission track age of 10.6 ± 1.2 Ma for the Moutsounas locality.

Consequently, no accurate age can be given for shearing. The minimum age must be at the younger age limit of the range of 19 to 14 Ma for M2 metamorphism (Keay, 1998,Keay et al., 2001 and Martin et al., 2006) and older than the zircon fission track age of 10.6 ± 1.2 Ma (Seward et al., 2009). These constraints make it likely that the normal faulting largely postdates the metamorphism M1, is constrained between ca. 14 and 10.6 Ma and occurred late during the deposition of the Pesulia Formation and is contemporaneous with initial stages of the Eromonisia Formation. However, a potential problem exists concerning the more precise chronostratigraphic indicators to limit its minimum timing of crystal fabric development relative to plastic deformation and dynamic recrystallization. Therefore, the time of shearing should be dated by future $^{40}Ar/^{39}Ar$ dating of single white grains grown during shearing.

The above mentioned large range of ages for the transition from M1 to M2 is explained by two models: (1) Buick and Holland (1989) invoked a progressive temperature increase during exhumation on the basis of thermobarometric data obtained on samples from different parts of the island. (2) Wijbrans and McDougall (1988) argued for partial thermal resetting of the HP–LT M1 paragenesis after its exhumation.

9.5. Stability of white mica and chlorite in sheared rocks

The new data from fine-grained sheared rocks reveal that earlier formed phengite-rich white mica potentially remains stable although later mechanical overprint during shearing. In contrast, chlorite is found to loose stability even within subgreenschist facies metamorphic conditions during shearing. A reason could be the interaction of older mineral assemblages with hydrous fluids, possibly hydrothermal fluids along the shear zone. This is believed to reset the chemical composition of chlorite. The final temperatures of chlorite thermometry need independent proof by other methods (e.g. de Caritat et al., 1993). We are taking, therefore, the temperatures of the chlorite thermometry only as approximate preliminary values. However, the data show that earlier microfabrics as those in microlithons (Type 1) yield higher temperatures than the subsequent sheared chlorites (Type 2). If the temperatures are approximately correct this would suggest that ductile shear in phyllosilicate-rich rocks could potentially occur at subgreenschist facies metamorphic conditions.

9.6. Tectonic interpretation of the Moutsounas shear zone

Since the discovery of the MCC of the lower unit at Naxos, the shear zone at the eastern boundary was considered as a normal fault toward the east (e.g. Lister et al., 1984 and Gautier et al., 1993; Jolivet et al., 1994) because of the juxtaposition of metamorphic rocks toward unmetamorphosed uppermost Oligocene to Miocene sediments (see also Vanderhaeghe et al., 2007). We consider the Moutsounas shear zone as a ductile oblique-slip low-angle normal fault with a motion of the hangingwall units toward the NNE. Later, the ductile oblique-slip fault was steepened due to subsequent E–W shortening. The E–W shortening is a late-stage process, which also resulted in the updoming of the MCC (Kruckenberg et al., 2011 and references therein) and in the formation of a synclinal structure containing Miocene sediments between the MCC in the east and the Naxos granodiorite in the west (Fig. 3b). The structures along the eastern upper margin of the

Moutsounas shear zone strongly correlate to contemporaneous structures like NNE shear and E-W shortening by folding (Urai et al., 1990,Vanderhaeghe, 2004 and Kruckenberg et al., 2011) within the interior of the Naxos MCC.

10. CONCLUSIONS

(1) We interpret the east-dipping Moutsounas shear zone as a lateral boundary of the Naxos MCC, which bounds tilted Miocene sediments exposed in the eastern hangingwall unit. Microstructural and textural analysis of samples and field observations reveals that the movement along the Moutsounas shear zone is associated with retrograde metamorphism, in which greenschist- and sub-greenschist metamorphic conditions overprinted the early high-temperature rocks (lower amphibolite facies).

(2) The Moutsounas shear zone formed in response to regional extension of Aegean continental crust. Extension on Naxos was accommodated, in part, by top-to-the-NNE ductile to ductile-brittle simple shear within greenschist-facies conditions. The microstructures and c-axis fabrics, taken together, disclose that ductile and brittle deformation was achieved by penetrative, non-coaxial deformation dominated by progressive simple shear.

(3) Based on the microstructure, texture and grain size of recrystallized quartz and calcite, we estimate that the differential stress and strain rate for the mylonitic rocks were respectively 20–77 MPa and 10^{-15}–10^{-13} s^{-1} ca. 350 °C for quartz and ca. 300 °C for calcite.

(4) Combined with published geochronological data, microstructures of samples and field observations, this indicates that the Moutsounas shear zone of the Naxos MCC is related to the main activity with a top-to-the-NNE shear with the main stage of updoming during migmatite formation and associated exhumation between ca. 15 and 11 Ma.

(5) Chlorite thermometry of the Moutsounas shear zone reveals two temperature groups, 300–360 °C, and 200–250 °C, are related to late stages of shearing. The lower temperature group is interpreted to result from late-stage hydrothermal overprint, which affected also older chlorite-bearing microfabrics. Chlorite in sheared rocks is likely accessible to late-stage resetting.

ACKNOWLEDGMENTS

We gratefully acknowledge critical but very constructive detailed reviews by Seth Kruckenberg and Janos Urai that helped us to considerably clarify ideas and presentation of this manuscript, and for the assistance of journal editor Thomas G. Blenkinsop This work is financially supported by the Austrian Science Fund (FWF)-Lise Meitner program, grant no. M-1343 to SC. Sincere thanks to Mr. R.-Y. Zhang and Mr. Y.-L. Dong for help during EBSD data acquisition.

REFERENCES

1. Altherr, R., Kreuzer, H., Wendt, I., Lenz, H., Wagner, G.A., Keller, J., Harre, W., Hohndorf, A., 1982. A late oligocene/early miocene high temperature belt in the Attic-Cycladic crystalline complex (SE Pelagonian, Greece). Geologisches Jahrbuch 23, 97–164.
2. Andriessen, P.A.M., 1991. K-Ar and Rb-Sr age determinations on micas of impure marbles of Naxos, Greece: the influence of metamorphic fluids and lithology on the blocking temperature. Schweizerische Mineralogische und Petrographische Mitteilungen 71, 89–99.
3. Andriessen, P.A.M., Banga, G., Hebeda, E.H., 1987. Isotopic age study of pre-Alpine rocks in the basal units on Naxos, Sikinos and Ios, Greek Cyclades. Geologie en Mijnbouw 66, 3–14.
4. Andriessen, P.A.M., Boelrijk, N.A.I.M., Hebeda, E.H., Priem, H.N.A., Verdurmen, E.A.T., Verschure, R.H., 1979. Dating the events of metamorphism and granitic magmatism in the Alpine Orogen of Naxos (Cyclades Greece). Contributions to Mineralogy and Petrology 69, 215–225.
5. Angelier, J., Glacon, G., Müller, C., 1978. Sur l'existance et la position tectonique du Miocène inférieur marin des l'archipel de Naxos (Cyclades, Grèce). Comptes Rendus Academie de Société de Paris 286, 21–24.
6. Angelier, J., Lyberis, N., Le Pichon, X., Barrier, E., Huchon, P., 1982. The tectonic development of the Hellenic arc and the Sea of Crete: a synthesis. Tectonophysics 86, 159–196.
7. Austin, N.J., Evans, B., 2007. Paleowattmeters: a scaling relation for dynamically recrystallized grain size. Geology 35, 343–346.

8. Avigad, D., 1998. High-pressure metamorphism and cooling on SE Naxos (Cyclades, Greece). European Journal of Mineralogy 10, 1309–1319.

9. Avigad, D., Garfunkel, Z., 1991. Uplift and exhumation of high-pressure metamorphic terrains: the example of the Cycladic blueschist belt (Aegean Sea). Tectonophysics 188, 357–372.

10. Bargnesi, E.A., Stockli, D., Mancktelow, N., Soukis, M., 2012. Miocene core complex development and coeval supradetachment basin evolution of Paros, Greece, insights from (UeTh)/He thermochronometry. Tectonophysics 595– 596, 165–182.

11. Barnhoorn, A., Bystricky, M., Burlini, L., Kunze, K., 2004. The role of recrystallization on the deformation behaviour of calcite rocks: larg strin torsion experiments on Carrara marble. Journal of Structural Geology 26, 885–903.

12. Barnhoorn, A., Bystricky, M., Burlini, L., Kunze, K., 2005. Post-deformational annealing of calcite rocks. Tectonophysics 403, 167–191.

13. Behrmann, J., 1985. Crystal plasticity and superplasticity in quartzite-a natural example. Tectonophysics 115, 101–129.

14. Behrmann, J.H., Seckel, C., 2007. Structures, flow stresses, and estimated strain rates in metamorphic rocks of the Small Cyclades Islands Iraklia and Schinoussa (Aegean Sea, Greece). Geotectonic Research 95, 1–11.

15. Bestmann, M., Kunze, K., Matthews, A., 2000. Evolution of a calcite marble shear zone complex on Thassos Island, Greece: microstructural and textural fabrics and their kinematic significance. Journal of Structural Geology 22, 1789–1807.

16. Böger, H., 1983. Stratigraphische und tektonische Verknuepfungen kontinentaler Sedimente des Neogens im Aegais-Raum. Geologische Rundschau 72, 771–813.

17. Bohnhoff, M., Rische, M., Meier, T., Endrun, B., Harjes, H.-P., Stavrakakis, G., 2004. A temporary seismic network on the Cyclades (Aegean Sea, Greece). Seismological Research Letters 75, 352–357.

18. Boronkay, K., Doutsos, T., 1994. Transpression and transtension within different structural levels in the central Aegean region. Journal of Structural Geology 16, 1555–1573.

19. Brichau, S., Ring, U., Ketcham, R.A., Carter, A., 2006. Constraining the longeterm evolution of the slip rate for a major extensional fault system in the central Aegean, Greece, using thermochronology. Earth and Planetary Science Letters 241, 293–306.

20. Brichau, S., Thomson, S., Ring, U., 2010. Thermochronometric constraints on the tectonic evolution of the Serifos detachment, Aegean Sea, Greece. International Journal of Earth Sciences 99, 379–393.

21. Brun, J.P., Sokoutis, D., 2007. Kinematics of the southern Rhodope core complex (North Greece). International Journal of Earth Sciences (Geol Rundsch) 96, 1079–1099.

22. Brunel, M., 1980. Quartz fabrics in shear zones mylonites: evidence of a major imprint due to late strain increments. Tectonophysics 64, 33–44.

23. Buick, I.S., 1991a. The late Alpine evolution of an extensional shear zone, Naxos, Greece. Journal of the Geological Society of London 148, 93–103.

24. Buick, I.S., 1991b. Mylonite fabric development on Naxos, Greece. Journal of Structural Geology 13, 643–655.

25. Buick, I.S., Holland, T.J.B., 1989. The PeTet path associated with crustal extension, in Evolution of Metamorphic Belts. In: Geological Society London, Special Publications, vol. 43, pp. 365–369.

26. Burkhard, M., 1993. Calcite twins, their geometry, appearance and significance as stress-strain markers and indicators of tectonic regime: a review. Journal of Structural Geology 15, 351–368.

27. Cao, S.Y., Liu, J.L., Hu, L., 2010. Orientation-related deformation mechanisms of naturally deformed amphibole in amphibolite mylonites from the Diancang Shan, SW Yunnan, China. Journal of Structural Geology 32, 606–622.

28. Cao, S.Y., Neubauer, F., Liu, J.L., Genser, J., Leiss, B., 2011a. Exhumation of the Diancang Shan metamorphic complex along the Ailao Shan-Red River belt, Yunnan, southwestern, China: evidence from 40Ar/39Ar thermochronology. Journal of Asian Earth Science 42, 525–550.

29. Cao, S.Y., Liu, J.L., Leiss, B., Vollbrecht, A., Genser, J., Neubauer, F., Zhao, C.Q., 2011b. Initiation of left-lateral deformation along the Ailao Shan-Red River shear zone: new microstructural, textural and geochronological constraints from the Diancang Shan metamorphic massif, SW Yunnan, China. International Geology Review 54, 348–367.

30. Cathelineau, M., 1988. Cation site occupancy in chlorites and illites as a function of temperature. Clay Minerals 23, 471–485.

31. Cathelineau, M., Nieva, D., 1985. A chlorite solid solution geothermometer. The Los Azufres (Mexico) geothermal system. Contributions to Mineralogy and Petrology 91, 235–244.

32. Coney, P., 1987. The regional tectonic setting and possible causes of Cenozoic extension in the North American Cordillera. In: Geological Society London, Special Publications, vol. 28, pp. 177–186.

33. Coney, P., Harms, T., 1984. Cordilleran metamorphic core complexes: Cenozoic extensional relics of Mesozoic compression. Geology 12, 550–554.

34. Davis, G.H., 1983. Shear-zone model for the origin of metamorphic core complexes. Geology 11, 342–347.

35. de Bresser, J., Ter Heege, J., Spiers, C., 2001. Grain size reduction by dynamic recrystallization: can it result in major rheological weakening? International Journal of Earth Science 90, 28–45.

36. de Bresser, J.H.P., Urai, J.L., Olgaard, D.L., 2005. Effect of water on the strength and microstructure of Carrara marble axially deformed at high temperatures. Journal of Structural Geology 27 (2), 265–281.

37. de Caritat, P., Hutcheon, I., Walshe, J.L., 1993. Chlorite geothermometry: a review. Clays and Clay Minerals 41, 219–239.

38. Duchêne, S., Aissa, R., Vanderhaeghe, O., 2006. Pressure-temperature-time evolution of metamorphic rocks from Naxos (Cyclades, Greece): constraints from thermobarometry and Rb/Sr dating. Geodinamica Acta 19, 301–321.

39. Doutsos, T., Kokkalas, S., 2001. Stress and deformation patterns in the Aegean region. Journal of Structural Geology 23, 455–472.

40. Egydio-Silva, M., Vauchez, A., Bascou, J., Hippertt, J., 2002. High-temperature deformation in the Neoproterozoic transpressional Ribeira belt, southeast Brazil. Tectonophysics 352, 203–224.

41. Ferrill, D.A., Morris, A.P., Evans, M.A., Burkhard, M., Groshong, R.H., Onasch, C.M., 2004. Calcite twin morphology: a low-temperature deformation geothermometer. Journal of Structural Geology 26, 1521–1529.

42. Floyd, M.A., Billiris, H., Paradissis, D., Veis, G., Avallone, A., Briole, P., McClusky, S., Nocquet, J.M., Palamartchouk, K., Parsons, B., 2010. A new velocity field for Greece: implications for the kinematics and dynamics of the Aegean. Journal of Geophysical Research 115, B10403.

43. Gautier, P., Ballévre, M., Brun, J.P., Jolivet, L., 1990. Extension ductile et bassins sédimentaires mio-liocénes dans les Cyclades (îles de Naxos et Paros). In: Comptes Rendus de l'Académie des Sciences. Série I. Mathématique, vol. 310, pp. 147–153.

44. Gautier, P., Brun, J.P., Jolivet, L., 1993. Structure and kinematics of upper Cenozoic extensional detachement on Naxos and Paros (Cyclades Islands, Greece). Tectonics 12, 1180–1194.

45. Grasemann, B., Schneider, D.A., Stöckli, D.F., Iglseder, C., 2012. Miocene bivergent crustal extension in the Aegean: evidence from the western Cyclades (Greece). Lithosphere 4, 23–29.

46. Gueydan, F., Mehl, C., Parra, T., 2005. Stress-strain rate history of a midcrustal shear zone and the onset of brittle deformation inferred from quartz recrystallized grain size. In: Geological Society London, Special Publications, vol. 243, pp. 127–142.

47. Hacker, B.R., Yin, A., Christie, J.M., Davis, G.A., 1992. Stress magnitude, strain rate, and rheology of extended middle continental crust inferred from quartz grain sizes in the Whipple Mountains, California. Tectonics 11, 36–46.

48. Hatzfeld, D., Besnard, M., Makropoulos, K., Voulgaris, N., Kouskouna, V., Hatzidimitriou, P., Panagiotopoulos, D., Karakaisis, G., Deschamps, A., LyonCaen, H., 1993. Subcrustal microearthquake seismicity and fault plane solutions beneath the Hellenic Arc. Journal of Geophysical Research 98, 9861– 9870.

49. Hejl, E., Riedl, H., Soulakellis, N., Van De Haute, P., Weingartener, H., 2003. Young Neogene tectonics ad relief development on the Aegean islands of Naxos, Paros and Ios (Cyclades, Greece). Mitteilungen Österreichischen Geologischen Gesellschaft 93, 105–127.

50. Hejl, E., Riedl, H., Weingartner, H., 2002. Post-plutonic unroofing and morphogenesis of the Attic-Cycladic complex (Aegea, Greece). Tectonophysics 349, 37–56.

51. Henjes-Kunst, F., Altherr, R., Kreuzer, H., Hansen, B.T., 1988. Disturbed U-Th-Pb systematics of young zircons and uranothorites: the case of the Miocene Aegean granitoids (Greece). Chemical Geology 73, 125–145.

52. Herwegh, M., de Bresser, J.H.P., ter Heege, J.H., 2005. Combining natural microstructures with composite flow laws: an improved approach for the extrapolation of lab data to nature. Journal of Structural Geology 27, 503–521.

53. Herwegh, M., Kunze, K., 2002. The influence of nano-scale second-phase particles on deformation of fine grained calcite mylonites. Journal of Structural Geology 24, 1463–1478.

54. Hirth, G., Teyssier, C., Dunlap, W.J., 2001. An evaluation of quartzite flow laws based on comparisons between experimentally and naturally deformed rocks. International Journal of Earth Science 90, 77–87.

55. Hirth, G., Tullis, J., 1992. Dislocation creep regimes in quartz aggregates. Journal of Structural Geology 14, 145–159.

56. Huet, B., Pourhiet, L.L., Labrousse, L., Burov, E.B., Jolivet, L., 2011. Formation of metamorphic core complex in inherited wedges: a thermomechanical modelling study. Earth and Planetary Science Letters 309, 249–257.

57. Hollenstein, C.M., Müller, M.D., Geiger, A., Kahle, H.G., 2008. Crustal motion and deformation in Greece from a decade of GPS measurements, 1993–2003.

58. Tectonophysics 449, 17e40. Jacobshagen, V., 1986. Geologie von Griechenland. Berlin (Gebrüder Borntraeger), p. 636. Jansen, J.B.H., 1973. Geological Map of Naxos-Nation. Inst. of Geology and Mining Research, Athens.

59. Jansen, J.B.H., 1977. The geology of Naxos. Geology and Geophysical Research [Athens, Greece] 19, 1–100.

60. Jansen, J.B.H., Schuiling, R.D., 1976. Metamorphism on Naxos: petrology and geothermal gradients. American Journal of Science 276, 1225–1253.

61. John, B.E., Howard, K.A., 1995. Rapid extension recorded by cooling-age patterns and brittle deformation, Naxos, Greece. Journal of Geophysical Research, 9969–9979.

62. Jolivet, L., Famin, V., Mehl, C., Parra, T., Aubourg, C., Hébert, R., Philippot, P., 2004. Strain localization during crustalescale boudinage to form extensional metamorphic domes in the Aegean Sea, In: Gneiss Domes in Orogeny. Geological Society of America 380, 185–210.

63. Jolivet, L., Faccenna, C., Piromallo, C., 2009. From Mantle to crust: stretching the Mediterranean. Earth and Planetary Science Letter 285, 198–209.

64. Jolivet, L., Lecomte, E., Huet, B., Denèle, Y., Lacombe, O., Labrousse, L., Le Pourhiet, L., Mehl, C., 2010. The north Cycladic detachment system. Earth and Planetary Science Letters 289, 87–104.

65. Jolivet, L., Brun, J.-P., Gautier, P., Lallemant, S., Patriat, M., 1994. 3Dkinematics of extension in the Aegean region from the early Miocene to the present: insights from the ductile crust. Bulletin de la Société Géologique Geologique de France 165 (3), 195–209.

66. Jolivet, L., Patriat, M., 1999. Ductile extension and the formation of the Aegean Sea. In: Durand, B., Jolivet, L., Horvàth, F., Séranne, M. (Eds.), The Mediterranean Basins: Tertiary Extension within the Alpine

Orogen, Geological Society Special Publication. Geological Society London, vol. 156, pp. 427–456.

67. Jowett, E.C., 1991. Fitting iron and magnesium into the hydrothermal chlorite geothermometer. In: GAC/MAC/SEG Joint Annual Meeting (Toronto, May 27.29, 1991). Program with Abstracts 16, A62. Toronto.

68. Keay, S., 1998. The Geological Evolution of the Cyclades, Greece: Constraints from SHRIMP U-Pb Geochronology. PhD thesis. Australian National University, Canberra, p. 341.

69. Keay, S., Lister, G., Buick, I., 2001. The timing of partial melting, Barrovianmetamorphism and granite intrusion in the Naxosmetamorphic core complex, Cyclades, Aegean Sea, Greece. Tectonophysics 342, 275–312.

70. Koch, P., 1983. Rheology and Microstructures of Experimentally Deformed Quartz Aggregates. Ph.D. thesis. University of California, Los Angeles. Koukouvelas, I.K., Kokkalas, S., 2003. Emplacement of the Miocene west Naxos pluton (Aegean sea, Greece): a structural study. Geological Magazine 140, 45–61.

71. Krabbendam, M., Urai, J.J., Vliet, L.J., 2003. Grain size stabilisation by dispersed graphite in a high-grade quartz mylonite: an example from Naxos (Greece). Journal of Structural Geology 25, 855–866.

72. Kranidiotis, P., MacLean, W.H., 1987. Systematics of chlorite alteration at the Phelps Dodge massive sulfide deposit, Matagami, Quebec. Economic Geology 82, 1898–1911.

73. Kruckenberg, S.C., Ferré, E.C., Teyssier, C., Vanderhaeghe, O., Whitney, D.L., Seaton, N.C.A., Skord, J.A., 2010. Viscoplastic flow in migmatites deduced from fabric anisotropy: an example from the Naxos dome, Greece. Journal of Geophysical Research 115, B09401.

74. Kruckenberg, S.C., Vanderhaeghe, O., Ferré, E.C., Teyssier, C., Whitney, D.L., 2011. Flow of partially molten crust and the internal dynamics of a migmatite dome, Naxos, Greece. Tectonics 30, TC3001.

75. Kruhl, J.H., 1996. Prism- and basal-plane parallel subgrain boundaries in quartz: a microstructural geothermobarometer. Journal of Metamorphic Geology 14, 581e589.

76. Kuhlemann, J., Frisch, W., Dunkl, I., Kázmér, M., Schmiedl, G., 2004. Miocene siliciclastic deposits of Naxos Island: geodynamic and environmental implications for the evolution of the southern Aegean Sea (Greece). In: Geological Society of America, Special Paper, vol. 378, pp. 51–65.

77. Law, R.D., 1986. Relationships between strain and quartz crystallographic fabrics in the Roche Maurice Quartzites of Plougastel, western Brittany. Journal of Structural Geology 8, 493–516.

78. Lee, J., Lister, G.S., 1992. Late Miocene ductile extension and detachment faulting, Mykonos, Greece. Geology 20, 121–124.

79. Leiss, B., Molli, G., 2003. "High-Temperature" texture in naturally deformed Carrara marble from the Alpi Apuane, Italy. Journal of Structural Geology 25, 649–658.

80. Leiss, B., Barber, D.J., 1999. Mechanisms of dynamic recrystallization in naturally deformed dolomite inferred from EBSP-analyses. Tectonophysics 303, 51–69.

81. Lewis, S., Holness, M., Graham, C., 1998. Ion microprobe study of marble from Naxos, Greece: Grain-scale fluid pathways and stable isotope equilibration during metamorphism. Geology 26, 935–938.

82. Le Pichon, X., Angelier, J., 1979. The Hellenic Arc and Trench system: a key to the Neotectonic evolution of the eastern Mediterranean area. Tectonophysics 60, 1–42.

83. Lister, G.S., Banga, C., Feenstra, A., 1984. Metamorphic core complexes of Cordilleran type in the Cyclades, Aegean Sea, Greece. Geology 12, 221–225.

84. Lister, G.S., Davis, G.A., 1989. The origin of metamorphic core complexes and detachment faults formed during Tertiary continental extension in the northern Colorado River region, U.S.A. Journal of Structural Geology 11, 65–94.

85. Lister, G.S., Snoke, A.W., 1984. S-C mylonites. Journal of Structural Geology 6, 617– 638.

86. Lister, G.S., Williams, P.F., 1979. Fabric development in shear zones: theoretical controls and observed phenomena. Journal of Structural Geology 1, 283–297.

87. Liu, J.L., Walter, J.M., Weber, K., 2002. Fluid-enhanced low-temperature plasticity of calcite marble: microstructures and mechanisms. Geology 30, 787–790.

88. Liu, J.L., Tang, Y., Tran, M.-D., Cao, S.Y., Zhao, Z.C., Zhao, Z.D., Chen, W., 2012. The nature of the Ailao ShaneRed River (ASRR) shear zone: constraints from structural, microstructural and fabric analyses of metamorphic rocks from the Diancang Shan, Ailao Shan and Day Nui Con Voi massifs. Journal of Asian Earth Sciences 47, 231–251.

89. Martin, L.A.J., Duchêne, S., Deloule, E., Vanderhaeghe, O., 2008. Mobility of trace elements and oxygen in zircon during metamorphism: Consequences for geochemical tracing. Earth and Planetary Science Letters 267, 161–174.

90. Martin, L., Duchêne, S., Deloule, E., Vanderhaeghe, O., 2006. The isotopic composition of zircon and garnet: a record of the metamorphic history of Naxos, Greece. Lithos 87, 174–192.

91. Massonne, H.J., Schreyer, W., 1987. Phengite geobarometry based on the limiting assemblage with K-feldspar, phlogopite, and quartz. Contributions to Mineralogy and Petrology 96, 212–224.

92. Massonne, H.J., Szpurka, Z., 1997. Thermodynamic properties of white micas on the basis of high pressure experiments in the systems K2OeMgOeAl2O3eSiO2 and K2OeFeOeAl2O3eSiO2eH2O. Lithos 41, 229e250.

93. McClusky, S., Balassanian, S., Barka, A., Demir, C., Ergintav, S., et al., 2000. Global Positioning System constraints on plate kinematics and dynamics in the eastern Mediterranean and Caucasus. Journal of Geophysical Research 105, 5695–5719.

94. Mehl, L., Hirth, G., 2008. Plagioclase preferred orientation in layered mylonites: evaluation of flow laws for the lower crust. Journal of Geophysical Research 113, B05202.

95. Mercier, J.C.C., Anderson, D.A., Carter, N.L., 1977. Stress in the lithosphere: Inferences from steady-state flow of rocks. Pure and Applied Geophysics 115, 199–226.

96. Okrusch, M., Bröcker, M., 1990. Eclogites associated with high-grade blueschist in the Cyclades archipelago, Greece: a review. European Journal ofMineralogy 2, 451–476.

97. Okudaira, T., Shigematsu, N., 2012. Estimates of stress and strain rate in mylonites based on the boundary between the fields of grain-size sensitive and insensitive creep. Journal of Geophysical Research 117, B03210.

98. Ord, A., Christie, J.M., 1984. Flow stresses from microstructures in mylonitic quartzites of the Moine thrust zone, Assynt area, Scotland. Journal of Structural Geology 6, 639–654.

99. Otani, M., Wallis, S.R., 2006. Quartz lattice preferred orientation patterns and static recrystallization: natural examples from the Ryoke belt, Japan.Geology 34, 561–564.

100. Passchier, C.W., Trouw, R.A.J., 2005. Microtectonics. Springer-Verlag, Berlin, p. 366.

101. Paterson, M.S., Luan, F.C., 1990. Quartzite rheology under geological conditions. In: Knipe, R.J., Rutter, E.H. (Eds.), Deformation Mechanisms, Rheology and Tectonics, Geological Society London Special Publication, vol. 54, pp. 299–307.

102. Pe-Piper, G., 2000. Origin of S-type granites coeval with I-type granites in the Hellenic subduction system, Miocene of Naxos, Greece. European Journal of Mineralogy 12, 859–875.

103. Pe-Piper, G., Kotopouli, C.N., Piper, D.J.W., 1997. Granitoid rocks of Naxos, Greece; regional geology and petrology. Geological Journal 32, 153–171.

104. Pieri, M., Burlini, L., Kunze, K., Stretton, I., Olgaard, D.L., 2001a. Rheological and microstructural evolution of Carrara marble with high shear strain; results from high temperature torsion experiments. Journal of Structural Geology 23, 1393–1413.

105. Pieri, M., Kunze, K., Burlini, L., Stretton, I., Olgaard, D.L., Burg, J.P., Wenk, H.R., 2001b. Texture development of calcite by deformation and dynamic recrystallisation at 1000 K during torsion experiments of marble to large strains. Tectonophysics 330, 119–140.

106. Poirier, P., 1980. Shear localization and shear in stability in materials in the ductile field. Journal of Structural Geology 2, 135–142.

107. Prior, D.J., Trimby, P.W., Weber, U.D., Dingley, D.J., 1996. Orientation contrast imaging of microstructures in rocks using forescatter detectors in the scanning electron microscope. Mineralogical Magazine 60, 859e869.

108. Reischmann, T., 1998. Pre-Alpine origin of tectonic units from the metamorphic core complex of Naxos, Greece, identified by single zircon Pb/Pb dating. Bulletin of the Geological Society of Greece 32, 101–111.

109. Riedl, H., 1982a. Vergleichende Untersuchungen zur Geomorphologie der Kykladen unter besonderer Berücksichtigung der Insel Naxos. Salzburger Exkursionsberichte 8, 9e54.

110. Riedl, H., 1982b. Die Altflächenentwicklung der Kykladen. Annales Geologiques des Payes Héllenicques 31, 33–84.

111. Ring, U., Glodny, J., Will, J., Thomson, S., 2010. The Hellenic subduction system: high-pressure metamorphism, exhumation, normal faulting, and large-scale extension. Annual Review of Earth and Planetary Sciences 38, 45–76.

112. Rösler, G., 1978. Relics of non-metamorphic sediments on central Aegean Islands. In: Closs, H., Roeder, D., Schmidt, K. (Eds.), Alps,

Apennines, Hellenides. In: Stuttgart, E. (Ed.), Schweizerbart, Inter-union Commission on Geodynamics Scientific Report, 38, pp. 480–481.

113. Romeo, B., Capote, R., Lunar, R., 2007. Crystallographic preferred orientations and microstructure of a Variscan marble mylonite in the Ossa-Morena Zone (SW Iberia). Journal of Structural Geology 29, 1353–1368.

114. Rutter, E.H., 1998. On the relationship between the formation of shear zones and the form of the flow law for rocks undergoing dynamic recrystallization. Tectonophysics 303, 147–158.

115. Rutter, E.H., 1995. Experimental study of the influence of stress temperature and strain on the dynamic recrystallization of Carrara marble. Journal of Geophysical Research 100, 24,651–24,663.

116. Rutter, E.H., Casey, M., Burlini, L., 1994. Preferred crystallographic orientation development during the plastic and superplastic flow of calcite rocks. Journal of Structural Geology 16, 1431–1446.

117. Schenk, O., Urai, J., Evans, B., 2005. The effect of water on recrystallization behavior and grain boundary morphology in calcite-observations of natural marble mylonites. Journal of Structural Geology 27, 1856–1872.

118. Schenk, O., Urai, J.L., van der Zee, W., 2007. Evolution of boudins under progressively decreasing pore pressure e a case study of pegmatites enclosed in marble deforming at high grade metamorphic conditions, Naxos, Greece. American Journal of Science 307, 1009–1033.

119. Schmid, S.M., Casey, M., 1986. Complete fabric analysis of some commonly observed quartz C-axis patterns, in mineral and rock deformation: laboratory Studies. Geophysical Monograph Series 36, 263–286.

120. Schmid, S.M., Boland, J.N., Paterson, M.S., 1977. Superplastic flow in fine grained limestone. Tectonphysics 43, 257–291.

121. Schmid, S.M., Paterson, M.S., Boland, J.N., 1980. High temperature flow and dynamic recrystallization in Carrara marble. Tectonophysic 65, 245–280.

122. Seward, D., Vanderhaeghe, O., Siebenaller, L., Thomson, S., Hibsch, C., Zingg, A., Holzner, P., Ring, U., Duchêne, S., 2009. Cenozoic tectonic evolution of Naxos Island through a multi-faceted approach of fission-track analysis. In: Geological Society London, Special Publications, vol. 321, pp. 179–196.

123. Siebenaller, L., Boiron, M.C., Vanderhaeghe, O., Hibsch, C., Jessell, M.W., Andremayer, A.S., France-Lanord, C., Photlades, A., 2013. Fluid record of rock exhumation across the brittleeductile transition during formation of a Metamorphic Core Complex (Naxos Island, Cyclades, Greece). Journal of Metamorphic Geology 31, 313–338.

124. Stipp, M., Stünitz, H., Heilbronner, R., Schmid, S., 2002. The eastern Tonale fault zone: a 'natural laboratory' for crystal plastic deformation of quartz over a temperature range from 250 to 700 C. Journal of Structural Geology 24, 1861–1884.

125. Stipp, M., Tullis, J., 2003. The recrystallized grain size piezometer for quartz. Geophysical Research Letters 30, 2088.

126. Stipp, M., Tullis, J., Scherwath, M., Behrmann, J.H., 2010. A new perspective on paleopiezometry: dynamically recrystallized grain size distributions indicate mechanism changes. Geology 38, 759–762.

127. Strumpf, N., 1997. Das Kernkristallin von Naxos (Kykladen, Griechenland). Berlin. Schriftenreihe für Geowissenschaften 5, 1–140.

128. Toy, V.G., Prior, D.J., Norris, R.J., 2008. Quartz fabrics in the Alpine Fault mylonites: Influence of pre-existing preferred orientations on fabric development during progressive uplift. Journal of Structural Geology 30, 602–621.

129. Twiss, R.J., 1980. Static theory of size variations with stress for subgrains and dynamically recrystallized grains, in U.S.G.S. Open-file Report 80-625, 665- 683.

130. Twiss, R.T., 1977. Theory and applicability of a recrystallized grain size paleopiezometer. Pure and Applied Geophysics 115, 227–244.

131. Urai, J.L., Schuiling, R.D., Jansen, J.B.H., 1990. Alpine deformation on Naxos (Greece). In: Geological Society London, Special Publication, vol. 54, pp. 509–522.

132. Urai, J.L., Feenstra, A., 2001. Weakening associated with the diaspore-corundum dehydration reaction in metabauxites; an example from Naxos (Greece). Journal of Structural Geology 23, 941–950.

133. Vanderhaeghe, O., 2004. Structural development of the Naxos migmatite dome. In: Whitney, D.L., Teyssier, C., Siddoway, C.S. (Eds.), Gneiss Domes in Orogeny, Geological Society of America Special Paper, 380, pp. 211–227.

134. Vanderhaeghe, O., Hibsch, C., Siebenaller, L., Duchêne, S., de St Blanquat, M., Kruckenberg, S., Fotiadis, A., Martin, L., 2007. Penrose conference e extending a continent e naxos field guide. In: Lister, G.,

Forster, M., Ring, U. (Eds.), Inside the Aegean Metamorphic Core Complexes. Journal of the Virtual Explorer vol. 27.

135. Walcott, C., White, S., 1998. Constraints on the kinematics of post-orogenic extension imposed by stretching lineations in the Aegean region. Tectonophysics 298, 155–175.

136. Walker, A.N., Rutter, E.H., Brodie, K.H., 1990. Experimental study of grain-size sensitive flow of synthetic, hot-pressed calcite rocks. Geological Society Special Publications 54, 259–284.

137. Warren, J.M., Hirth, G., 2006. Grain size sensitive deformation mechanisms in naturally deformed peridotites. Earth and Planetary Science Letters 248, 438– 450.

138. Wawrzenitz, N., Krohe, 1998. Exhumation and doming of the Thasos metamorphic core complex (S Rhodope, Greece). Structural and geochronological constraints. Tectonophysics 285, 301–332.

139. Wenk, H.R., Takeshita, T., Bechler, E., Erskine, B.G., Matthies, S., 1987. Pure shear and simple shear calcite textures. Comparison of experimental, theoretical and natural data. Journal of Structural Geology 9, 731–745.

140. Wernicke, B., 1985. Uniform sense simple shear of the continental lithosphere. Canadian Journal of Earth Sciences 22, 108–125.

141. Wernicke, B., Axen, G.J., 1988. On the role of isostasy in the evolution of normal fault systems. Geology 16, 848–851.

142. Whitney, D.L., Teyssier, C., Rey, P., Buck, R.W., 2013. Continental and oceanic core complexes. Geological Society of America Bulletin 125, 273–298.

143. Wijbrans, J.R., McDougall, I., 1986. 40Ar/39Ar dating of white micas from an Alpine high-pressure metamorphic belt on Naxos Greece): the resetting of the argon isotopic system. Contributions to Mineralogy and Petrology 93, 187–194.

144. Wijbrans, J.R., McDougall, I., 1988. Metamorphic evolution of the Attic Cycladic metamorphic belt on Naxos (Cyclades, Greece) utilizing 40Ar/39Ar age spectrum measurements. Journal of Meramorphic Geology 6, 571–594.

145. Wortel, M.J.R., Spakman, W., 2000. Subduction and slab detachment in the MediterraneaneCarpathian region. Science 290, 1910–1917.

5

CHAPTER

Characterization and Origin of Aluminous A-type Granites from the Lachlan Fold Belt, Southeastern Australia

P.L. King[1,*], A.J.R. White[2], B.W. Chappell[1] and C.M. Allen[1]

[1]*Department of Geology, Australian National UniversityCanberra, A.C.T. 0200, Australia*

[2]*VIEPS, Department of Earth Sciences, Melbourne UniversityParkville, VIC. 3052, Australia*

ABSTRACT

The metaluminous to weakly peraluminous A-type granites of the Lachlan Fold Belt are a distinctive group of igneous rocks, on the basis of chemical and mineralogical criteria. Those granites that contain ~65–72% SiO_2 can be distinguished from other types on the basis of higher abundances of $Fe_{total}/(Fe_{total} + Mg)$, high field strength elements, trivalent rare earth elements, Ga and Zn. Mineralogically, they contain Fe-rich hydrous mafic minerals and primary ilmenite, and hence are reduced relative to the Ni–NiO buffer. However, the extremely felsic A-type granites (SiO_2> ~72%) have the same chemical and mineralogical characteristics as felsic, fractionated I-type granites. Recent analyses indicate that the Lachlan Fold Belt A-type granites have Sc, F, alkali element, trace transition element and H_2O contents similar to those of other unfractionated I-type granites. Rb–Sr and Nd–Sm isotopic compositions are highly variable, probably reflecting source region heterogeneity. The metaluminous to weakly peraluminous A-type granites of the Lachlan Fold Belt are distinct from peralkaline rocks in terms of chemical composition, petrography and field associations, although these rocks have been grouped together as a single type in current classification schemes. We propose that the metaluminous to weakly peraluminous A-type granites, such as those of the Lachlan Fold Belt, should be defined as 'aluminous A-type' granites and should not be grouped with peralkaline granites. The Lachlan Fold Belt aluminous A-type granites have relatively high calculated zircon saturation temperatures. We suggest that these granites were produced by high-temperature partial melting of a felsic infracrustal source.

Key words: A-type granite; aluminous A-type; peralkaline; Australia

INTRODUCTION

Rock classification schemes are useful, not only for characterizing rocks, but also because they potentially bring us closer to understanding fundamental rock-forming processes. One of the most common methods of defining granite types is in terms of mineralogical and/or chemical compositions, which are measurable and non-genetic parameters (e.g. cordierite granite, peraluminous granite). A chemical–tectonic approach to granite classification has been taken by some researchers, where chemical parameters are taken as indicative of the tectonic environment in which the granite formed (e.g. Pearce *et al.*, 1984). Others have inferred magma source compositions from chemical and mineralogical features, for example the I- and S-type classification of Chappell & White (1974, 1992). The proliferation of classification schemes has been reviewed by Clarke (1992) and Pitcher (1993), and reflects both the difficulty of choosing definitive criteria for classification and the complexity of granites.

The term 'A-type' was proposed in an abstract by Loiselle & Wones (1979)to distinguish 'mildly alkaline' rocks (high K_2O + Na_2O) from typical 'calc-alkaline' (I-type) rocks. Other distinctive features that those workers proposed were high ratios of $Fe_{total}/(Fe_{total} + Mg)$ and F/H_2O ratios, high abundances of high field strength elements (HFSE) and trivalent rare earth elements (REE^{3+}), and low abundances of Fe–Mg trace elements (Cr, V, Ni, Cu), Ba, Sr and Eu. Loiselle & Wones (1979) included peralkaline, metaluminous and weakly peraluminous rocks in their A-type group. In addition, they noted that A-type granites are characterized by emplacement in a rift zone or stable continental block, and are thus 'anorogenic'.

The characterization and origin of A-type granites have been the subject of much debate. Previous workers have noted that the original definition is problematic for three main reasons. First, A-type classification is inconsistent with other classification schemes (Sylvester, 1989; Creaser *et al.*, 1991). This has led to some workers proposing further subdivisions of the A-type group, and those subdivisions use different criteria [compareWhalen & Currie (1990) and Poitrasson *et al.* (1995) with Eby (1992)]. Second, A-type rocks overlap compositionally with felsic I-type rocks (Whalen *et al.*, 1987; Chappell & White, 1992). Third, the tectonic definition of 'anorogenic' A-type granites is confusing and also it does not always correspond to the geochemical definition (e.g. Maniar & Piccoli, 1989; Greenberg, 1990; Rogers & Greenberg, 1990; below). However, despite the extensively published views that the term A-type is poorly defined, many workers find it a useful way to describe rocks (e.g. Hogan *et al.*, 1992).

We consider that the current A-type definitions do not adequately define a distinct group of igneous rocks world-wide and that the popular A-type definitions (Loiselle & Wones, 1979; Eby, 1990, 1992) are too broad. In this paper, we present new data on aluminous (metaluminous to weakly peraluminous) granites from the Lachlan Fold Belt (LFB) of southeastern Australia that have previously been defined as 'A-type' (Collins *et al.*, 1982; Whalen *et al.*, 1987; Chappell *et al.*, 1991). The wealth of data on the LFB granites (Chappell & White, 1992) makes it an ideal place to assess the distinctive characteristics that might also be used to identify A-type granites elsewhere. We use these characteristics to distinguish the LFB A-type granites from some I-type granites, particularly those that are fractionated (where the term 'fractionated' refers to fractional crystallization of dominantly quartz and feldspars). We also compare the LFB A-type granites with some peralkaline granites, and argue that rocks with peralkaline associations are very different from aluminous A-type rocks and that the two types should not be grouped together. Finally, we provide a petrogenetic model for the LFB aluminous A-type granites.

GEOLOGIC SETTING

Granites, dominantly of Silurian and Devonian age, make up ~20% of the area of the LFB. On the basis of geochemical arguments, a small proportion of these rocks (0.6%) have been referred to as 'A-type' (Collins*et al.*, 1982; Whalen *et al.*, 1987; Chappell *et al.*, 1991). Those rocks are listed in Table 1 and illustrated in Fig. 1. We will not consider the A-type granites that have only limited geochemical and petrographic data.

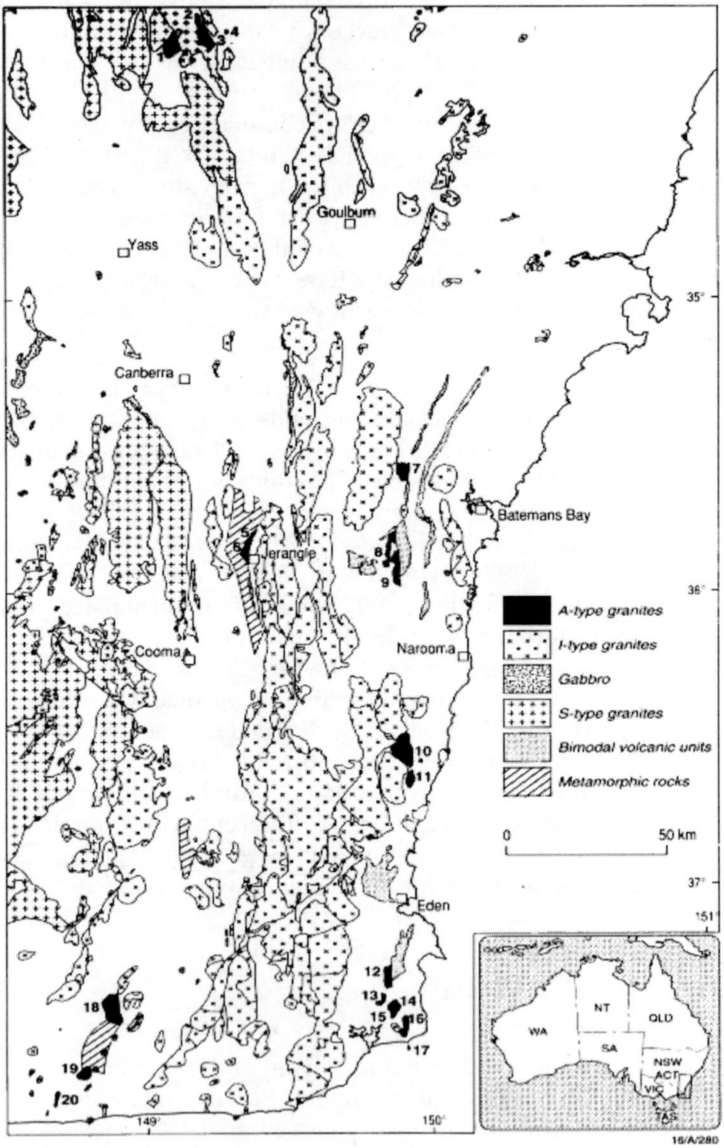

Figure 1. Map showing the location of the I-, S- and A-type granites in the Lachlan Fold Belt, along with andalusite to sillimanite grade metamorphic rocks, gabbros and bimodal rift complexes. The A-type granites have a wide distribution and make up ~0.6% of the granites exposed in the belt. The names of the granite units are given in Table 1.

Table 1. List of A-type granites from the Lachlan Fold Belt

Batholith	Suite	Unit	No. in Fig. 1	Area (km²)	Amphibole type*	No. of analyses	% SiO_2 mean	p.p.m. Zr mean	Reference
Metaluminous to weakly peraluminous rocks									
Wyangala	Yewrangara	Yewrangara	1	38	fhast P	7	73·91*	248	Yacopetti (1987)
	Spring Road	Spring Road	2	1·7	hb O	1	70·01	390	Yacopetti (1987)
	Blackmans Ck	Blackmans Ck	3	30	hb O	1	76·83	165	Yacopetti (1987)
	Burnt Hill	Burnt Hill	4	0·2	?				Yacopetti (1987)
Bega†	Wangrah	Danswell Ck	5	4·4	hb–fed P	5	71·25	464	Richards (1967), King (1992)
	Wangrah	Wangrah	6	17·5	hb–fed P	8	74·00*	240	Richards (1967), King (1992)
	Wangrah	Dunskeig	6	2·5	hb–fed P	3	76·60	120	Richards (1967), King (1992)
	Wangrah	Eastwood	6	4	hb–fed P	2	75·87	166	Richards (1967), King (1992)
	Monga	Monga	7	24	fact P	4	72·41*	343	Wyborn & Owen (1986)
	Mongamula		8	?	fact O				Wyborn & Owen (1986)
	Coondella Ck		9	27	fact O				Wyborn & Owen (1986)
	Mumbulla	Mumbulla	10	56	none	6	77·23	173	Collins (1977), Collins *et al.* (1982)
	Mumbulla	Dr George Mt	11	13	none	2	77·17	164	Collins (1977), Collins *et al.* (1982)
Gabo Island	Gabo Island	Watergums	12	28	fhast–r P	3	73·00	466	Collins (1977), Collins *et al.* (1982)
	Gabo Island	Naghi	13	3·5	fhast O	1	73·62	496	Collins (1977), Collins *et al.* (1982)
	Gabo Island	Nagha	14	5·3	fhast O	2	73·42	502	Collins (1977), Collins *et al.* (1982)
	Gabo Island	Carlyle	15	4·6	fhast O				Collins (1977), Collins *et al.* (1982)
	Gabo Island	Howe Range	16	14·5	fhast O	2	72·40	523	Collins (1977), Collins *et al.* (1982)
	Gabo Island	Gabo Island	17	1·6	fhast P	1	73·06	464	Collins (1977), Collins *et al.* (1982)
Bonang	Ellery	Ellery	18	64	fhast P	3	71·91	311	Dodd (1981)
	Murrungowar	Murrungowar	19	19·5	none	2	69·76	335	Dodd (1981)
		Mt Raymond	20	5·8	r O				Dodd (1981)
Wagga‡	Mt Arthur	Mt Arthur	?	?	?	?	75·87	191	Pogson (pers. comm., 1991)
	unassigned	Mt Foster	?	?	?	?			Pogson (pers. comm., 1991)
	unassigned	Mt Harris	?	?	?	?			Pogson (pers. comm., 1991)
	Boohlahbone	Boohlahbone	—	?	none				Barron *et al.* (1982)
Peralkaline rock associations									
Wagga	Narraburra	Narraburra	—	23	arf P, r O	10	77·73	428	Wormald & Price (1988), Wormald (1991)
	Narraburra	Fernleigh	—	1	arf O, r O	2	77·64	294	Wormald (1991)
	Narraburra	Wattle Park§	—	?	arf P	1	77·53	440	Wormald (1991)
	Narraburra	Boginderra	—	?	arf P, r P	4	74·60	593	Wormald (1991)
	Narraburra	Pinehurst	—	?	arf P, r O	1	75·67	600	Wormald (1991)
	Narraburra	Solitary Hill	—	0·5	arf O	1	75·76	239	Wormald (1991)
	Narraburra	Chellington	—	1·5	altered	1	66·78	181	Wormald (1991)
Western Victoria	unassigned	Victoria Valley	—	?	arf, r O	2	76·81	778	Atkinson (1976)

hast, ferrohastingsite; fact, ferroactinolite; hb, (ferro)hornblende; fed, ferroedenite; r, riebeckite; fed, ferroedenite; arf, arfvedsonite; O, composition determined optically; P, composition determined from electron microprobe analysis (some analyses performed as part of this work, others by the references cited). Amphibole classification is after Leake (1978).

*The Yewrangara, Wangrah and Monga units have a range in SiO_2 >4%. The other units are much more homogeneous, on the basis of available analyses.

†Reference to the A-type granites of the Bega Batholith is found in Lewis *et al.* (1994).

‡The Mount Foster and Mount Arthur granites were classified as A-type by Chappell *et al.* (1991); however, further work indicates that these are not definitely A-type. The Mount Arthur Granite is also known as the Milmiland Granite (Geological Survey of N.S.W.).

§The Wattle Park pluton was given that name by Chappell *et al.* (1991); however, this pluton was named Avondale by Wormald (1991). As there is another pluton named Avondale in the LFB, Wattle Park is preferred.

See Fig. 1 for the locations of these units.

The LFB provides a somewhat controversial geological setting for the A-type granites [see papers in Fergusson & Glen (1992)]. This is partly because it is a fold belt at a continental margin with no precise evidence for the nature of the heat source. For example, distinctive rock types found in modern subduction zone environments are not present. On the basis of geochemical arguments put forward by Chappell (1994) relating to the origin of the S–type granites, the widely distributed Ordovician turbidite deposits (that provide the matrix to all of the granites not associated with volcanic rocks) were apparently deposited on continental crust. Chappell also proposed that an older basement may have been thinned by crustal extension near the end of the Cambrian. This model is consistent with deposition of marine Ordovician sediments on a depressed, thinned basement followed by enhanced heat flow in the Silurian to Devonian causing massive crustal melting to produce the granites.

Because of the restricted extent of A-type granites in the LFB, it is difficult to generalize about their field associations, although most occur in two linear belts at the margins of batholiths or near major pre-existing faults (King *et al.*, 1992). Some A-type granites are associated with bimodal volcanic (rift) complexes with metaluminous A-type rhyolite (Fig. 1), suggesting emplacement in an extensional setting. Other A-type granites intrude pre-existing andalusite to sillimanite grade regional metamorphic rocks and/or are found near rare gabbro (Fig. 1). Like many other granites in the LFB, the A-type granites are mostly shallow-level intrusions, and in some cases, comagmatic rhyolitic volcanic rocks are preserved (e.g. Boohlahbone Suite; Barron *et al.*, 1982). Near-surface crystallization of these granites is also supported by the observation of aplites, miarolitic cavities, some pod pegmatites and graphic intergrowths seen in thin section (Burnham & Ohmoto, 1980).

In the LFB, A-type granites show evidence for emplacement both during and after other granite magmatism and deformation (King *et al.*, 1992). This differs from the common inference that A-type granites world-wide are anorogenic. For example, unfoliated A-type plutons intrude foliated S-type granites, indicating emplacement after a major episode(s) of magmatism and deformation (Yewrangara and Blackman's Creek). Evidence for emplacement pre- or syn-deformation comes from A-type granites that are strongly foliated parallel to the regional deformation (Bonang Complex). Finally, emplacement of both I- and A-type granites late in the magmatic sequence is evidenced by U–Pb ion microprobe zircon dating, as noted by Williams *et al.* (1991), Williams (1992) and King *et al.*(1992). Those workers found that the Watergums A-type granite and nearby I-type granites had magmatic ages of 395 Ma; ~20 Ma younger than the majority of granites in the adjacent Bega Batholith.

Such complexities imply that although A-type granites commonly form late in a tectonic and magmatic cycle, that is not always the case; therefore 'anorogenic' cannot be used as a general term to characterize these rocks. Also, the term 'anorogenic' is used with various meanings. Some workers have applied the term anorogenic magmatism to magmatism postdating major episodes of granitic magmatism (e.g. Bonin *et al.*, 1978). Other workers define orogeny to include both magmatic and structural activity. Furthermore, it can be difficult to constrain an orogeny to a specific time period, owing to lateral variations in the intensity of structural activity within a terrane, some of which may result from orogenic activity in adjacent terranes (e.g. the LFB and New England Fold Belt; Chappell, 1994).

PETROGRAPHY

In the LFB, A-type granites are commonly pink, owing to abundant brick red K-feldspar, but cream–white K-feldspar is also present (Wangrah, Ellery and Monga suites and the A-type granites of the Wyangala Batholith). Plagioclase is normally zoned (commonly An_{40-15}) and no cores with higher anorthite contents have been identified. As with other near-surface granites, quartz occurs as embayed bipyramidal crystals in some rocks and quartz-feldspar intergrowths are common. In many of the granites, quartz and feldspar proportions are close to near minimum-temperature melt compositions for pressures of 100–200 MPa. All A-type granites of the LFB contain annite and the majority contain small amounts of amphibole. Amphibole compositions are summarized in Table 1. The crystal habits of mafic minerals tend to be specific to individual plutons. Both annite and amphibole are observed as anhedral and euhedral crystals and are found interstitially, individually, and in mineral aggregates. Annite rarely pseudomorphs Fe-rich amphibole.

Notably, in the LFB, the aluminous A-type granites never contain pyroxene or olivine, although some Ba-rich A-type volcanic rocks contain pyroxene. This differs from many A-type and peralkaline rocks described elsewhere, e.g. Pikes Peak batholith (Barker *et al.*, 1975), North America–Labrador–Greenland (Anderson, 1983), Padthaway Ridge, South Australia (Turner *et al.*, 1992). Collins *et al.* (1982) inferred that dark brown material in the Gabo Island Suite pseudomorphs fayalite; our electron microprobe analyses indicate that this is either stilpnomelane or a smectite–Fe oxide mixture and there is no firm evidence for the identity of the original mineral.

Ubiquitous accessory minerals include euhedral zircon and apatite (both interpreted as melt-precipitated), whereas allanite and xenotime occur in some granites. Oxide minerals are ilmenite with lesser magnetite, the significance of which is discussed below. Minor deuteric alteration, typical of intrusive rocks, is common, with secondary minerals including chlorite, muscovite, fluorite, epidote, sphene, siderite, magnetite, hematite, stilpnomelane and amorphous iron oxides.

Enclave types in the A-type granites of the LFB have abundances that are suite-specific. Granites of the Gabo Island and Mumbulla suites contain rare enclaves, xenocrysts and phenocrysts. In contrast, the Wangrah Suite contains phenocrysts, rapakivi texture and microgranular enclaves, with considerable textural heterogeneity at all scales.

CHEMICAL COMPOSITIONS

Chemical compositions of 13 selected LFB A-type granites and the average composition are listed in Table 2. Major element data were obtained by X-ray spectrometry using fused discs (Norrish & Hutton, 1969). The trace elements Rb, Sr, Ba, Zr, Nb, Y, V, Ni, Cu, Zn, Ga, As, Sn and Pb were measured on pelletized powder samples using the X-ray spectrometric methods described by Chappell (1991). Instrumental neutron activation analysis (Chappell & Hergt, 1989) was used to determine Cs, Hf, REE, Sc, Cr, Co, Th and U contents. Wet chemistry techniques from Peck (1964)were used to determine FeO content. Comprehensive chemical data for the Gabo Island and Mumbulla suites were previously given by Collins *et al.*(1982). On the basis of petrographic investigations and the techniques described by Chappell & White (1992), no altered samples are included in the data presented.

Table 2. Chemical analyses of A-type granites from the Lachlan Fold Belt

	Average of 55* A-type samples	±1σ	AB412	GI1	AB239	WB140	BG20	BG15	AB422	VB203	AB421	AB118	AB238	AB401	WB121
wt %															
SiO_2	73.83	±2.45	70.45	73.60	70.95	72.06	69.85	71.92	72.53	75.87	75.32	77.00	76.38	76.67	76.69
TiO_2	0.28	0.10	0.54	0.38	0.50	0.25	0.52	0.24	0.37	0.13	0.15	0.15	0.18	0.09	0.06
Al_2O_3	12.78	0.66	13.26	12.72	13.17	13.19	13.77	13.54	13.08	12.29	12.71	11.83	12.14	12.10	12.60
Fe_2O_3	0.84	0.46	0.93	1.99	1.61	0.56	0.72	0.64	0.89	0.88	0.51	0.40	0.81	0.60	0.34
FeO	1.54	0.84	3.32	0.82	2.28	1.82	3.14	2.17	1.71	0.50	1.16	1.05	0.78	0.38	0.62
MnO	0.05	0.02	0.08	0.04	0.05	0.05	0.05	0.06	0.05	0.03	0.06	0.04	0.02	0.02	0.03
MgO	0.29	0.27	0.61	0.20	0.46	0.19	0.91	0.27	0.48	0.14	0.23	0.04	0.09	0.05	0.04
CaO	1.04	0.55	1.93	0.83	1.46	1.19	2.09	1.52	1.31	0.70	0.90	0.61	0.29	0.53	0.54
Na_2O	3.42	0.33	3.39	3.62	3.48	3.52	2.83	3.25	3.32	3.80	3.36	3.08	3.16	3.25	4.16
K_2O	4.67	0.43	3.98	4.11	4.16	5.27	4.65	4.90	4.88	4.60	4.61	4.98	5.13	5.12	4.60
P_2O_5	0.07	0.06	0.18	0.08	0.11	0.04	0.19	0.08	0.12	0.02	0.05	0.02	0.05	<0.01	<0.01
H_2O^+				0.73	1.11	0.58	0.77	0.69		0.47		0.47	0.71	0.47	0.50
H_2O^-				0.29	0.21	0.14	0.13	0.13		0.20		0.18	0.16	0.18	0.11
CO_2				0.29	0.18	0.01	0.15	0.39		0.02		0.07	0.02	0.07	0.02
rest			0.23	0.27	0.25	0.22	0.26	0.20	0.17	0.17	0.12	0.20	0.17	0.09	0.10
			98.80	100.03	99.98	99.19	100.03	100.00	98.90	99.82	99.18	100.10	100.09	98.90	100.41
Trace elements (p.p.m.)															
F				1390	1120	750	730	1320	1040	910	670	1290	230		2090
Rb	197	±49	146	166	181	136	125	209	228	168	216	290	230	187	284
Cs			5	12	6	3	6	9	13	10	10	11	7	7	9
Sr	90	55	128	161	110	112	250	133	87	58	56	50	34.5	14.0	13.5
Ba	480	200	505	745	725	685	1005	530	215	575	148	655	650	61	58
Zr	301	137	535	496	453	350	325	314	313	191	190	187	152	116	107
Hf			13	12	12	9	8	8	8	7	5	6	5	5	5
Nb	24	7	26	27	17	36	21	36	21	22	20	18	12	11	43
Y	76	25	75	81	63	56	14	53	64	74	77	100	61	125	75
La			64	54	75	76	60	55	58	42	28	58	35	24	14
Ce			145	136	159	163	124	126	121	95	63	132	80	64	37
Nd			63	68	65	63	40	49	50	47	34	62	34	47	18
Sm			14.2	15.0	15.1	14.6	7.4	12.0	10.2	10.0	8.6	16.4	9.3	15.3	6.7
Eu			2.85	2.40	2.10	2.05	2.55	1.95	1.40	1.65	1.15	1.60	0.80	0.71	0.30
Gd				12.9	13.6	12.5						15.3	9.5		8.2
Tb			2.9	2.4	2.1	2.2	0.9	1.9	2.8		2.5	2.0	1.7	4.8	2.0
Ho			2.5	3.3	2.7	2.3	0.6	1.9	2.3	2.1	2.9	4.2	2.4	4.4	2.8
Yb			7.8	8.7	6.0	5.2	1.5	5.4	7.0	7.0	8.6	10.6	5.8	12.8	7.8
Lu			1.15	1.36	0.87	0.69	0.23	0.71	0.88	1.00	1.28	1.57	0.80	1.72	1.06
Sc	11	5	13	17	11	6	8	9	7	6	6	12	5	4	1
V	8	9	23	7	15	5	31	5	12	2	5	2	2	<1	<1
Cr	2	3	7	2	2	<1	10	<1	5	<1	1	<1	<1	<1	<1
Co	2	2		2	5	2	10	2		3		1	2		3
Ni	1	2	3	<1	3	<1	7	3	3	<1	1	<1	<1	<1	<1
Cu	4	4	2	5	7	<1	2	2	2	7	3	8	9	1	<1
Zn	84	24	81	139	85	72	70	87	52	92	52	106	53	28	55
Ga	22	2	22.2	21.6	19.2	23.0	18.6	23.2	21.2	20.6	19.6	20.0	17.0	22.4	27.9
As	2	2	1.5	1.5	1.5	1.0	1.5	1.5	1.5	2.0	1.0	3.0	1.0	1.0	2.5
Sn	8	3	5	7	8	7	5	9	11	9	11	8	8	7	18
Pb	28	7	19	30	19	20	29	30	28	25	36	33	30	28	32
Th			17	22	23	20	21	22	32	18	17	22	17	25	30
U			3.8	6.0	4.1	1.3	2.1	3.8	6.6	2.6	7.1	5.1	4.0	7.0	6.8

Sample	Unit	Suite	100 000 sheet	Grid ref.	Latitude	Longitude	ANU no.
AB412′	Danswell Ck	Wangrah	Michelago (8726)	125334	35° 49·22′S	149° 21·11′E	51924
GI1	Watergums	Gabo Island	Eden (8823)	513684	37° 17·81′S	149° 50·10′E	39551
AB239	Monga	Monga	Araluen (8826)	622596	35° 34·35′S	149° 53·57′E	40348
WB140	Yewrangara	Yewrangara	Crookwell (8729)	933216	34° 07·71′S	149° 05·74′E	48343
BG20	Murrungowar	Murrungowar	Murrungowar (8622)	505357	37° 36·80′S	148° 42·31′E	30250
BG15	Ellery	Ellery	Bendock (8623)	570619	37° 22·57′S	148° 46·39′E	30245
AB422	Wangrah	Wangrah	Michelago (8726)	120259	35° 53·28′S	149° 20·90′E	51928
VB203	Mt Arthur	Mt Arthur	Canonba (8335)	418683	31° 01·08′S	147° 26·26′E	48203
AB421	Eastwood	Wangrah	Michelago (8726)	106278	35° 52·27′S	149° 19·94′E	51922
AB118	Mumbulla	Mumbulla	Bega (8824)	571500	36° 33·64′S	149° 52·35′E	34617
AB238	Monga	Monga	Araluen (239)	613597	35° 34·31′S	149° 52·97′E	40347
AB401	Dunskeig	Wangrah	Michelago (8726)	111250	35° 53·78′S	149° 20·32′E	51929
WB121	Yewrangara	Yewrangara	Crookwell (8729)	903210	34° 09·06′S	149° 03·80′E	48324

*Chappell & White (1992) presented an average for 43 A-type granites in their work; we include additional analyses from the Wangrah Suite in this average.

OXYGEN FUGACITY AND VOLATILE CONTENTS

Ilmenite, typically found within mafic minerals, is the most common Fe–Ti oxide mineral in the LFB A-type granites. Our data indicate that the hematite and $MnTiO_3$ components in ilmenite are very low (generally <2%), which could indicate that ilmenite was the sole primary Fe–Ti oxide mineral (Czamanske *et al.*, 1981). In most cases, magnetite has textural relations that suggest it exsolved from primary ilmenite. An exception to this is the Gabo Island Suite, which has relatively high whole-rock Fe_2O_3/FeO contents (Collins *et al.*, 1982) and may contain primary magnetite. We suggest that ilmenite is the primary Fe–Ti oxide mineral found in the majority of LFB A-type granites and that magnetite is a subsolidus product. If ilmenite is primary, then the LFB A-type granites evolved under reduced conditions relative to the Ni–NiO (NNO) buffer. The occurrence of Fe-rich mafic minerals supports the hypothesis that A-type magmas are relatively reduced (Wones, 1989). This probably reflects the bulk composition of the source. The occurrence of primary ilmenite is similar to many other A-type granites (e.g. the North American Pikes Peak and Wolf River batholiths; Barker *et al.*, 1975; Anderson, 1980), althoughAnderson (1983) noted that oxygen fugacity may range from below to above the NNO buffer in other areas.

A-type magmas have been inferred to have low H_2O contents, on the basis of crystallization of annite interstitial to abundant feldspar and quartz. The stabilization of annite late in the crystallization sequence has been attributed to progressive enrichment in H_2O in residual melt upon crystallization of quartz and feldspars. Alternately, Skjerlie & Johnston (1992) have implicated high F contents for stabilizing interstitial biotite; however, evidence against this suggested role for F in the LFB A-type granites is discussed below. The presence of inferred primary, melt-precipitated amphibole in some of the LFB A-type granites suggests that initial H_2O contents may not be as low as in those granites where the only mafic mineral is interstitial annite. Unfortunately, most current experimental data do not apply specifically to the mildly alkaline A-type granite–granodiorite compositions found in the LFB (compare with Naney, 1983; Merzbacher & Eggler, 1984; Johnson & Rutherford, 1989). Clemens*et al.* (1986) conducted phase equilibria experiments on the amphibole-bearing LFB Watergums Granite at 100 MPa and 725–1000°C and did not find stable amphibole. They suggested that the H_2O content of that granite was between 2.4 and 4.3% at 100 MPa (inferred emplacement conditions), on the basis of field and textural arguments. We concur with their argument that some of the LFB A-type granites had magmatic H_2O contents at levels commonly cited for I-type granites, thus it is not appropriate to apply the term 'anhydrous' to these granites.

Many workers have proposed that A-type granites have anomalously high F contents (e.g. Loiselle & Wones, 1979; Collins *et al.*, 1982; Whalen *et al.*, 1987). In the LFB, fluorite is observed interleaved with annite in the Gabo Suite (Collins *et al.*, 1982) and as an accessory mineral in other A-type granites. We consider that the abundance of fluorite in these rocks has been overemphasized and that some of it could be secondary or subsolidus in origin, on the basis of further petrographic investigations and whole-rock and mineral chemistry.

In the LFB A-type granites F contents are comparable with, or lower than those of other felsic granites (Fig. 2). Previously cited F enrichment in the LFB A-type rocks (Collins *et al.*, 1982) resulted from comparison of F contents in the A-type rocks with the relatively unfractionated I-type granites. The LFB A-type granites have F abundances ranging from 880 to 1430 p.p.m., whereas strongly fractionated, felsic I- and S-type granites have higher F contents, up to 7000 p.p.m. (Fig. 2). Less felsic I- and S-type granites generally contain <1000 p.p.m. F.

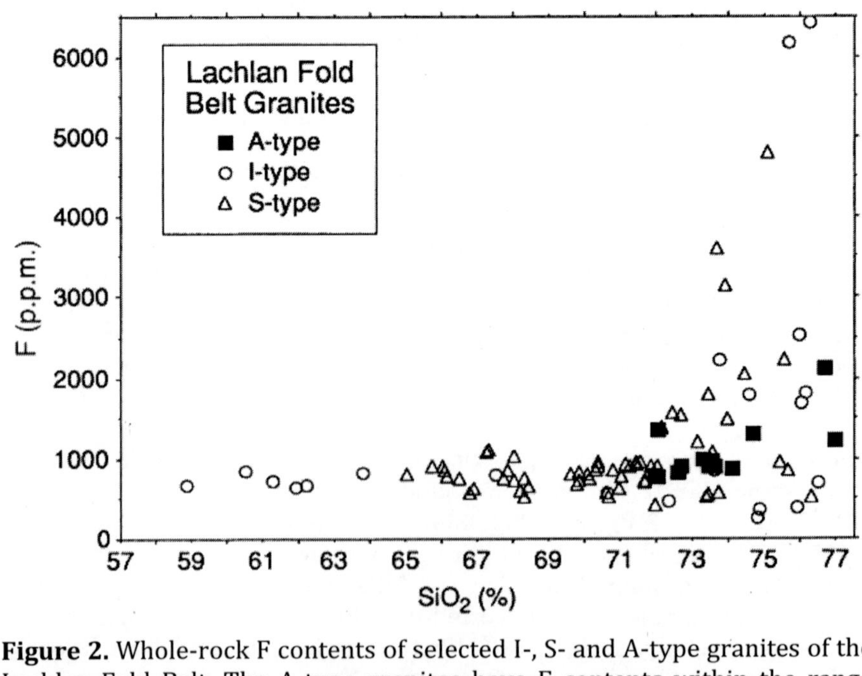

Figure 2. Whole-rock F contents of selected I-, S- and A-type granites of the Lachlan Fold Belt. The A-type granites have F contents within the range observed for the S- and I-type granites, but do not reach the F contents of the most felsic, fractionated granites.

P. Blevin (personal communication, 1992) has shown that annite in LFB A-type granites contains lower halogen contents than annite with similar Fe/(Σoctahedral cations) contents from I-type granites (generally < ~1.5%). This differs from some A-type granites world-wide in which biotite has F >1.5% (Anderson, 1983). Fluorine contents for amphiboles from the LFB A-type granites are comparable with those for fractionated and unfractionated I-type granites from the LFB and North Queensland (Champion, 1991; Fig. 3).

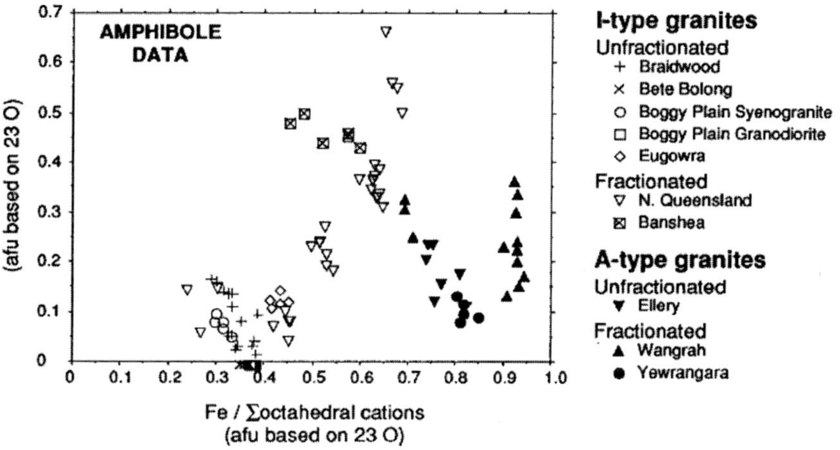

Figure 3. F vs Fe/(Σoctahedral cations) contents for amphiboles from selected I- and A-type granites of the Lachlan Fold Belt and North Queensland. F contents for the A-type granites are within the range of values observed for the I-type granites. It should be noted that the A-type granites contain amphibole with higher Fe/(Σoctahedral cations) contents, even compared with some of the fractionated Fe-rich granites from North Queensland. The data for the unfractionated granites are from a representative thin section of that granite. The data for the fractionated granites are from at least two thin sections from different phases of the granite. The data for the North Queensland granites are courtesy of Champion (1991), and include analyses of different plutons in that area. Data for the Banshea granite are from Budd (1992).

Taken together, these findings indicate that the F contents of unfractionated A-type granites are generally less than those observed for the most fractionated granites. Therefore, high F content, like low H_2O, does not discriminate A-type from I-type granites in the LFB.

CALCULATED CRYSTALLIZATION TEMPERATURES

Watson (1979) and Watson & Harrison (1983) showed experimentally that the partition coefficient $D^{Zr}{}_{zircon}$) is a function of the parameter M = (Na + K + 2Ca)/ (Si × Al), and temperature. If a metaluminous rock composition is that of a liquid that was just saturated in Zr, then its temperature can be calculated from the measured major element composition and Zr content. This temperature calculation cannot be used for peralkaline rocks, where different processes control Zr saturation (Watson, 1979). Also, because many granites do not correspond to melt compositions, and zircon inheritance is a widespread feature (Chen & Williams, 1990; Williams *et al.*, 1991), calculated temperatures correspond to crystallization temperatures only for the most felsic, fractionated rocks. However, it is those felsic granites that compare most closely with the A-type granites in composition. In Table 3, we present calculated zircon saturation temperatures for the I-type granites from the LFB that have been classified as unfractionated and fractionated by Chappell & White (1992). A notable feature of those data is that the temperatures calculated for the A-type granite suites are generally higher than those of the I-type granites. Below, we discuss some of the factors that contribute to the large standard deviations observed in calculated temperature values and the importance of using the suite concept to examine the meaning of calculated zircon saturation temperatures.

Ion microprobe analysis (Williams, 1992) indicates rare zircon age inheritance for the only A-type granite of the LFB examined by that technique (Watergums pluton, analysis GI1 in Table 2). This suggests that most of the zircon in the Watergums Granite was a near-liquidus phase, which is supported by the euhedral form of zircons. The occurrence of melt-precipitated zircon along with rare zircon showing calculated age inheritance could imply that the melt was just oversaturated in Zr so that the calculated temperature of 885°C is a little above that of the magma, but close to the true value within the limitations of the method. The Watergums Granite is typical of the whole Gabo Island Suite, for which nine analyses from five units give calculated temperatures of 893±10°C.

As granites undergo fractional crystallization Zr contents generally decrease. This happens because the melt becomes more felsic and because the temperature falls, resulting in decreased Zr solubility and precipitation and removal of zircon (Watson, 1979). One can deduce that if felsic I-type rocks have low Zr contents then they were derived by fractional crystallization from more mafic liquids or they were felsic primary melts.

Table 3. Some chemical data and calculated Zr saturation temperatures for the A-type Yewrangara Granite, average A-type granite, average unfractionated I-type granite, and average fractionated I-type granite

	Yewrangara Granite							Average A-type granite	Unfract. I-type granite	Fract. I-type granite
	WB140	WB139	WB120	WB141	WB142	WB122	WB121			
Count:								55	131	103
wt %										
SiO_2	72·06	73·67	73·46	72·68	74·10	74·73	76·69	73·83	72·90	76·17
FeO_{total}	2·42	2·09	2·03	1·90	1·88	1·40	0·93	2·29	1·96	1·00
MgO	0·19	0·16	0·12	0·21	0·12	0·10	0·04	0·29	0·66	0·12
CaO	1·19	1·02	1·09	1·02	0·94	0·82	0·54	1·04	1·63	0·61
Na_2O	3·52	3·61	3·72	4·13	3·45	3·73	4·16	3·42	3·27	3·37
K_2O	5·27	5·20	5·13	4·89	5·14	4·94	4·60	4·67	4·42	4·92
Trace elements (p.p.m.)										
F	750	860	880	860	840	1350	2090			
Ga	23·0	23·4	24·0	24·2	23·2	24·8	27·8	22	16	19
Rb	136	147	152	137	144	181	284	197	219	424
Cs	2·8	2·9	4·2	3·3	2·2	4·8	9·1			
Sr	112	88	86	78	82	50	13·5	90	147	31
Ba	685	550	515	365	495	280	58	480	488	99
La	76	63	62	44	52	38	14			
Lu	0·69	0·70	0·67	0·68	0·65	0·74	1·06			
Zr	350	281	287	275	263	176	107	301	151	116
Temp.	850	832	831	829	828	791	751	839	781	764

Temp., the calculated zircon saturation temperature in °C. Unfract., unfractionated granites (Rb < 270 p.p.m.); fract., fractionated granites (Rb > 270 p.p.m.); both defined by Chappell & White (1992). Complete analyses of WB140, WB121 and the LFB A-type average are given in Table 2.

The behavior of zircon in A-type granites can be examined using the Yewrangara A-type granite, which displays compositional trends consistent with fractional crystallization (Table 3). The status of the least felsic sample (WB140) as far as Zr saturation is concerned, and therefore temperature, is uncertain because it possibly represents a primary unfractionated melt composition. The other rocks have decreasing Zr contents that correlate with indicators of increasing fractionation, which is consistent with the hypothesis that zircon was precipitating and being removed from the evolving melt. Therefore, the calculated zircon saturation temperatures are realistic within the limits of the method. For samples WB120–WB122 in Table 3, the compositions are distinctly A-type, and the calculated temperatures are higher than for felsic I-type granites. The A-type character of the most felsic granite (WB121) is only apparent through its field association with the other rocks of the Yewrangara pluton. This granite has chemical features that overlap with fractionated I-type granites and its calculated temperature is likewise in the range of I-type granites. Similarly, in individual suites, the I-type granites that are fractionated have lower Zr, and hence lower calculated temperatures, than the unfractionated rocks. For this reason, it is necessary to examine suites when comparing calculated temperatures; the least fractionated rocks that represent liquid compositions record the highest calculated temperature of the granite suite.

We conclude that the A-type granites of the LFB formed from melts at temperatures significantly higher than those that produced other felsic granites. The genetic significance of this will be considered below.

COMPOSITIONAL COMPARISON WITH I-TYPE GRANITES

Many felsic, highly fractionated I-type granites have compositions that overlap those of the intrinsically felsic A-type granites. When dealing with the more felsic granites, there are difficulties with most genetic schemes, mainly because such rocks tend to be close to haplogranite (near minimum-temperature melt) compositions. Thus, felsic granites typically have converging major element compositions and similar mineral compositions, with quartz and two feldspars occurring in subequal amounts (Table 3). This immediately leads to problems; for instance, classification schemes dependent on mineralogy become meaningless for the more felsic rocks (SiO_2 > 72%). The so-called 'alkali-feldspar granites' are a good example of this: Pitcher (1993) suggested that these are A-type rocks, whereas other workers have demonstrated that some are I-type (MacKenzie *et al.*, 1988, 1990). To use chemical classification schemes when dealing with felsic rocks it is necessary to use distinctive granite associations where the granite type can be most readily identified using the more mafic members (Whalen *et al.*, 1987; Chappell & White, 1992).

In the case of discriminating the A-type granites from other granites we are principally concerned with I-type rocks. A-type granites can be easily discriminated from S-type granites because the latter have much higher P_2O_5, lower Na_2O contents, and show increasing P_2O_5 with fractionation, which contrast with A-type trends (Chappell & White, 1992). The felsic S-type granites are always peraluminous and become strongly so with fractionation (two-mica granites), thus mineralogical criteria alone will distinguish the two types.

Collins *et al.* (1982) and Whalen *et al.* (1987) compared the Gabo Island and Mumbulla A-type suites with other granites, and in doing so created classification criteria that many workers use to distinguish A-type character. Landenberger & Collins (1996) similarly compared data from I- and A-type granites from the Chaelundi Complex, New South Wales, Australia. We now have a larger database of LFB granites (1939 analyses with SiO_2 >57%) that allows us to better describe the compositional effects of fractionation in the LFB. As fractionation changes the compositional features used for determining granite type, it is useful to undertake this kind of comparison. Average chemical data for the LFB A-type granites, and unfractionated and fractionated I-type granites are presented in Table 3.

Relative to typical I-type granites, LFB A-type granites have higher $Fe_{total}/(Fe_{total} + Mg)$, although compositional convergence occurs for the most felsic rocks. This feature could be a function of the relatively reduced nature of A-type magmas. The LFB A-type granites are not unusually 'alkaline' compared with other granites with similar silica contents; as $Na_2O + K_2O$ increases with SiO_2 content in any granite suite these elements are not diagnostic (Fig. 4; Table 3). The LFB A-type granites are only clearly more K_2O rich than the I-type granites at SiO_2 contents of 65–70%. At higher silica contents, the A-types are weakly peraluminous as a result of fractionation. The perception that A-type granites are 'extremely alkaline' seems to have arisen from the convenient association of the letter 'a' with alkaline, although A-type granites were originally defined as 'mildly' alkaline (Loiselle & Wones, 1979) or 'somewhat' alkaline (Collins *et al.*, 1982).

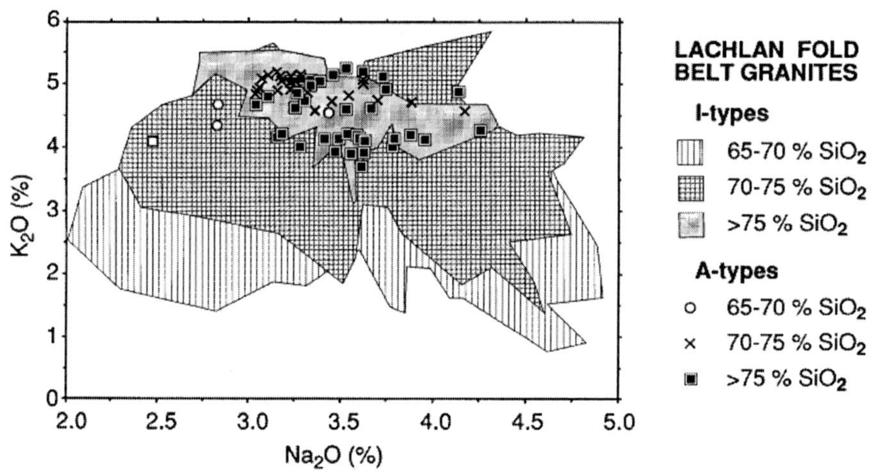

Figure 4. K_2O vs Na_2O for I-type and A-type granites of the Lachlan Fold Belt. Both K_2O and Na_2O increase with increasing SiO_2 content for both granite types; thus high $K_2O + Na_2O$ is a characteristic of both I- and A-type felsic granites. It should also be noted that the A-type rocks are more potassic than sodic. Three extreme outliers of the I-type granites have not been included in the fields for the I-type granites. These are from the Urialla and Jungle Creek plutons (both 70–75% SiO_2; Na_2O contents of 2.02% and 2.64%, and K_2O contents of 3.78% and 2.25%, respectively) and the Coles Bay pluton (Freycinet, >75% SiO_2, Na_2O and K_2O contents of 4.91% and 3.71%, respectively).

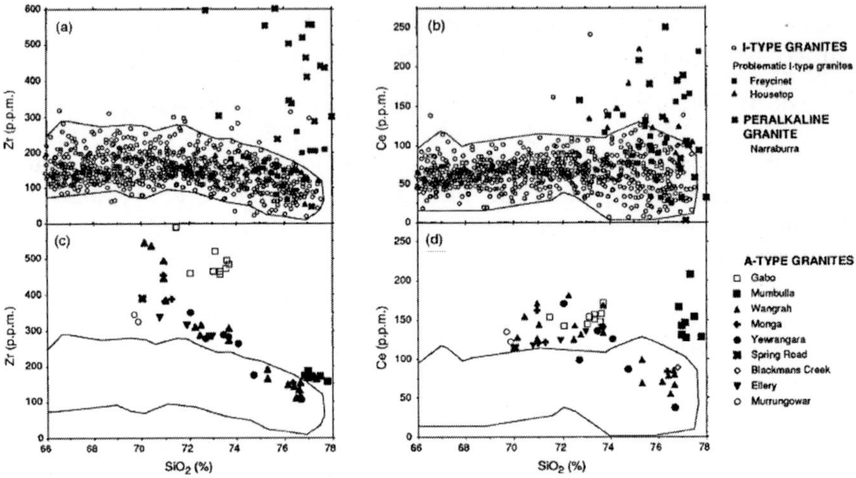

Figure 5. Fig. 5. (a) Zr vs SiO_2 for I-type, peralkaline and A-type granites of the Lachlan Fold Belt. The outline arbitrarily encloses ~98% of the I-type granites. The I-type granites that have Zr contents >300 p.p.m. include the Burrinjuck and Yeoval granites, which belong to the Boggy Plain Supersuite (discussed in text; Wyborn *et al.,* 1987). Also, the Gobondery and Yackandandah granites each have one analysis with >300 p.p.m. Zr. None of those granites are considered to be A-type, as other criteria are not satisfied. It should be noted that the peralkaline Narraburra Granite (Wormald & Price, 1988; Wormald, 1991) has extremely variable Zr contents, with some granites having very high contents. Two peralkaline rocks from the Narraburra Suite were not plotted in the diagram (from the Boginderra Granite;Wormald, 1991). Those rocks have Zr contents of 638 and 799 p.p.m., and SiO_2 contents of 73.95 and 75.48%. (b) Ce vs SiO_2 for I-type, peralkaline and A-type granites of the Lachlan Fold Belt. The outline arbitrarily encloses ~99.4% of the I-type granites. The I-type granites that plot out of the outlined area include the Burrinjuck and Jackson granites of the Boggy Plain Supersuite (discussed in text; Wyborn *et al.*, 1987); also the Jindera, Mumbedah and Butmaroo granites. Those granites are not considered to be A-type because other criteria are not satisfied. One analysis from the Jindera Granite was not plotted on the diagram (SiO_2 = 70.67%, Ce = 422 p.p.m.).

Ba contents appear to be suite-specific for both the A- and I-type granites in the LFB. For example, the Gabo Suite (average 73% SiO_2) and Mumbulla Suite (average 77% SiO_2) granites have relatively high average Ba contents (767 and 575 p.p.m., respectively). In contrast, granites of the Wangrah and Yewrangara Suites have Ba contents that decrease from ~820 p.p.m. to ~40 p.p.m., while SiO_2 increases from ~70% to 77%. Ba contents in granites may reflect differences in source regions or pyroxene-dominated fractionation; however, we currently do not fully understand the variations that are observed. Sr and Ca levels are generally lower in the A-type than the I-type granites (Table 3).

Zirconium content is relatively high (commonly >300 p.p.m.) in the less felsic A-type granites (65–72% SiO_2) and decreases with fractionation in both the A- and I-type granites (Fig. 5; Tables 2 and Table 3). Y and Nb are also relatively abundant in the less felsic A-type granites, but show no clear trends with fractionation, while in the I-type granites those elements generally increase with fractionation (Table 3). Zn decreases with fractionation in both granite types, but average Zn content for A-type granites is higher than that observed for the most felsic I-type granites. Thus, Zn is an indicator of A-type character, but not a complete discriminant. Ga does not vary greatly with fractionation in the A-type granite suites from the LFB, except in the Yewrangara Suite, where it increases (Table 3). However, Ga contents increase with fractionation in many I-type granites, which results in Ga/Al trends that can be readily discerned from the majority of LFB A-type granites (Whalen *et al.*, 1987).

The REE behave similarly with fractionation in both A- and I-type granites, although the more mafic A-type granites have flatter REE^{3+} patterns and much higher REE contents than typical I-type granites (Tables 2 and 3). The LREE (La–Sm) decrease with fractionation in both types. The most mafic rocks do not have a strong negative Eu/Eu* anomaly (e.g. Gabo Island and Mumbulla suites; Collins *et al.*, 1982). However, Eu decreases as the rocks become more felsic, resulting in a strong negative Eu/Eu* anomaly (e.g. WB121), indicative of feldspar fractionation. The HREE (Gd–Yb) increase slightly, or remain constant with fractionation in both the I- and A-type granites. Sample BG20 (Murrungowar) has strikingly different REE geochemistry from the other LFB A-type granites, with a steeper chondrite-normalized slope, and lower concentrations of HREE. We attribute these differences to BG20 being the most mafic rock analyzed with annite as the sole mafic mineral (amphibole is present in all of the other less felsic compositions).

The A-type granites of the Gabo and Mumbulla suites contain high abundances of Sc (14–18 p.p.m.), and high Sc was cited as a diagnostic feature of A-type granites by Collins *et al.* (1982). However, the Sc content of other A-type granites is variable and generally lies within the range observed for I-type granites of the same SiO_2 content. Within a particular SiO_2 range, the average value for Sc in the A-type granites is slightly higher than the average for the I-type granites and so, for this element, the average value is misleading and high Sc contents are not diagnostic for LFB A-type granites.

Low abundances of trace transition elements have also been cited as characteristic of LFB A-type granites by Collins *et al.* (1982), but this feature is also a reflection of the inherently felsic nature of A-type granites; felsic I-type granites also contain low trace transition element abundances (Table 3). Low trace transition element content cannot be used as a criterion for identifying A-type granites.

Application of chemical criteria to A-type rocks

As shown above, the more mafic (~65–72% SiO_2) LFB A-type granites are distinguished chemically from I-type granites by relatively high $Fe_{total}/(Fe_{total} + Mg)$, HFSE (Zr, Nb, Y), REE^{3+}, Ga and Zn and relatively flat REE patterns (Table 3). These chemical features are slightly different from those proposed by Collins *et al.* (1982) and Whalen *et al.* (1987). As the concentrations of the 'diagnostic element' contents change with fractionation in I- and A-type granites, it is difficult to assign meaningful fields on geochemical plots (compare with Whalen *et al.*, 1987). In some cases, rocks may plot in the A-type field of one chemical discrimination diagram, but will not plot in the A-type field of another chemical discrimination diagram, owing to fractionation (e.g. the 'aluminous subsolvus A-type' rocks from Corsica; Poitrasson *et al.*, 1994, 1995). In the absence of a close relationship with unfractionated rocks, distinction of strongly fractionated A-type granites from similarly fractionated I-type granites is difficult. At present, classification of very felsic rocks can only be made confidently when there is an association with less evolved rocks within the same suite. Thus a small proportion of very felsic and fractionated rocks are ambiguous in terms of I- or A-type character as a result of their prolonged evolution. Compositional separation between the two types is very clear in less evolved rocks, and mineralogical criteria can be used to discriminate the types. Therefore, the A-type concept is valid.

Granites problematic for classification

As noted above, the extremely felsic granites are difficult to classify using chemical data. Altered, felsic granites are even more problematic for classification and, as an example, we present data on the Freycinet and Housetop granites of Tasmania. These are felsic, strongly fractionated (Rb > 270 p.p.m.) granites that are difficult to classify as either fractionated I- or A-type granites. They were designated I-type by Chappell *et al.* (1991)simply because that is by far the more common type of granite throughout the belt, although chemical compositions overlap with those of A-type granites for some elements (e.g. Zr and Ce; Fig. 5). The Tasmanian rocks appear to be altered, on the basis of petrographic evidence, such as an abundance of albitic plagioclase, and compositional evidence (e.g. most have extremely low CaO abundances <1%). Current data indicate that these rocks have variable LREE contents, reflecting the presence or absence of accessory minerals in a given sample (Camacho, 1987). The A-type granites differ from these unusual granites in their lack of intense alteration. Furthermore, the A-type granites tend to have LREE trends that display a strong negative correlation with increasing SiO_2 content. More work is necessary to expand the current data set of altered felsic granites in the LFB, but we suggest that some of the granites regarded as 'A-type' world-wide may in fact be altered, and possibly fractionated, felsic I-type granites.

COMPARISON WITH PERALKALINE GRANITES

The diagnostic features originally attributed to the A-type group byLoiselle & Wones (1979) are clear in peralkaline rocks. Peralkaline rocks are unequivocally 'A-type' in this sense, whereas the metaluminous to weakly peraluminous A-type rocks are more ambiguous. We contend that there are distinctive features of the peralkaline granites that are not shared with the aluminous A-type granites, such as indeed their peralkalinity, consequent distinctive mineralogy, characteristic field associations and general chemical compositions.

Many of the peralkaline rocks are hypersolvus (with alkali feldspar the only feldspar) and the mafic minerals include fayalite, hedenbergite, aegirine, ferrorichterite, riebeckite and arfvedsonite. These features indicate that the peralkaline magmas had an anhydrous origin. In general, the peralkaline rocks are associated with voluminous basalt, syenite and/or anorthosite. In contrast, the LFB aluminous A-type granites do not include any hypersolvus rocks and those mafic minerals are not observed, excepting riebeckite, which is rare (Table 1). We argue above that the LFB A-type granites have 'normal' H_2O contents and these petrographic observations support this proposal. In addition, the LFB aluminous A-type granites are only associated with minor basalt and gabbros and never associated with syenite or anorthosite.

Previous workers have discussed the differences between peralkaline and aluminous A-type rocks. Whalen & Currie (1990) studied large metaluminous–peralkaline complexes, emphasizing the Topsails Complex, Newfoundland. They suggested that those rock associations have a different petrogenesis from other smaller, highly fractionated A-type granites. We contend that the metaluminous–peralkaline complexes differ from the smaller metaluminous–weakly peraluminous granites of the LFB and that the latter rocks need not be highly fractionated. In other regions, metaluminous A-type rocks and the larger peralkaline A-type rock associations have been grouped together. For example, Eby (1992) divided A-type granites from around the world into A_1 and A_2 groupings on the basis of tectonic affiliations and the Y/Nb and Yb/Ta ratios of the rocks. If a wider group of trace elements are examined [e.g. the figures of Eby (1992)], then compositions of the A_1, A_2, peralkaline and metaluminous rocks overlap. This is due to the inherently fractionated nature of many of the peralkaline granites, which leads to great compositional variation within a suite. However, if the most mafic member of a suite is examined, the peralkaline rocks typically contain more Ba, Sr, HFSE and REE^{3+} and strong negative Eu/Eu* anomalies relative to metaluminous A-type granites. We submit that these differences result from different petrogenetic schemes for the peralkaline rocks (fractionation from a mafic source; below) and the aluminous rocks (partial melting of an infracrustal source; below).

In Fig. 5, we show some of the variation in Eby's A_2 group by comparing the peralkaline Narraburra Suite with the LFB aluminous A-type granites. The Narraburra Suite rocks have much more variable Zr contents than those observed for the LFB aluminous A-type granites and some of them have substantially higher Zr contents. World-wide, peralkaline granites commonly contain 1000 p.p.m. Zr (Eby, 1990), but contents as great as ~5000 p.p.m. have been reported (e.g. Habd-Aldyaheen complex, Arabian Peninsula; Radain *et al.*, 1981). This is consistent with experimental evidence that Zr is more soluble in peralkaline than metaluminous melts (Watson, 1979). Zr content increases with fractional crystallization in the peralkaline rocks if the magma reaches the granite stage of fractionation at high enough temperatures and the solubility of Zr has not been exceeded.

The chemical trends observed for many peralkaline rocks could be achieved by Al_2O_3 depletion of either an I- or A-type parent magma or basaltic magma by extreme crystal fractionation of plagioclase and clinopyroxene (Tuttle & Bowen, 1958) or Ca-bearing alkali feldspar (Carmichael & McKenzie, 1963). These mechanisms appear consistent with peralkaline rock compositions (H_2O-poor, high Ba, Sr and HFSE, strong negative Eu/Eu*) and field associations (common mafic rocks), and have been proposed by many workers (e.g. Baker, 1974; Walsh *et al.*, 1979;Turner *et al.*, 1992).

The differences between the metaluminous to weakly peraluminous A-type rocks and the peralkaline rocks are such that they should not be grouped together in rock classification schemes. In grouping these two different types of granite together, Loiselle & Wones (1979) had access to many fewer data than now available. The name 'peralkaline' should be retained for a group of rocks that have long been regarded as distinctive. The 'aluminous A-type granites' of the LFB and elsewhere are different from peralkaline rocks, and as the two types are likely to have different origins, it is inappropriate to group them together. We recommend using the term 'aluminous A-type' in reference to the LFB rocks and similar rocks elsewhere. Retaining the 'A' term is useful because of the extensive literature on those rocks using that name, even though some 'A' characteristics are not distinctive in the LFB and probably elsewhere (e.g. extremely 'alkaline', 'anhydrous' and 'anorogenic'). Perhaps 'A' should now refer to 'anomalous' in the LFB, in reference to the anomalously high HFSE contents of the aluminous A-type granites (compare with Hogan *et al.*, 1992).

ISOTOPIC WORK

Isotopic compositions were measured using a multi-collector VG354 thermal ionization mass spectrometer at the Centre for Isotope Studies, North Ryde, New South Wales. For each analysis, a 100 mg sample was dissolved in an open beaker on a hot plate employing HF, HNO_3 and $HClO_4$. Sr was separated using a one-step ion exchange column in a BioRad AG50W-X8 resin and dilute HCl. Sr isotopic compositions were corrected for mass fractionation with an $^{86}Sr/^{84}Sr$ value of 0.1194. Nd was extracted using a single column pass process (see Korsh & Gulson, 1986). Both Sr and Nd isotopic ratios for a single dissolution were collected at least 50 times such that 2σ values were at most 0.0015%. Procedural blanks were 790 pg of Nd and 250 pg of Sm. For Rb–Sr analyses, 10 runs of a standard (SRM987) gave $^{84}Sr/^{86}Sr$ = 0.056489±5, $^{86}Sr/^{88}Sr$ = 0.120182±196, $^{87}Sr/^{86}Sr$ = 0.710262±3. For Nd isotopic analyses, 27 dissolutions of Onion's standard gave $^{146}/^{144}Nd$ = 0.721396±320 (2σ), $^{145}Nd/^{144}Nd$ = 0.348407±2, $^{143}Nd/^{144}Nd$ = 0.511112±2.

Isotopic ratios are potentially useful indicators of granite source regions; however, in the case of the Sm–Nd and Rb–Sr systems, the A-type granites are ambiguous, as ratios overlap with those of I-type and peralkaline rocks (Table 4, Fig. 6). None the less, no sedimentary component is indicated by isotopic data. The A-types have relatively high initial $^{143}Nd/^{144}Nd$ ratios with ε_{Nd} values (at 400 Ma) varying from +5.01 to -3.15; there is a range from mantle to crustal values. This variation is consistent with Nd isotope heterogeneity in granites from eastern Australia and from Corsica (Chappell *et al.*, 1990; Turner *et al.*, 1992;Poitrasson *et al.*, 1994, 1995). In both cases, isotope heterogeneity has been interpreted to reflect magma source region heterogeneity. These results illustrate some of the difficulties in applying Nd isotopes to elucidate the genesis of extremely felsic magmas, such as the A-type granites.

Table 4. Rb–Sr and Sm–Nd isotopic data for selected A-type granites from the Lachlan Fold Belt

	AB412	GI1	AB239	WB140	BG20	BG15	AB422	VB203	AB421	AB118	AB238	AB401	WB121
$^{87}Rb/^{86}Sr$	3·309		4·778	3·522	1·449	4·562	7·617	8·323	11·231	13·416	19·510	39·443	63·068
$^{87}Sr/^{86}Sr$	0·724632		0·736103	0·725114	0·716215	0·733704	0·745348	0·750039	0·763997	0·780686	0·815897	0·909173	1·067880
Sr_T*	0·70579		0·70888	0·70505	0·70796	0·70772	0·70196	0·70263	0·70002	0·70426	0·70476	0·6845	0·70864
$^{147}Sm/^{144}Nd$	0·11601	0·14602	0·11856	0·12091	0·08441	0·11893	0·11419	0·11854	0·14442	0·15022	0·14056	0·18877	0·18412
$^{143}Nd/^{144}Nd$	0·512527	0·512431	0·512331	0·512496	0·512223	0·512369	0·512540	0·512690	0·512500	0·512355	0·512350	0·512779	0·512681
ε_{Nd0}†	−2·17	−4·04	−5·99	−2·77	−8·10	−5·25	−1·91	1·01	−2·69	−5·52	−5·62	2·75	0·84
Initial Nd_T*	0·51222	0·51205	0·51202	0·51218	0·512000	0·51206	0·51224	0·51238	0·51212	0·51196	0·51198	0·51228	0·51220
ε_{SrT}*	97·7		142·1	87·2	128·9	125·4	43·0	52·6	15·3	76·0	83·1	−206·8	138·5
ε_{NdT}*	1·95	−1·46	−2·00	1·10	−2·37	−1·28	2·3	5·01	−0·03	−3·15	−2·76	3·15	1·48
Model Nd ‡	813	1370	1139	904	969	1084	779	583	1191	1640	1436	1589	1845

*The subscript T refers to the value calculated at 400 Ma.
†The subscript o refers to the initial value.
‡The CHUR values used were $^{147}Sm/^{144}Nd$ = 0·1966 and $^{143}Nd/^{144}Nd$ = 0·512638 (Jacobsen & Wasserburg, 1980).

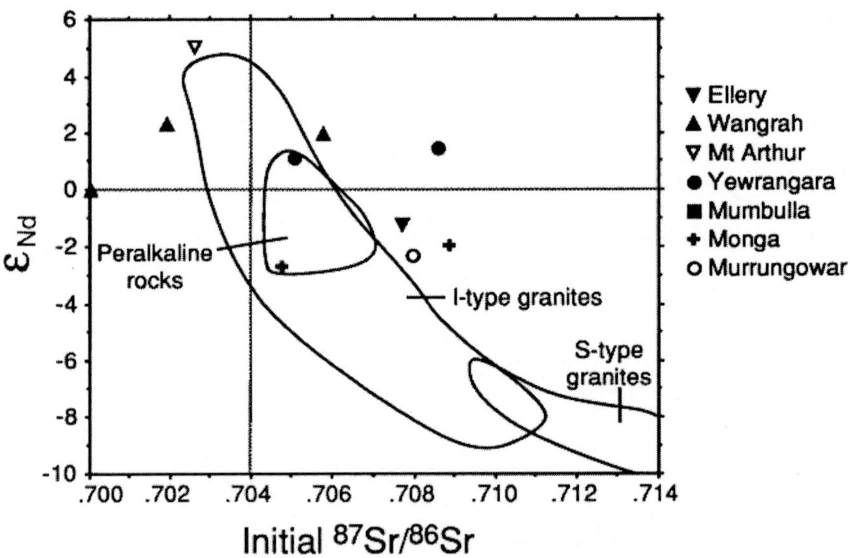

Figure 6. Plot of ε_{Nd} vs initial $^{87}Sr/^{86}Sr$ (both calculated for 400 Ma) showing the LFB A-type granites. Also shown are the fields for I- type and S-type granites from the LFB [data from Chappell *et al.*(1990)] and the field for peralkaline rocks from the Padthaway Ridge, South Australia (Turner *et al.*, 1992). It should be noted that the A-type granites overlap isotopically with the I-type and peralkaline rocks. The sample AB401 (Dunskeig pluton, Wangrah Suite) has been excluded from this plot, as explained in the text.

Calculated initial $^{87}Sr/^{86}Sr$ ratios (at 400 Ma) are in the range 0.7002–0.70888, excluding sample AB401. Sample AB401 has extremely low initial $^{87}Sr/^{86}Sr$ that is thought to reflect an inaccurate age correction owing to the very high Rb/Sr ratio. We do not consider that the Sr initial ratios are meaningful in the case of the LFB A-type granites, owing to their relatively low Sr contents and thus their high sensitivity to changes of Rb/Sr or Sr isotopic compositions.

PETROGENESIS

The LFB metaluminous to weakly peraluminous A-type granites have chemical and mineralogical attributes which must reflect a somewhat different petrogenesis from typical I-type granites. Previously, the distinctive features of the A-type granites, such as high HFSE and Ga, were explained in terms of high F, low H_2O and high alkali element contents in the source (e.g. Collins *et al.*, 1982; Clemens *et al.*, 1986; Whalen *et al.*, 1987; Peterson *et al.*, 1991; Skjerlie & Johnston, 1992). Our data indicate that other petrogenetic models are necessary to explain the LFB A-type granites and possibly other aluminous A-type granites world-wide.

Granites, as felsic rocks of 'minimum-temperature melt composition', represent either products of fractional crystallization of quartz, feldspars and mafic minerals from a more mafic melt, or primary melts derived from a source containing quartz and feldspars. Within the constraints imposed by these two broad mechanisms, the distinctive features of the aluminous A-type granites could reflect either a characteristic parent magma composition and/or source rock, or differences in the physical and/or chemical conditions of melting, magma separation or crystallization. We briefly review some of the arguments previously made for the genesis of I-type and peralkaline rocks, in the context of evaluating the potential genetic schemes for the aluminous A-type rocks.

Involvement of a mafic magma

Fractional crystallization, generally of a mantle-derived magma, to produce I-type granites or peralkaline compositions has been proposed by many workers (e.g. Bowen, 1928; Baker, 1974; Walsh *et al.*, 1979). An example of such a process in the LFB is the I-type Boggy Plain Supersuite (Wyborn *et al.*, 1987). Those rocks do not share the chemical features of the LFB A-types, except at their most felsic compositions, where we have seen that there is overlap between I- and A-type compositions. At the more mafic compositions, where A-type granite chemical features are distinctive, the fractionated granites can be easily distinguished from the Boggy Plain Supersuite rocks. For example, Zr contents are generally lower in the Boggy Plain Supersuite because they are derived from a mafic magma that was undersaturated with respect to Zr (exceptions are listed in Fig. 5). Once Zr becomes saturated in a magma, high Zr contents cannot be reached via fractional crystallization, especially if zircon is one of the fractionated minerals and temperature falls during fractionation. A few granites do have high Zr contents in the Boggy Plain Supersuite (>300 p.p.m.; Fig. 5), but they are not considered to be A-type as other criteria are not satisfied.

One argument against a mechanism of fractional crystallization of a mafic magma to produce the aluminous A-type granites of the LFB is that some of the least evolved rocks have Eu, Rb, Sr and Ba contents that preclude extensive feldspar fractionation from a mafic source. The composition of the very felsic Mumbulla Suite rocks [high SiO_2, ~77%; high Ba, 500–655 p.p.m.; moderate Rb, 225–261 p.p.m.; and low to moderate Eu, 1.21–1.68 p.p.m. (Collins *et al.*, 1982)] indicates that there has been no appreciable feldspar fractionation. The most mafic members of the Yewrangara and Wangrah Suites (WB140 and AB412, Table 2) similarly have compositions that were probably not produced by fractionation from a mafic source, but appear to be the result of partial melting.

Magma mixing is a process that has been advanced by some workers to produce A-type granites (e.g. Bédard, 1990). However, as many of the intrinsically felsic A-type granites of the LFB are weakly peraluminous, very little strongly metaluminous mafic magma could have been added to the granite magma, especially considering that the mafic magmas have higher concentrations of Al and $\Sigma(Ca + Na + K)$. Also, magma mixing seems unlikely in the LFB, on the basis of the occurrence of bimodal A-type rock associations and the restricted miscibility of the two magma types (Sparks & Marshall, 1986).

Partial melting

Many workers have suggested that A-type granites are produced by combined partial melting-fractionation from source regions of slightly different composition than those for I-type granites (e.g. Anderson & Cullers, 1978; Cullers *et al.*, 1981; Collins *et al.*, 1982; Anderson, 1983;Clemens *et al.*, 1986; Whalen *et al.*, 1987; Wormald & Price, 1988; Creaser*et al.*, 1991; Landenberger & Collins, 1996). Proposed source compositions that have been favored are lower crustal, including tonalite, granodiorite, peraluminous granulite, charnockite and granulitic residuum from melting of I-type granites. On the whole, the researchers listed above have discussed granites with similar compositions to the LFB A-type granites. The general approach has been to try and account for an H_2O-poor, F-rich source. These models, or any others that invoke a distinctive source for the A-type granites, have to be reconciled with the fact that A-type granites may be adjacent to or intrusive into other granite types and, in the LFB, have a very limited distribution. This means that if the sources for A-type granites are distinctive, then those source regions had to be either very localized and/or separated vertically from adjacent granite source regions,

Synthesis

In assessing the possible genetic schemes for the LFB A-type granites, the most important single feature is their high Zr contents relative to the I-type granites, which, as we have noted above, is related to higher calculated zircon saturation temperatures. Zr is an ideal discriminant for the aluminous, unfractionated LFB A-type granites. In addition, its behavior in granite magmatic systems has been characterized experimentally (Watson, 1979; Watson & Harrison, 1983). We suggest that the other HFSE that are relatively abundant in the A-type granites have partition coefficients that are similarly dependent on temperature.

The critical observation that A-type magmas existed at higher calculated zircon saturation temperatures than other felsic granites implies that they were derived from a fertile felsic source region that required a high temperature before an extractable magma was produced. We favor a felsic infracrustal source that could overlap in composition with I-type sources, on the basis of similarities between the two rock types. The major difference between petrogenetic schemes for the aluminous A-type magmas and the I-type magmas is that different physical conditions prevailed. Limited availability of H_2O and relatively low oxygen fugacity during partial melting, and therefore high temperatures, may be all that is required to produce aluminous A-type granites. As melt viscosity is reduced at high temperatures, an extractable melt may be generated at smaller degrees of partial melting than needed for other granite types. The necessarily high temperatures required to produce an extractable magma may have been initiated by mantle upwelling or mafic magma influx into a localized area. Although we cannot rule out magma mixing as a factor in petrogenesis, there is no direct evidence to support it, therefore we do not favor it.

CONCLUSIONS

1. The term 'A-type' as proposed by Loiselle & Wones (1979) is too broad to succinctly define rocks world-wide. We recommend using the term 'aluminous A-type' in reference to the LFB rocks and similar rocks elsewhere. The peralkaline rocks should be known as such, and differ from the metaluminous to weakly per- aluminous A-type granites such as those of the LFB. This identification of 'aluminous A-type' rocks has petrogenetic implications.
2. Aluminous A-type granites commonly have Fe-rich mafic minerals and primary ilmenite. The more mafic (~65–72% SiO_2) granites are distinguished chemically from I-type granites by relatively high

$Fe_{total}/(Fe_{total} + Mg)$, HFSE (especially Zr, Nb, Y), REE^{3+}, Ga and Zn. The Rb–Sr and Nd–Sm isotopic compositions for the LFB A-type granites overlap with those of peralkaline and I-type granites from elsewhere.

3. More evolved A-type granites are difficult or impossible to distinguish from fractionated I-type granites. To characterize any kind of felsic granite it is necessary to compare suites of rocks in which less evolved rocks are also present. New data indicate that some features of LFB aluminous A-type granites, which were previously considered diagnostic of the type, are actually similar to those of many felsic I-type granites; for example, $Na_2O + K_2O$, Sc and trace transition element contents.

4. The LFB aluminous A-type granites generally have reduced compositions relative to the NNO buffer, which most probably reflects the character of the source region. Our data also show that those granites do not have high F contents compared with fractionated granites. There is no evidence that H_2O contents are lower than for felsic I-type granites; thus the term 'anhydrous' in reference to the A-type rocks is misleading. These intensive parameters have a key role in the petrogenesis of this granite type.

5. Calculated zircon saturation temperatures for aluminous A-type granites are higher than those of other granite types. We suggest that the other HFSE that are relatively abundant in the aluminous A-type granites have partition coefficients that, like Zr, are dependent on temperature.

6. The term anorogenic is unclear and thus should not be used for characterizing A-type granites. Geochronological evidence suggests that LFB aluminous A-type granites can be emplaced at any time during a tectonic–magmatic episode. In the LFB, aluminous A-type granites are found adjacent to I-type granites of similar age and have similar field relations.

7. We favor a petrogenetic scheme in which the aluminous A-type granites were derived by small degrees of partial melting of a felsic, infracrustal source region. The source region was probably relatively reduced, with water and F contents similar to those of I-type source regions. The A-type source regions could be either very localized or separated from other granite source regions vertically in the crust. The necessarily high temperatures required to produce an extractable magma may have been initiated by mantle upwelling or mafic magma influx into a localized area.

ACKNOWLEDGEMENTS

This manuscript benefited from constructive reviews by John Hogan, John Foden and an anonymous reviewer. We appreciate Tony Ewart's careful editorial assistance. We are grateful for helpful conversations and data from Phillip Blevin, Ian Williams, David Champion, Dennis Pogson, and Anthony Budd. Discussions with Phillip Candela, the late Robert Hill, Cal Barnes and Doone Wyborn were also useful. John Holloway and Lee Silver are thanked for providing facilities during the final stages of manuscript production. Nick Ware, Tony Phimphisane and Geoff Denton are thanked for help with analytical work. Figure 1 was drafted by the Cartographic Services Unit at AGSO. Financial support from the Australian Research Council Grant A39232908 is acknowledged.

REFERENCES

1. Anderson J. L. Mineral equilibria and crystallization conditions in the late Precambrian Wolf River rapakivi massif, Wisconsin. American Journal of Science 1980;280:289-332.

2. Anderson J. L. Proterozoic anorogenic granite plutonism of North America. Geological Society of America Memoir 1983;161:133-154.

3. Anderson J. L., Cullers R. L. Geochemistry and evolution of the Wolf River Batholith: a late Precambrian rapakivi massif in north Wisconsin, U.S.A. Precambrian Research 1978;7:287-324.

4. Atkinson P. 1976. Geology and geochemistry of the Victoria Valley Batholith, Grampians Ranges, south-eastern Australia. Unpublished B.Sc. Honours Thesis, La Trobe University, Melbourne, Vic.

5. Baker P. E. Peralkaline acid volcanic rocks of oceanic islands. Bulletin of Volcanology 1974;38:737-754.

6. Barker F., Wones D. R., Sharp W. N., Desborough G. A. The Pikes Peak Batholith, Colorado Front Range, and a model for the origin of the gabbro–anorthosite–syenite–potassic granite suite. Precambrian Research 1975;2:97-160.

7. Barron L. M., Scheibner E., Suppel D. W. The Mount Hope Group and its comagmatic granites on the Mount Allen 1:100 000 Sheet, New South Wales. Quarterly Notes New South Wales Geological Survey 1982:24.

8. Bédard J. Enclaves from the A-type granite of the Mégantic Complex, White Mountain magma series: clues to granite magmagenesis. Journal of Geophysical Research 1990;95((B11)):17797-17819.

9. Bonin B., Grelou-Orsini C., Vialette Y. Age, origin and evolution of the anorogenic complex of Evisa (Corsica): a K–Li–Rb–Sr study. Contributions to Mineralogy and Petrology 1978;65:425-432.

10. Bowen N. L. The Evolution of the Igneous Rocks Princeton, NJ: Princeton University Press; 1928.

11. Budd A. R. Internal differentiation and mineralization of the Banshea pluton, New South Wales. Unpublished B.Sc. Honours Thesis, Australian National University, Canberra, A.C.T.

12. Burnham C. W., Ohmoto H. Late-stage processes of felsic magmatism. Mining Geology, Special Issue 1980;8:1-11.

13. Google Scholar

14. Camacho A. 1987. Geochemistry of some fractionated western Tasmanian granites. Unpublished M.Sc. Thesis, La Trobe University, Melbourne, Vic.

15. Carmichael I. S. E., McKenzie W. S. Feldspar–liquid equilibria in pantellerites: an experimental study. American Journal of Science 1963;261:382-396.

16. Champion D. C. 1991. Felsic granites of far north Queensland. Unpublished Ph.D. Thesis, Australian National University, Canberra, A.C.T.

17. Chappell B. W. Trace element analysis of rocks by X-ray spectrometry. Advances in X-ray Analysis 1991;34:263-276.

18. Chappell B. W. Lachlan and New England: fold belts of contrasting magmatic and tectonic development. Journal and Proceedings, Royal Society of New South Wales 1994;127:47-59.

19. Chappell B. W., Hergt J. M. The use of known Fe content as a flux monitor in neutron activation analysis. Chemical Geology 1989;78:151-157.

20. Chappell B. W., White A. J. R. Two contrasting granite types. Pacific Geology 1974;8:173-174.

21. Chappell B. W., White A. J. R. I- and S-type granites in the Lachlan Fold Belt. Transactions of the Royal Society of Edinburgh: Earth Sciences 1992;83:1-26.

22. Chappell B. W., Williams I. S., White A. J. R., McCulloch M. T. Granites of the Lachlan Fold Belt. ICOG 7 Field Guide Excursion A-2. Bureau of Mineral Resources, Geology & Geophysics Record 1990;1990/48.

23. Chappell B. W., English P. M., King P. L., White A. J. R., Wyborn D. Granites and related rocks of the Lachlan Fold Belt (1:1 250 000 scale map) Canberra: Bureau of Mineral Resources, Geology and Geophysics; 1991.

24. Chen Y. D., Williams I. S. Zircon inheritance in mafic inclusions from the Bega Batholith granites, southeastern Australia: an ion microprobe study. Journal of Geophysical Research 1990;95B:17787-17796.

25. Clarke D. B. Granitoid Rocks London: Chapman and Hall; 1992.

26. Clemens J. D., Holloway J. R., White A. J. R. Origin of an A-type granite: experimental constraints. American Mineralogist 1986;71:317-324.

27. Collins W. J. 1977. Gabo Island Granite Suite. Unpublished B.Sc. Honours Thesis, Australian National University, Canberra, A.C.T.

28. Collins W. J., Beams S. D., White A. J. R., Chappell B. W. Nature and origin of A-type granites with particular reference to southeastern Australia. Contributions to Mineralogy and Petrology 1982;80:189-200.

29. Creaser R. A., Price R. C., Wormald R. J. A-type granites revisited: assessment of a residual-source model. Geology 1991;19:163-166.

30. ↵ Cullers R. L., Koch R. J., Bickford M. E. Chemical evolution of magmas in the Proterozoic Terrane of the St. Francois Mountains, Southeastern Missouri. 2. Trace element data. Journal of Geophysical Research 1981;86:10388-10401.

31. Czamanske G. K., Ishihara S., Atkin S. A. Chemistry of rock-forming minerals of the Cretaceous–Paleocene batholith in Southwestern Japan and implications for magma genesis. Journal of Geophysical Research 1981;86:10431-10469.

32. Dodd S. J. 1981. The deformation, metamorphic and intrusive history of the Kuark Metamorphic Belt, Southeast Gippsland. Unpublished B.Sc. Honours Thesis, La Trobe University, Melbourne, Vic.

33. Eby G. N. The A-type granitoids: a review of their occurrence and chemical characteristics and speculations on their petrogenesis. Lithos 1990;26:115-134.

34. Eby G. N. Chemical subdivision of the A-type granitoids: petrogenetic and tectonic implications. Geology 1992;20:641-644.

35. Fergusson C. L., Glen R. A., editors. Tectonophysics. 1992. The Palaeozoic eastern margin of Gondwanaland: tectonics of the Lachlan Fold Belt, southeastern Australia and related orogens; p. 214.

36. Greenberg J. K. Anorogenic granite associations as products of progressive continental evolution. In: Gower C. F., Rivers T., Ryan B., editors. Mid-Proterozoic Laurentia–Baltica. Geological Association of Canada, Special Paper. Vol. 38. 1990. p. 447-457.

37. Hogan J. P., Gilbert M. C., Weaver B. L. A-type granites and rhyolites: is A for ambiguous? EOS, Transactions, American Geophysical Union (newsletter) 1992;73:508.

38. Jacobsen S. C., Wasserburg G. J. Sm–Nd isotopic evolution of chondrites. Earth and Planetary Science Letters 1980;50:139-155.

39. Johnson M. C., Rutherford M. J. Experimentally determined conditions in the Fish Canyon Tuff, Colorado, magma chamber. Journal of Petrology 1989;30:711-737.

40. King P. L. 1992. A-type granites from the Lachlan Fold Belt: a case study and reassessment. Unpublished B.Sc. Honours Thesis, Australian National University, Canberra, A.C.T.

41. King P. L., Chappell B. W., White A. J. R., Williams I. S. A-type granites from the Lachlan Fold Belt, eastern Australia. EOS, Transactions, American Geophysical Union 1992;73:346.

42. Korsch M. J., Gulson B. L. Nd and Pb isotopic studies of an Archaean layered mafic–ultramafic complex, Western Australia, and implications for mantle heterogeneity. Geochimica et Cosmochimica Acta 1986;50:1-10.

43. Landenberger B., Collins W. J. Derivation of A-type granites from a dehydrated charnockitic lower crust: evidence from the Chaelundi Complex, Eastern Australia. Journal of Petrology 1996;37:145-170.

44. Abstract/FREE Full Text

45. Leake B. E. Nomenclature of amphiboles. American Mineralogist 1978;63:1023-1052. Lewis P. C., Glen R. A., Pratt G. W., Clarke I. Bega-Mallacoota 1:250 000 Geological Sheet SJ/55–4, SJ/55–8: Explanatory Notes Sydney: Geological Survey of New South Wales; 1994.

46. Loiselle M. C., Wones D. R. Characteristics and origin of anorogenic granites. Geological Society of America, Abstracts 1979;11:468.↵ MacKenzie D. E., Black L. P., Sun S-S. Origin of alkali-feldspar granites: an example from the Poimena Granite, northeastern Tasmania, Australia. Geochimica et Cosmochimica Acta 1988;52:2507-2524.

47. MacKenzie D. E., Sun S-S., Black L. P. Reply to N. C. Higgins' Comment on Origin of alkali-feldspar granites: an example from the Poimena Granite, northeastern Tasmania, Australia'. Geochimica et Cosmochimica Acta 1990;54:2313-2322.

48. Maniar P. D., Piccoli P. M. Tectonic discrimination of granitoids. Geological Society of America Bulletin 1989;101:635-643.

49. Merzbacher C., Eggler D. H. A magmatic geohygrometer: application to Mount St. Helens and other dacitic magmas. Geology 1984;12:587-590.

50. Naney M. T. Phase equilibria of rock-forming ferromagnesian silicates in granitic systems. American Journal of Science 1983;283:993-1033.

51. Norrish K., Hutton J. T. An accurate X-ray spectrographic method for the analysis of a wide range of geological samples. Geochimica et Cosmochimica Acta 1969;33:431-453.

52. Pearce J. A., Harris N. B. W., Tindle A. G. Trace element discrimination diagrams for the tectonic interpretation of granitic rocks. Journal of Petrology 1984;25:956-983.

53. Peck L. C. Systematic analysis of silicates. US Geological Survey, Bulletin 1964:1170.

54. Peterson J. W., Chako T., Kuehner S. M. The effects of fluorine on the vapor-absent melting of phlogopite + quartz: implications for deep-crustal processes. American Mineralogist 1991;76:470-476.

55. Pitcher W. S. The Nature and Origin of Granite London: Blackie; 1993.

56. Poitrasson F., Pin C., Duthou J-L., Platevoet B. Aluminous subsolvus anorogenic granite genesis in the light of Nd isotopic heterogeneity. Chemical Geology 1994;112:199-219.

57. Poitrasson F., Duthou J-L., Pin C. The relationship between petrology and Nd isotopes as evidence for contrasting anorogenic granite genesis: example of the Corsican Province (SE France). Journal of Petrology 1995;36:1251-1274.

58. Radain A. A. M., Fyfe W. S., Kerrich R. Origin of peralkaline granites of Saudi Arabia. Contributions to Mineralogy and Petrology 1981;78:358-366.

59. Richards D. N. 1967. The Geology of the Jerangle District. Unpublished B.Sc. Honours Thesis, Australian National University, Canberra, A.C.T.

60. Rogers J. J. W., Greenberg J. E. Late-orogenic, post-orogenic, and anorogenic granites: distinction by major-element and trace-element chemistry and possible origins. Journal of Geology 1990;98:291-310.

61. Skjerlie K. P., Johnston A. D. Vapor-absent melting at 10 kbar of a biotite- and amphibole-bearing tonalitic gneiss: implications for the generation of A-type granites. Geology 1992;20:263-266.

62. Sparks R. S. J., Marshall L. A. Thermal and mechanical constraints on mixing between mafic and silicic magmas. Journal of Volcanology and Geothermal Research 1986;29:99-124.

63. Sylvester P. J. Post-collisional alkaline granites. Journal of Geology 1989;97:261-280.

64. Turner S. P., Foden J. D., Morrison R. S. Derivation of some A-type magmas by fractionation of basaltic magma: an example from the Padthaway Ridge, South Australia. Lithos 1992;28:151-179.

65. Tuttle O. F., Bowen N. L. Origin of granite in the light of experimental studies in the system NaAlSi3O8-KAlSi3O8-SiO2-H2O. Geological Society of America Memoir 1958;74:153.

66. Walsh J. M. N., Beckinsale R. D., Shelhorn R. R., Thorpe R. S. Geochemistry and petrogenesis of Tertiary granitic rocks from the island of Mull, NW Scotland. Contributions to Mineralogy and Petrology 1979;71:99-116.

67. CrossRefWeb of ScienceGoogle Scholar

68. Watson E. B. Zircon saturation in felsic liquids: experimental data and applications to trace element geochemistry. Contributions to Mineralogy and Petrology 1979;70:407-419.

69. Watson E. B., Harrison T. M. Zircon saturation revisited: temperature and composition effects in a variety of crustal magma types. Earth and Planetary Science Letters 1983;64:295-304.

70. Whalen J. B., K. L. Currie. The Topsails igneous suite, western Newfoundland; fractionation and magma mixing in an 'orogenic' A-type granite suite. Ore-bearing Granite Systems; Petrogenesis and Mineralizing Processes. Geological Society of America, Special Papers 1990;246:287-300.

71. Whalen J. B., Currie K. L., Chappell B. W. A-type granites: geochemical characteristics, discrimination and petrogenesis. Contributions to Mineralogy and Petrology 1987;95:407-419.

72. Williams I. S. Some observations on the use of zircon U–Pb geochronology in the study of granitic rocks. Transactions of the Royal Society of Edinburgh: Earth Sciences 1992;83:447-458.

73. Williams I. S., Chappell B. W., McCulloch M. T., Crook K. A. W. Inherited and detrital zircons—clues to the early growth of crust in the Lachlan Fold Belt. Geological Society of Australia, Abstracts 1991;29:58.

74. Wones D. R. Significance of the assemblage titanite + magnetite + quartz in granitic rocks. American Mineralogist 1989;74:744-749.

75. Wormald R. J. 1991. The petrology and geochemistry of Mid to Late Palaeozoic magmatism in the Temora region, New South Wales. Unpublished Ph.D. Thesis, La Trobe University, Melbourne, Vic.

76. Wormald R. J., Price R. C. Peralkaline granites near Temora, southern New South Wales: tectonic and petrological implications. Australian Journal of Earth Sciences 1988; 35:209-221.

77. Wyborn D., Owen M. Araluen, N.S.W.—1:100 000 Geological Map Commentary Canberra, A.C.T.: Australian Government Publishing Service; 1986.

78. Wyborn D., Turner B. S., Chappell B. W. The Boggy Plain Supersuite: a distinctive belt of I-type igneous rocks of potential economic significance in the Lachlan Fold Belt. Australian Journal of Earth Sciences 1987;34:21-43.

79. Yacopetti C. M. 1987. The A-type granites of the Lachlan Fold Belt. Unpublished B.Sc. Honours Thesis, Australian National University, Canberra, A.C.T.

6

CHAPTER

Metamorphism Under Stress: The Problem of Relating Minerals to Depth

Stefan M. Schmalholz* and Yuri Podladchikov

Institute of Earth Sciences, University of Lausanne, CH-1015 Lausanne, Switzerland

STRESS AND METAMORPHISM—THE DEPTH PROBLEM

Metamorphism and rock deformation often occur together. In a paper in this issue of *Geology*, Wheeler (2014, p. 647) presents theoretical results that show that a differential stress driving the deformation could have a dramatic effect on metamorphic reactions. If this is true, then Wheeler's result has considerable consequences for geological applications, because it is commonly assumed that the metamorphic reactions are controlled by the lithostatic pressure (the weight of the overlying rock) which can be directly related to depth. According to this common practice, metamorphic minerals are used to estimate the burial depth of the tectonic unit in which the minerals have been found (e.g., Jolivet et al., 2003). If such depth estimates could indeed be considerably inaccurate, then this inaccuracy could help to answer some open geodynamic questions. For example, the depth of the metamorphic sole beneath ophiolites, which is constrained by the thickness of the observed overlying rock (<~15 km), is usually much less than the depth estimated from the metamorphic minerals of the sole (>~30 km; Moulas et al., 2013, and references therein). However, the potential inaccuracy of depth estimates based on minerals would question current paradigms in geology.

PARADIGMS UNDER THREAT

The discovery of micrometer-scale ultrahigh-pressure (UHP) minerals, such as coesite and diamond, in exhumed crustal rocks has revolutionized both metamorphic petrology and geodynamics (Chopin, 1984, 2003; Smith, 1984). The plate tectonics paradigm, with an old (not recycled into the mantle) continental crust floating horizontally on a denser mantle, had to be refined because the UHP (>28 kbar) minerals indicate the subduction of continental crust deep into the mantle (>100 km), and also their subsequent exhumation to the surface. Moreover, chronological studies show that the vertical motions during exhumation have the same velocities as the horizontal motion of continental plates, that is, several centimeters per year (e.g., Rubatto and Hermann, 2001). UHP metamorphism became an exciting research field because careful microscopic observation alone could have enormous consequences for the geodynamic history of the corresponding region. The formation at large mantle depth became widely accepted as the unique cause for UHP metamorphism in crustal rocks, and has been proposed as a new paradigm (Hacker and Gerya, 2013). The fundamental assumption of this UHP paradigm is that the pressure indicated by UHP minerals must be converted to depth by applying the lithostatic formula for pressure.

NON-HYDROSTATIC THERMODYNAMICS

The relationship between stress, pressure, and metamorphic reactions can be quantified with thermodynamic calculations. In these calculations, borrowed from gases and fluids and applied to solids without any modification (e.g., Connolly, 2009), it is commonly assumed that the stress state at every point is isotropic, and is fully represented by a single scalar "thermodynamic pressure" that controls the metamorphic reactions. This fluid-like version of thermodynamic theory is often referred to as hydrostatic thermodynamics. However, the relation between a single thermodynamic pressure and the stress in a solid is not so simple. For example, rock deformation experiments have shown that quartz was partially converted to coesite at a confining pressure of only 12.5 kbar although a thermodynamic pressure of 28 kbar is required (Hirth and Tullis, 1994). Hirth and Tullis (1994)suggested that it was the magnitude of the largest stress (28 kbar) in the loading direction that controlled the reaction from quartz to coesite, and neither the confining pressure nor the mean stress (18 kbar). Also, the observation of coesite in both garnet and omphacite in exhumed rock (Chopin, 1984; Smith, 1984) indicates pressure variations >10 kbar on the mineral scale according to the well-accepted "pressure vessel" model that explains the preservation of coesite during the exhumation (Gillet et al., 1984; Perrillat et al., 2003; O'Brien and Ziemann, 2008). Furthermore, a differential stress of up to 17 kbar on the mineral scale has been indicated by observed coherent mineral lamellae (e.g., Pryer and Robin, 1996). However, the influence of pressure variations and differential stress on metamorphic reactions is commonly neglected in hydrostatic thermodynamic calculations.

Different approaches exist to account for non-hydrostatic effects in thermodynamic relationships between solids under stress (Grinfeld, 1991, see p. 125–132). A more microscopic approach concerning grain-scale stress variation considers partial Gibbs energy (chemical potentials) of the solid phases as functions of the normal stresses that act locally on mineral interfaces (Gibbs, 1906; McLellan, 1980; Wheeler, 2014). The more macroscopic approach of Dahlen (1992) starts from the same local relationships ("closely following" McLellan, 1980) but ends with approximate partial Gibbs energies as a function of fluid pressure for a fluid-saturated rock, even if the reaction between solid phases does not involve fluids. Dahlen's approach is similar to considering the "pore fluid" pathway in the sense of Wheeler (2014). Consequently, Dahlen predicts zero sensitivity to differential stress, while Wheeler shows that if other pathways are active then the sensitivity may be dramatic.

The experiments of Hirth and Tullis (1994) contradict the approaches of bothWheeler (2014) and Dahlen (1992). With the impact of differential stress on metamorphic reactions suggested by the results of Wheeler (2014), Hirth and Tullis (1994) should see coesite in the quartz stability field also for a largest stress of <28 kbar, which was not the case. But according to Dahlen (1992) they should not see coesite at all. However, both approaches can be easily defended. Wheeler (2014) suggests only the possibility of forming coesite subject to the chosen reaction pathway. In the absence of such a pathway in the experiments, coesite would not form at lower stress. The approach of Dahlen (1992) could be justified because it predicts correctly the 99 vol% quartz in the experiments and only fails to predict the 1 vol% of coesite, which can be considered as error due to the approximations done by Dahlen (1992). Alarmingly, the current practice in petrology (fluid pressure = total pressure = lithostatic pressure) is in agreement with Dahlen (1992), meaning that small coesite inclusions in UHP rocks should not be used as a proxy for any of these pressures. In fact, the non-hydrostatic solid-solid reaction in Hirth and Tullis (1994) experiments can be explained by the classical Gibbs (1906) theory developed for the non-hydrostatic solid-fluid case. The suggestion of Moulas et al. (2013), that coesite formed under a nearly hydrostatic stress that is close to the value of the largest stress in the experiments, makes the coesite crystals thermodynamically indistinguishable from a fluid, and, hence, the solid-solid reaction between quartz and coesite reduces to the solid-fluid case considered by Gibbs (1906).

BACK TO THE FIELD

The reliability of depth estimates based on minerals is only under threat if the thermodynamic pressure can deviate significantly from the lithostatic pressure (for example, >20%). To quantify such deviation is difficult. One possible way is to compare the metamorphic record with the structural one. For example, Pleuger and Podladchikov (2014) performed a structural restoration for the Adula nappe (Western Alps, Switzerland), which suggests that the nappe was formed at a depth of ~60 km. However, the Adula nappe includes UHP rocks that indicate a maximum lithostatic depth of ~110 km. With regard to this "depth problem" a geologist can currently choose between two types of solutions: first, the structural record is more or less complete and metamorphic reactions occurred at a depth less than the lithostatic depth, or, second, the lithostatic depth estimate is accurate but the structural record is incomplete, that is, the structures that indicate the 50 km exhumation from mantle depth have not been identified. Most geologists seem to choose the second solution. Nevertheless, in a new textbook on Alpine geology (Pfiffner, 2014), the first solution has been proposed for the Adula nappe.

A 50 km depth difference corresponds to a lithostatic pressure of ~1400 MPa. The result of Wheeler (2014) based on non-hydrostatic thermodynamics suggests that under a differential stress of 140 MPa the UHP minerals could, in principle, form at a depth of 60 km given a suitable reaction pathway. An alternative possibility, to form the UHP minerals in 60 km depth using classical hydrostatic thermodynamics, would be that stresses deviated significantly from the lithostatic pressure at 60 km depth and that, hence, the pressure controlling the metamorphic reaction was ~1400 MPa larger than the lithostatic pressure (Schmalholz and Podladchikov, 2013; Schmalholz et al., 2014).

POTENTIAL APPLICATIONS AND OPPORTUNITIES

If minerals might not be reliable indicators of maximum burial, then one might fear that metamorphic petrology could lose one of its main applications. However, if tectonic pressure and differential stress could indeed considerably influence metamorphic reactions, then this could also create new applications. For example, if observed variations in metamorphic facies were due to pressure variations at the same depth, then these facies variations would contain important information on pressure, stress, and strength variations in rock. Then, there would be a need for a more accurate quantification of these variations, and also for a more detailed comparison of these variations with the deformation structures. Particularly, the regions without (U)HP metamorphism would be as important for geodynamic reconstructions as the (U)HP regions. Moreover, pressure variations can actually be considered in hydrostatic thermodynamic calculations. For example, Tajčmanová et al. (2014) explained the chemical zoning preserved in reaction rims as a result of mineral-scale pressure variations, and not, as usually done, as a result of sluggish kinetics of metamorphic reactions. The kinetics of metamorphic reactions (that is, the non-equilibrium thermodynamics) is yet another unsolved problem.

REFERENCES

1. Chopin, C., 1984, Coesite and pure pyrope in high-grade blueschists of the Western Alps - a 1st record and some consequences: Contributions to Mineralogy and Petrology, v. 86, p. 107–118
2. Chopin, C., 2003, Ultrahigh-pressure metamorphism: Tracing continental crust into the mantle: Earth and Planetary Science Letters, v. 212, p. 1–14
3. Connolly, J.A.D., 2009, The geodynamic equation of state: What and how: Geochemistry Geophysics Geosystems, v. 10
4. Dahlen, F.A., 1992, Metamorphism of nonhydrostatically stressed rocks: American Journal of Science, v. 292, p. 184–198.
5. Gibbs, J.W., 1906, On the equilibrium of heterogeneous substances, in The Scientific Papers of J. Willard Gibbs, volume 1: London, New York and Bombay, Longmans, Green and Company, p. 55–354.
6. Gillet, P., Ingrin, J., and Chopin, C., 1984, Coesite in subducted continental crust—P-T history deduced from an elastic model: Earth and Planetary Science Letters, v. 70, p. 426–436.

7. Grinfeld, M., 1991, Thermodynamic Methods in the Theory of Heterogeneous Systems: Harlow, UK, Longman Scientific & Technical, 399 p.

8. Hacker, B.R., and Gerya, T.V., 2013, Paradigms, new and old, for ultrahigh-pressure tectonism: Tectonophysics, v. 603, p. 79–88.

9. Hirth, G., and Tullis, J., 1994, The brittle-plastic transition in experimentally deformed quartz aggregates: Journal of Geophysical Research. Solid Earth, v. 99, B6, p. 11731–11747.

10. Jolivet, L., Faccenna, C., Goffe, B., Burov, E., and Agard, P., 2003, Subduction tectonics and exhumation of high-pressure metamorphic rocks in the Mediterranean orogens: American Journal of Science, v. 303, p. 353–409.

11. McLellan, A.G., 1980, The Classical Thermodynamics of Deformable Materials: Cambridge, UK, Cambridge University Press, 338 p.

12. Moulas, E., Podladchikov, Y.Y., Aranovich, L.Y., and Kostopoulos, D., 2013, The problem of depth in geology: When pressure does not translate into depth: Petrology, v. 21, p. 527–538.

13. O'Brien, P.J., and Ziemann, M.A., 2008, Preservation of coesite in exhumed eclogite: Insights from Raman mapping: European Journal of Mineralogy, v. 20, p. 827–834.

14. Perrillat, J.P., Daniel, I., Lardeaux, J.M., and Cardon, H., 2003, Kinetics of the coesite-quartz transition: Application to the exhumation of ultrahigh-pressure rocks: Journal of Petrology, v. 44, p. 773–788.

15. Pfiffner, O.A., 2014, Geology of the Alps: Chichester, UK, Wiley Blackwell, 376 p.

16. Pleuger, J., and Podladchikov, Y., 2014, A purely structural restoration of the NFP20- East cross section and potential tectonic overpressure in the Adula nappe (Central Alps): Tectonics, v. 33, p. 656–685.

17. Pryer, L.L., and Robin, P.Y.F., 1996, Differential stress control on the growth and orientation of flame perthite: A palaeostress-direction indicator: Journal of Structural Geology, v. 18, p. 1151.

18. Rubatto, D., and Hermann, J., 2001, Exhumation as fast as subduction?: Geology, v. 29, p. 3–6.

19. Schmalholz, S.M., and Podladchikov, Y.Y., 2013, Tectonic overpressure in weak crustal-scale shear zones and implications for the exhumation of high-pressure rocks: Geophysical Research Letters, v. 40, p. 1984–1988.

20. Schmalholz, S.M., Duretz, T., Schenker, F.L., and Podladchikov, Y.Y., 2014, Kinematics and dynamics of tectonic nappes: 2-D numerical modelling and implications for high and ultra-high pressure tectonism in the Western Alps: Tectonophysics.

21. Smith, D.C., 1984, Coesite in clinopyroxene in the Caledonides and its implications for Geodynamics: Nature, v. 310, p. 641–644.

22. Tajčmanová, L., Podladchikov, Y., Powell, R., Moulas, E., Vrijmoed, J.C., and Connolly, J.A.D., 2014, Grain-scale pressure variations and chemical equilibrium in high-grade metamorphic rocks: Journal of Metamorphic Geology, v. 32, p. 195–207.

23. Wheeler, J., 2014, Dramatic effects of stress on metamorphic reactions: Geology, v. 42, p. 647–650.

7
CHAPTER

Clay Smear: Review of Mechanisms and Applications

Peter J. Vrolijk[a], Janos L. Urai[b], Michael Kettermann[b]

[a] *ExxonMobil Upstream Research Company, S1.2A.478, 22777 Springwoods Village Parkway, Spring, TX 77389, USA*

[b] *Structural Geology, Tectonics and Geomechanics, Energy and Mineral Resources Group, RWTH Aachen University, Lochnerstrasse 4-20, D-52056 Aachen, Germany*

HIGHLIGHTS

- We review four decades of research into clay smear processes and applications.
- Review includes critical evaluations of outcrop, laboratory experiments, numerical models, and subsurface flow studies.
- Clay smear arises from a multitude of fault processes in clastic sedimentary rocks.
- Although long recognized as central to flow problem, processes that result in clay smear holes are poorly studied.
- Future clay smear models applied to fluid flow will benefit from consideration of failure modes.

ABSTRACT

Clay smear is a collection of fault processes and resulting fault structures that form when normal faults deform layered sedimentary sections. These elusive structures have attracted deep interest from researchers interested in subsurface fluid flow, particularly in the oil and gas industry. In the four decades since the association between clay-smear structures and oil and gas accumulations was introduced, there has been extensive research into the fault processes that create clay smear and the resulting effects of that clay smear on fluid flow. We undertake a critical review of the literature associated with outcrop studies, laboratory and numerical modeling, and subsurface field studies of clay smear and propose a comprehensive summary that encompasses all of these elements. Important fault processes that contribute to clay smear are defined in the context of the ratio of rock strength and *in situ* effective stresses, the geometric evolution of fault systems, and the composition of the faulted section. We find that although there has been progress in all avenues pursued, progress has been uneven, and the processes that disrupt clay smears are mostly overlooked. We highlight those research areas that we think will yield the greatest benefit and suggest that taking these emerging results within a more process-based framework presented here will lead to a new generation of clay smear models.

Keywords: Clay smear; Fault process; Sedimentary rocks; Experiments; Fluid flow

1. INTRODUCTION

Clay smear is among the family of structures that deflect fluid flow in sedimentary basins. Clay smear processes were originally conceived for (Smith, 1966 and Weber et al., 1978) and continue to be applied to problems of cross-fault flow when porous and permeable rocks, specifically sandstones and shales, are cut by normal faults. Within this conceptual framework a series of stacked sandstone and clay-rich shale beds are offset with a normal displacement, and during that offset clays from shale beds may become entrained within the fault zone and redistributed along its length between the footwall and hanging wall beds (Fig. 1). While this general description is accurate, it is too imprecise to be of much practical use. Our goals in this paper are to critically review and summarize the state of knowledge for the fault processes that occur within the ellipse in Fig. 1, to evaluate how these processes affect cross-fault flow and whether current models of these processes adequately perform their task, and to speculate on how the next level of understanding into these problems might be achieved. Although much of this literature is based on quantitative analysis, the breadth of this review requires us to approach the subject in a qualitative manner informed by the quantitative background.

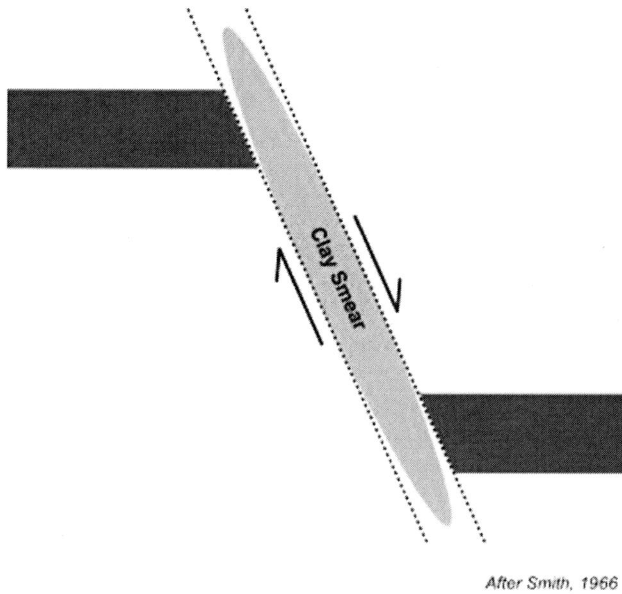

Figure 1. Most general definition of clay smear as conceptualized by Smith (1966) in which a fault zone (shear zone) contains clay derived from a faulted and offset shale bed. Many processes occur within yellow 'Clay Smear' ellipse, and a primary goal of review is to provide process framework that leads to more successful predictive models. Note that for an infinitely thin fault zone, there is no clay smear.

The central theme in this analysis is the geometry of all the components that comprise a fault zone, including:

- The geometry of the fault zone margins
- The heterogeneously deformed high shear-strain intervals within the fault zone that accommodate the vast majority of the fault displacement
- The distribution and occurrence of less deformed components within the fault zone, including fault relays and lenses
- The geometry and folding of beds bounding the fault in the footwall and hanging wall

It is this geometric arrangement that exerts the primary influence on cross-fault flow given the permeability of each component, just as the distribution, geometry, and properties of sandstones within fluvial channels is a primary concern of sedimentologists and stratigraphers when evaluating sub-surface flow. A fault zone must have finite thickness for clay smear to exist (Fig. 1); a fault surface, (i.e. an infinitely thin fault zone), contains no clay smear. Moreover, this geometric pattern is the product of the deformation caused by faulting the stratigraphic section. In theory, a forward model of the fault history based on geomechanical principles could yield the same geometric products. Although there have been attempts to relate clay smear to other flow problems, like flow along the fault zone (e.g., Caine and Minor, 2009), our discussion is restricted to summarizing clay smear processes and their effect on cross-fault flow.

Research into clay smear includes both basic and applied components. Basic research explores the initiation and evolution of normal faults, including localization phenomena, in clastic sedimentary rocks (Mandl, 1988) while the applied research approach focuses on the effects of normal faults on subsurface fluid flow. We begin this review with the field outcrop, laboratory experimental and numerical studies (Fig. 2) at the basic research end of the spectrum and then address applications of clay smear concepts in subsurface flow environments. While some might view basic and applied research areas as separate endeavors, we find tremendous benefit in considering them together. Indeed, we think the research community likely stands at a point where both ends of the research spectrum could improve with a new level of integration, how more process-based fault models will yield better cross-fault flow predictive models and how better subsurface flow studies will be more effective at testing improved predictive flow models.

1.1. Readers' guide

This is a comprehensive review of decades of research in many different technical areas that may go beyond the needs of some readers while others may require the detail provided in order to accept our conclusions. We chose this approach because we felt a comprehensive, critical evaluation of the literature is lacking. To accommodate the needs of different types of readers, we offer the following suggestions.

Curious reader: for readers who wish to learn something about this general area, we recommend Sections 1, 2, 3.4 and 4 and the following figures in addition to those cited in these sections (Fig. 4, Fig. 21, Fig. 22, Fig. 27, Fig. 28 and Fig. 29).

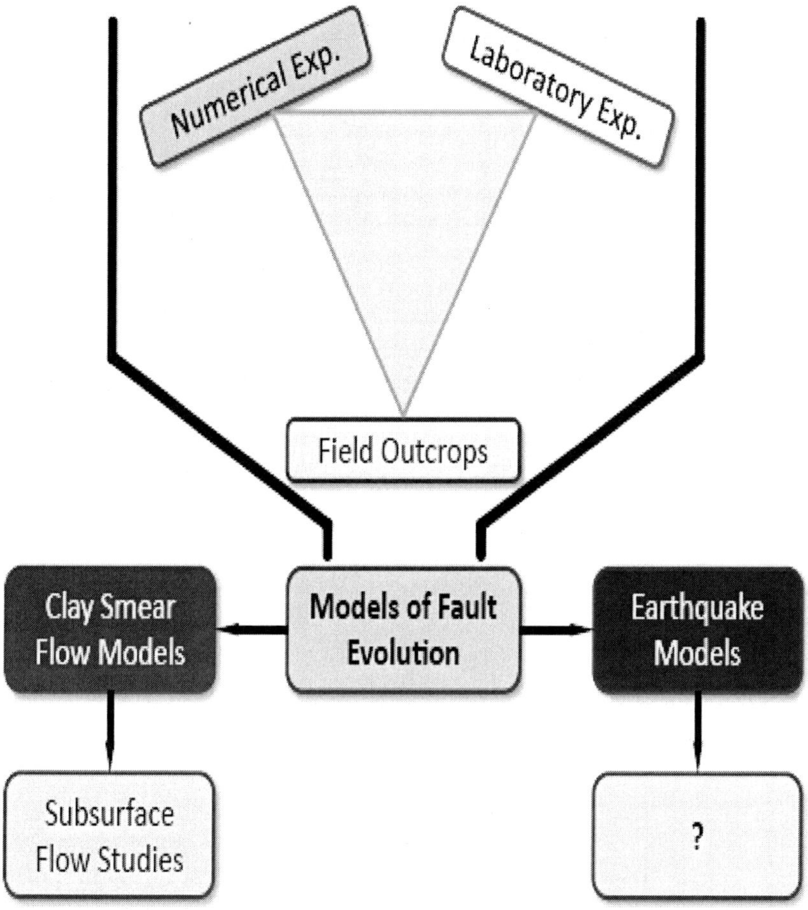

Figure 2. Clay smear process framework developed through studies of field outcrops, analytical & numerical experiments, and laboratory experiments, all of which contribute to models of fault evolution, which are in turn used to develop subsurface flow models. Clay smear flow models used to predict cross-fault flow and thus tested by subsurface flow studies. Less well developed is how models of fault evolution lead to earthquake prediction models, and data to test those models are beyond the scope of this work.

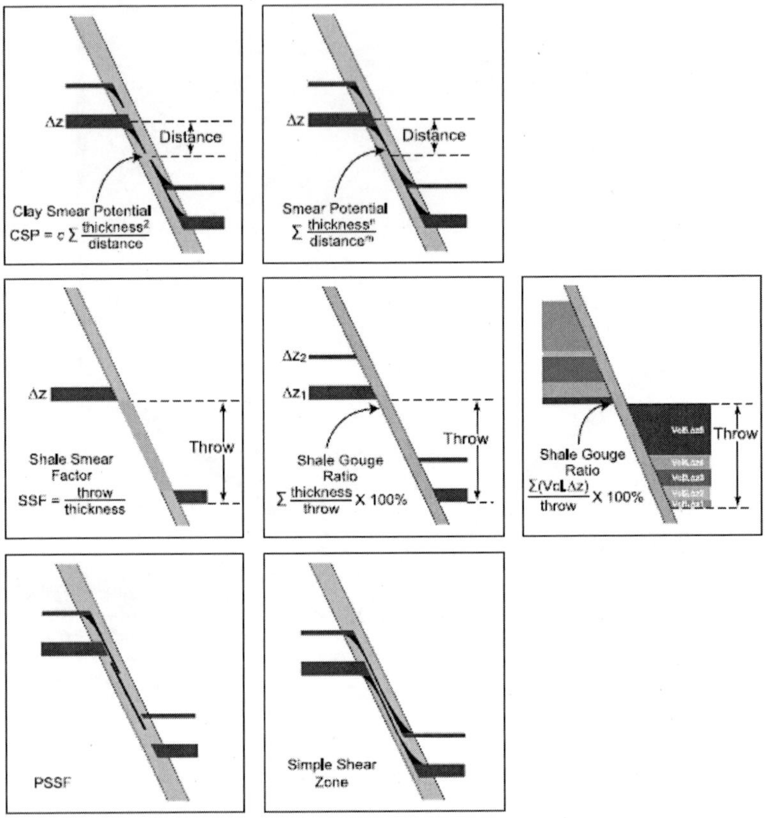

Figure 3. Summary of various algorithmic approaches to modeling effects of clay smear. Top row: Clay Smear Potential (left) and Smear Potential (right) in which exponents for thickness and offset distance variables are adjustable values (*n* and *m* are set to 2 and 1, respectively, for CSP); for CSP, c is a calibration constant intended to account for rheological and stress-dependency effects. Middle row: Shale Smear Factor (left) and two variations of Shale Gouge Ratio in middle and right-hand figures. In basic SGR (middle) stratigraphic section discretized into binary shale and sandstone lithologies and thickness represents only shale component of section. In right-hand SGR case, V_{shale} treated as continuous variable allowing sandstone with interpreted clay component contributing to modeled clay smear. Bottom row: Probabilistic Shale Smear Factor (left) reflects attempt to increase chance of holes forming with increasing SSF value, and Simple Shear Zone (right) represents end-member fault behavior where all lithologies become proportionately attenuated in fault zone according to original stratigraphic thicknesses. See text for further discussion and references for different algorithms. Figure modified from Yielding et al. (1997) and Childs et al. (2007).

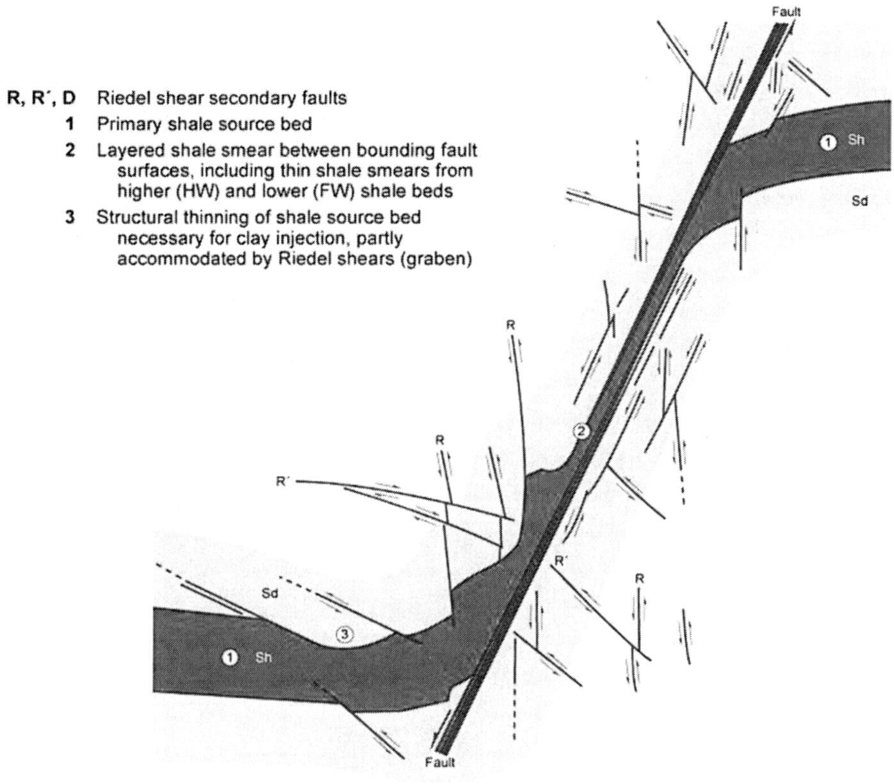

Figure 4. Summary of fault zone elements developed in the Hambach mines, modified after Lehner and Pilaar (1997), including a layered fault zone structure comprising clay smears from multiple source beds; secondary shears interpreted as Riedel R-, R'-, and D-shears; hanging wall and footwall folds; and structural thinning of clay source bed, particularly in hanging wall. Note that figure is intended to be dimensionless, but drawing made from outcrops that are typically meters to 10's of meters in scale. Figure reprinted from Lehner and Pilaar (1997) with permission from Elsevier.

Figure 5. Layered fault gouges from normal faults in (A) Hambach mines, and (B) Miri outcrops (after van der Zee and Urai, 2005). Hambach mine outcrop (A) maintains continuous sand layer between clay smears derived from different mudstone beds whereas sand becomes boudinaged into small phacoids in amalgamated matrix of clay smears in Miri outcrop (B). Fault offset in Hambach is several meters, while in Miri fault offset is 10's of meters.

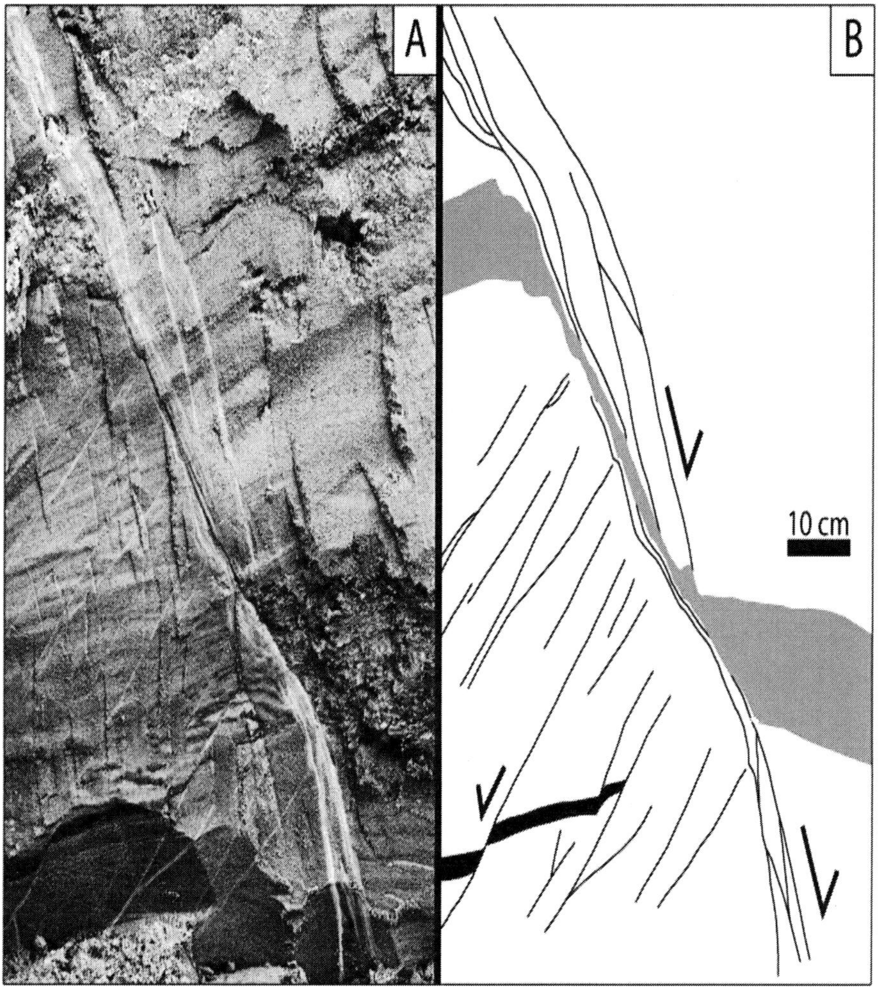

Figure 6. Full displacement of clay bed in Hambach mine, Germany. Note fault refraction at base of outcrop which may contribute to folding of clay bed (different orientation of clay bed in HW versus FW). Light bands in sandstone are deformation bands (secondary faults) that help accommodate overall fault strain. Asperity at fault bend at bottom of outcrop contributes to abrasion of HW clay bed, resulting in incorporation of more clay into fault.

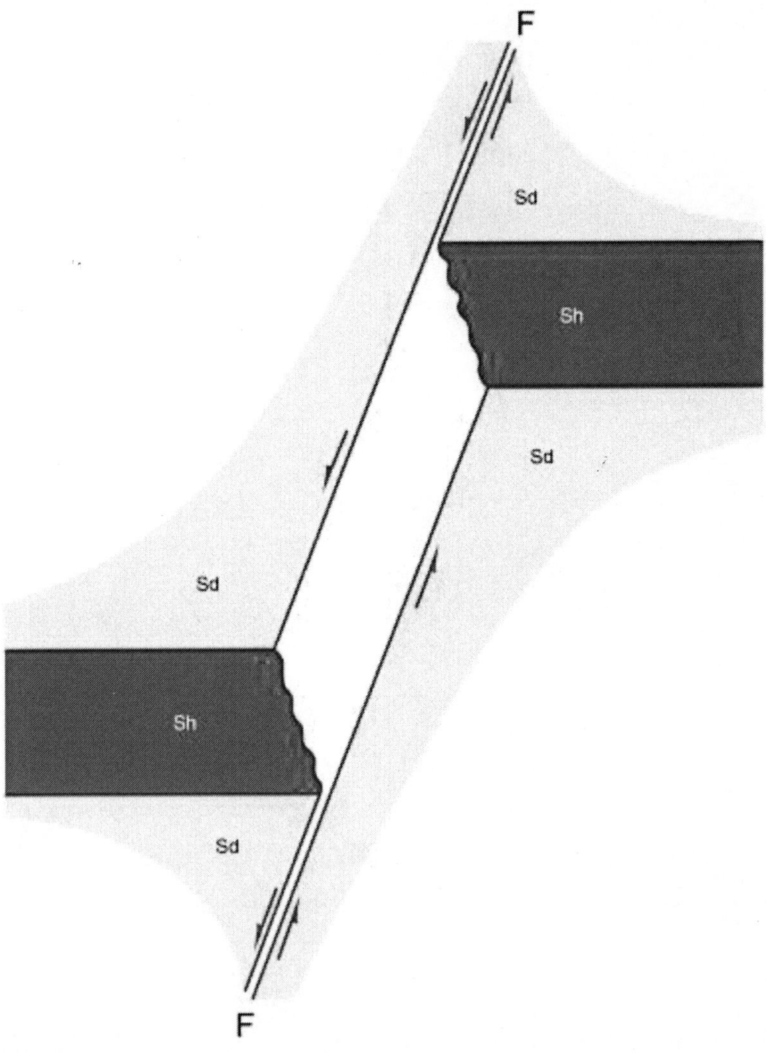

Figure 7. Schematic of 'pull-apart' geometry of overlapping fault segments developed in releasing step geometry, after Lehner and Pilaar (1997), in which dilatant volume created by this fault geometry is necessary component of clay-injection process. Note that figure that figure is intended to be dimensionless, but drawing made from outcrops that are typically meters in scale. Figure reprinted from Lehner and Pilaar (1997) with permission from Elsevier.

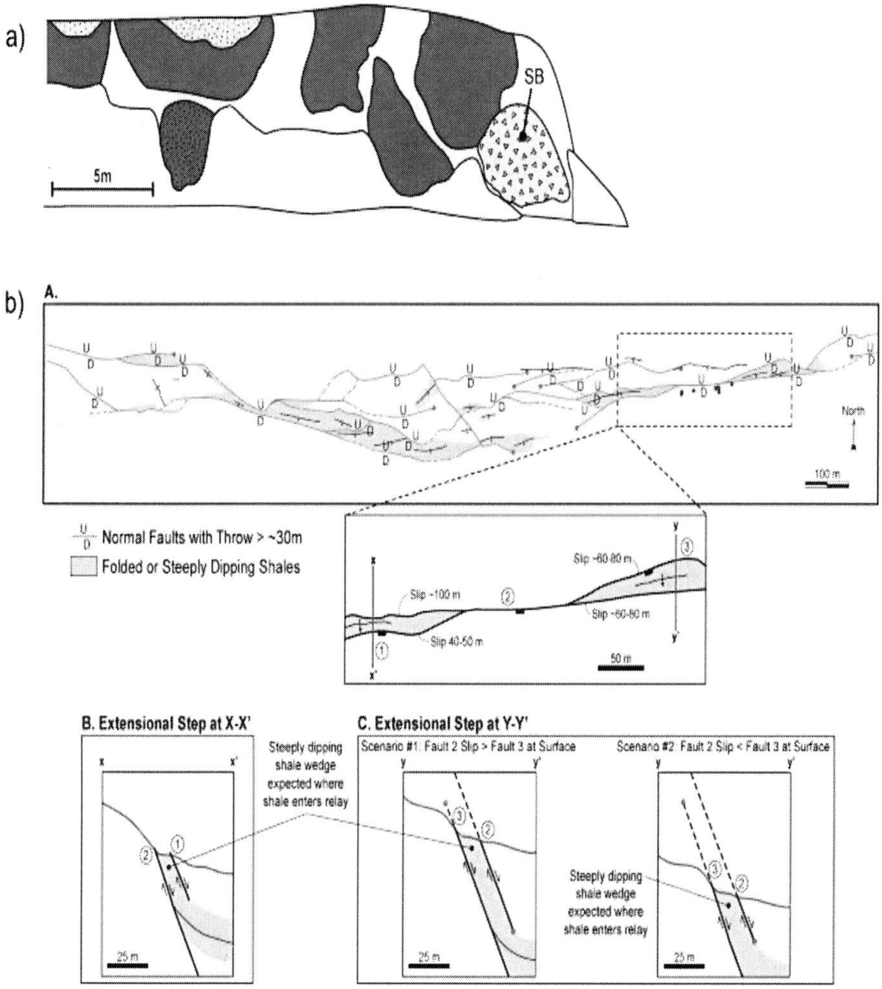

Figure 8. Plan view illustrations of fault lenses developed on scale of meters to 100's of m. (A) modified after Childs et al. (1997). Note in this Lancashire mine exposure that both shale- (green) and sandstone-dominated (stippled) lenses develop on fault surface (with minor sandstone breccia lens: SB). (B) Reproduced fromVrolijk et al., 2005a and Vrolijk et al., 2005b: mappable shale lenses develop along normal fault with 100's of meters of throw. Cross-sections illustrate geometry of overlapping fault segments (no vertical exaggeration). Note that observations along strike like these are essential for evaluating clay smear continuity for application to subsurface flow problems. Fig. 8a reprinted from Childs et al. (1997) with permission from Elsevier.

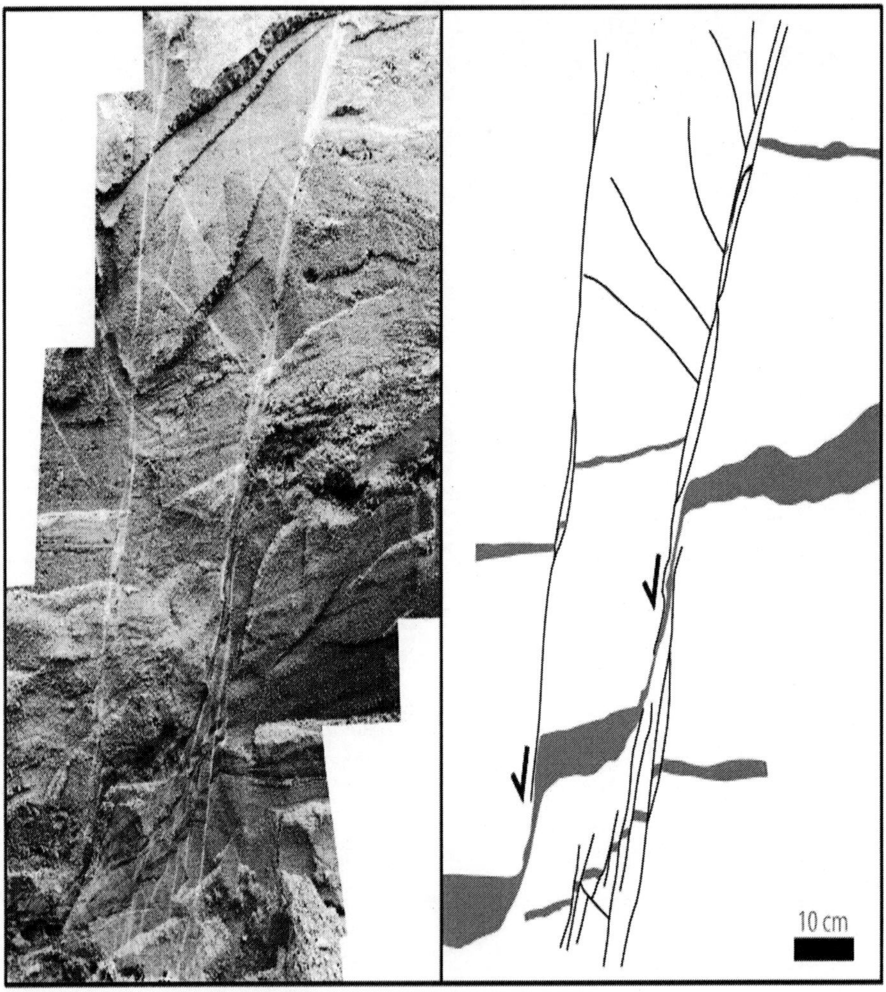

Figure 9. Outcrop figure from Hambach mine, Germany, in which dark clay bed is smeared along two primary fault zones defined by white deformation bands in sandstone, forming fault relay structure. Between bounding shear zones multiple secondary faults develop (base of outcrop), resulting in formation of small fault lenses of relatively undeformed clay bounded by clay-filled shear zones.

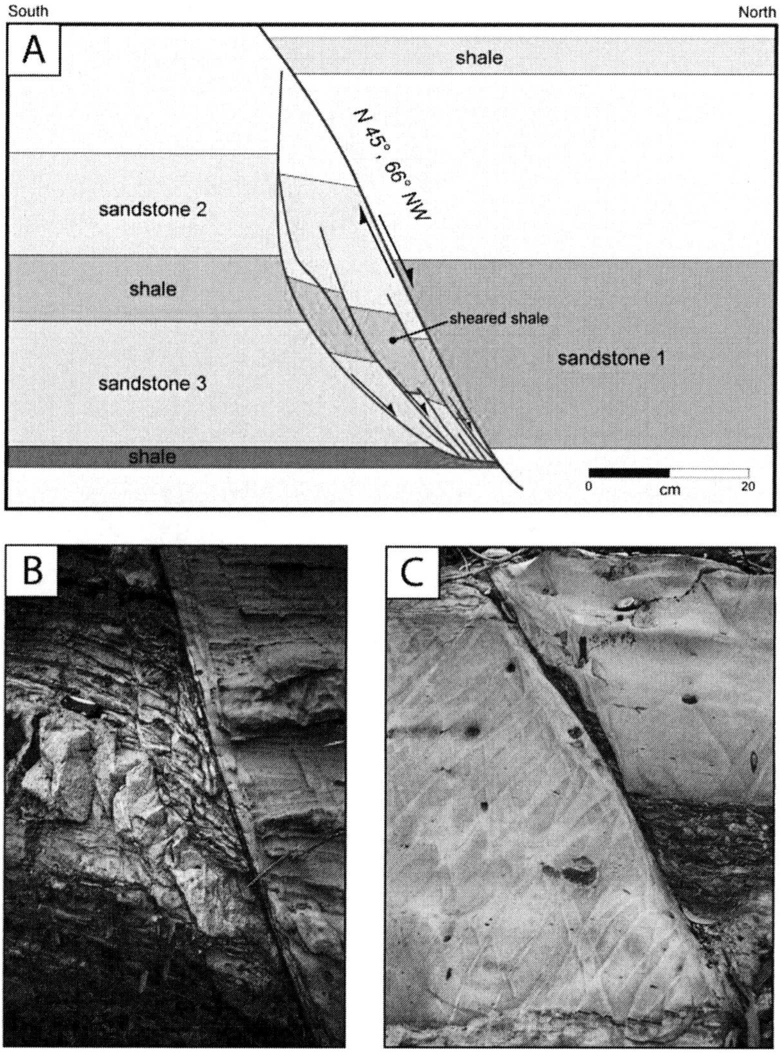

Figure 10. Fault relays developed in Miri outcrops at bed scale illustrate early stages of shale incorporation into fault zone as result of fault relay development and bed telescoping. (A) Field drawing after Burhannudinnur and Morley (1997) showing progressive rotation and shearing of shale bed approaching main fault surface, possibly characterizing a fault outcrop (B) described by van der Zee et al. (2003). (C) Characteristic shale wedges described by van der Zee and Urai (2005) as result of fault relay evolution; see van der Zee and Urai (2005) for reconstruction of fault structure. Fig. 10c reprinted from van der Zee and Urai (2005) with permission from Elsevier.

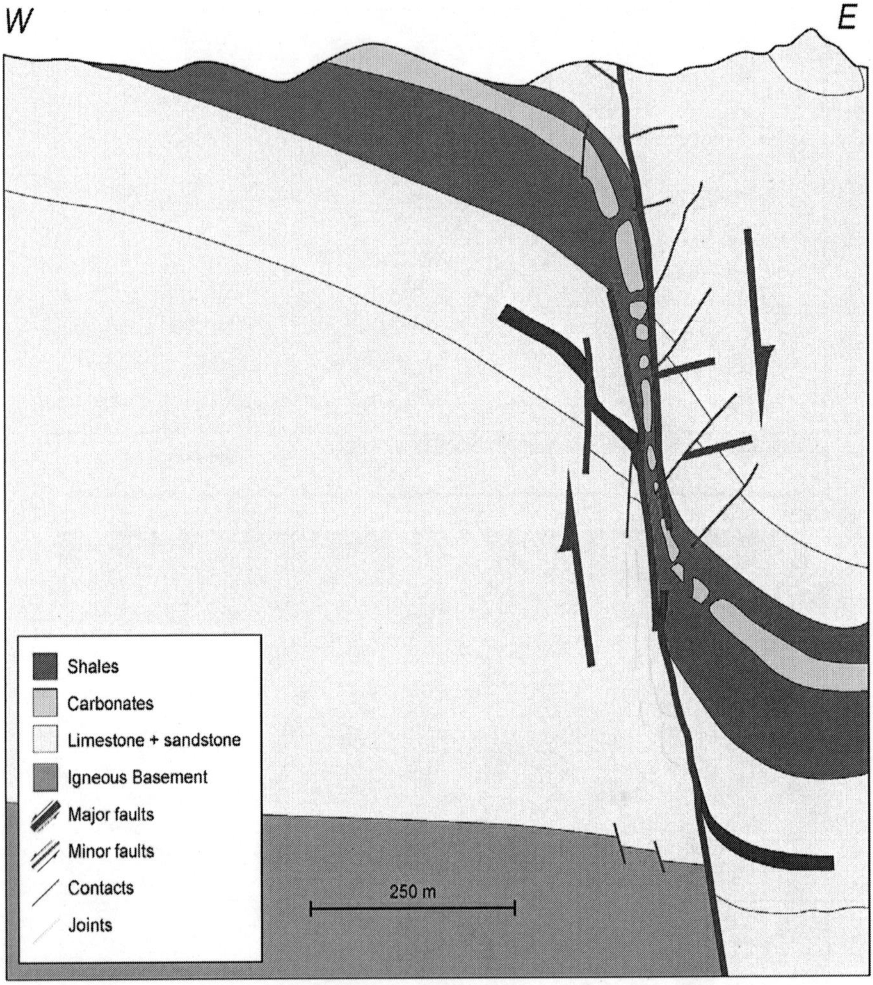

Figure 11. Summary of fault zone structures documented by Aydin and Eyal (2002). Important fault structures defined at this scale include stratigraphically layered fault zone, with bed thickness in fault zone roughly proportional to original source bed thickness; minor faulting of footwall and hanging wall beds immediately adjacent fault; boudinage of carbonate-dominated interval contained within shale interval; and secondary (Riedel?) faults developed mostly in hanging wall. Note proximity of a rigid basement in footwall, making this a good comparison for sandbox experiments introduced in later Experimental Studies section (4.2.1).

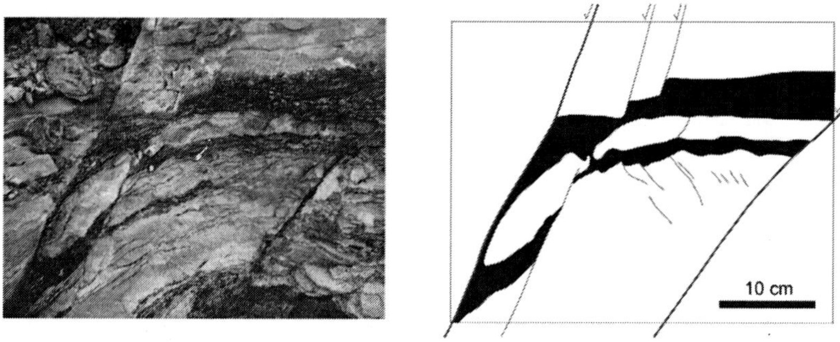

Figure 12. Example of how folding and secondary fault development (synthetic to main fault surface) interact to both entrain shale into fault zone and begin to dismember and boudinage intervening sandstone layers. Photo of outcrop from Miri presented in van der Zee et al. (2003).

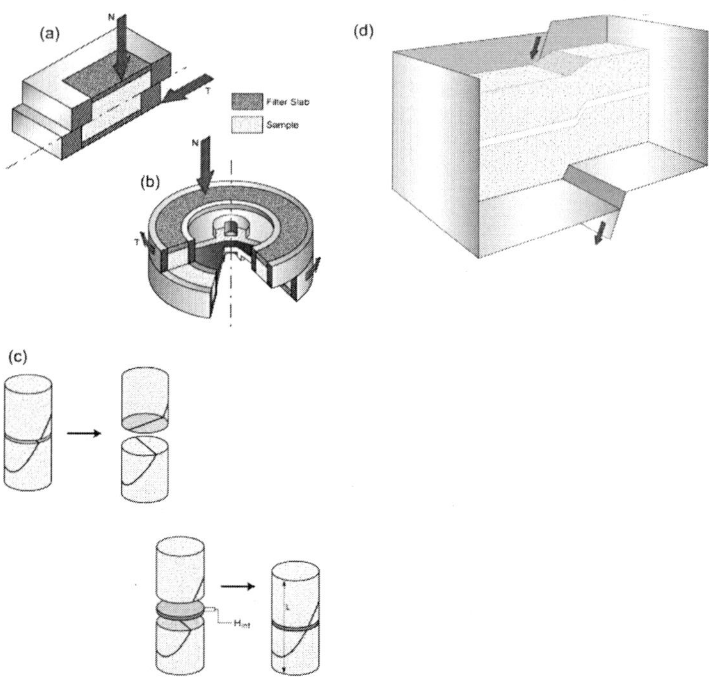

Figure 13. Schematic illustration of different experimental configurations to model claysmear. (A) Direct shear (red arrows indicate stresses); (B) Ringshear (red arrows are applied stresses); (C) Triaxial sample, illustrating how a pre-cut fault cylinder is separated into two, a sandwich of siltstone inserted to simulate fine-grained smear material, and sample reassembly; (D) Sandbox deformed over rigid basement fault with red arrows indicating displacement imposed by apparatus.

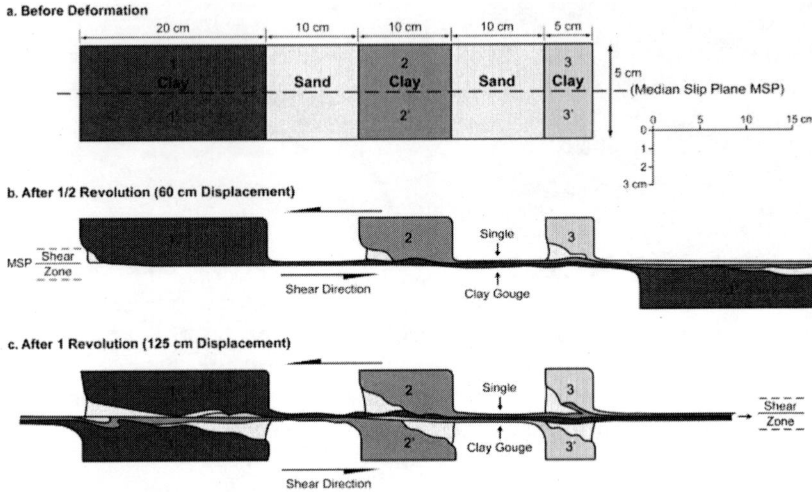

Figure 14. Development of multilayered composite clay gouge in sand-clay sample deformed in ring-shear apparatus (redrawn after Weber et al., 1978). Note how layers in clay gouge are correlated with their source layer, and sand wedges (squeezing blocks) which move into clay layer and are associated with transport of clay into shear zone. Copyright 1978, Offshore Technology Conference. Reproduced with permission of OTC. Further reproduction prohibited without permission.

Figure 15. Top view of a clay smear formed in ringshear experiment (Sperrevik et al., 2000) after brushing away top sand layer. Dark block at bottom of photo (inner ring) is source clay layer.

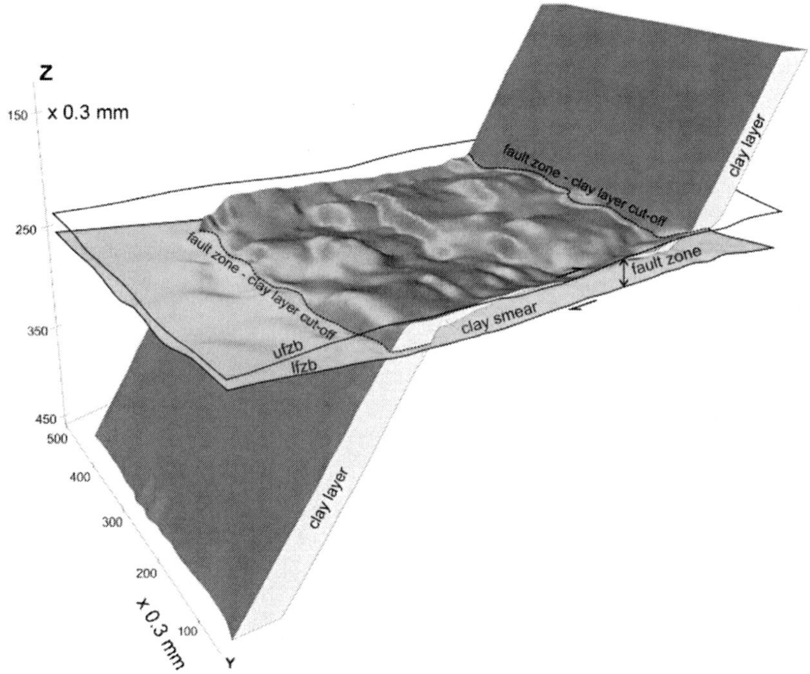

Figure 16. Three dimensional structure of clay smear formed in a large modified direct shear experiment (Çiftçi et al., 2013), showing sheared clay layer and shear zone boundaries. Note that although simulating normal fault processes, direct shear system creates horizontal displacement. ufzb and lfzb are upper and lower fault zone boundaries, respectively.

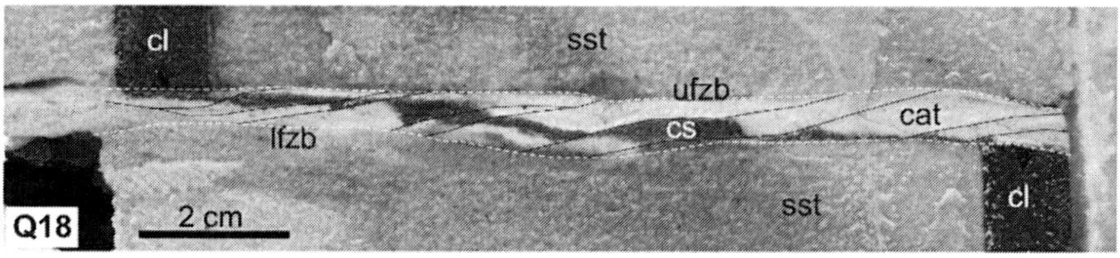

Figure 17. Photograph of sawcut of deformed sandstone–clay sample (Çiftçi et al., 2013). Clay layer is 12 mm thick. Note secondary shear zones dissecting sheared clay in wider shear zone. Sheared clay layer has highly variable thickness. sst = sandstone; cl = clay; ufzb and lfzb = upper and lower fault zone boundary, respectively; cat = cataclasite; cs = clay smear.

Figure 18. Lateral clay injection in sandbox experiment with very soft clay analog under relatively high overburden stress. Note almost complete injection of clay layer under squeezing block into dilatant fault. Experiment by P.D. Richard and J.L. Urai (2001).

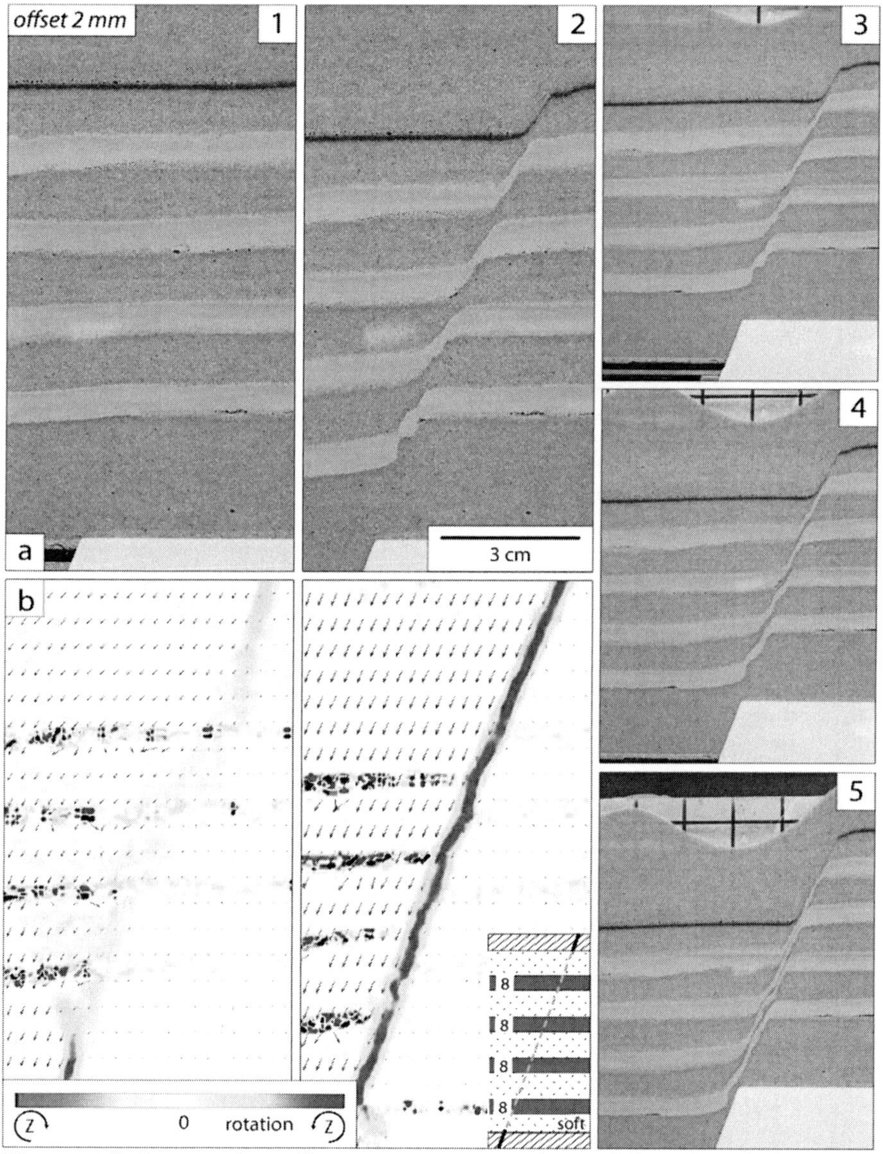

Figure 19. Stages of sandbox experiment with sand (gray) and clay (tan) deforming in shear and having similar strengths. First two stages are from experiment discussed in Fig. 16 of Schmatz et al. (2010b), showing PIV derived strain rate contours to illustrate lateral migration of zone of shearing. Last three stages show how progressive deformation of initially layered sand-clay shear zone results in one coherent sand-clay mixture band. Figure reprinted from Schmatz et al. (2010b) with permission from Elsevier.

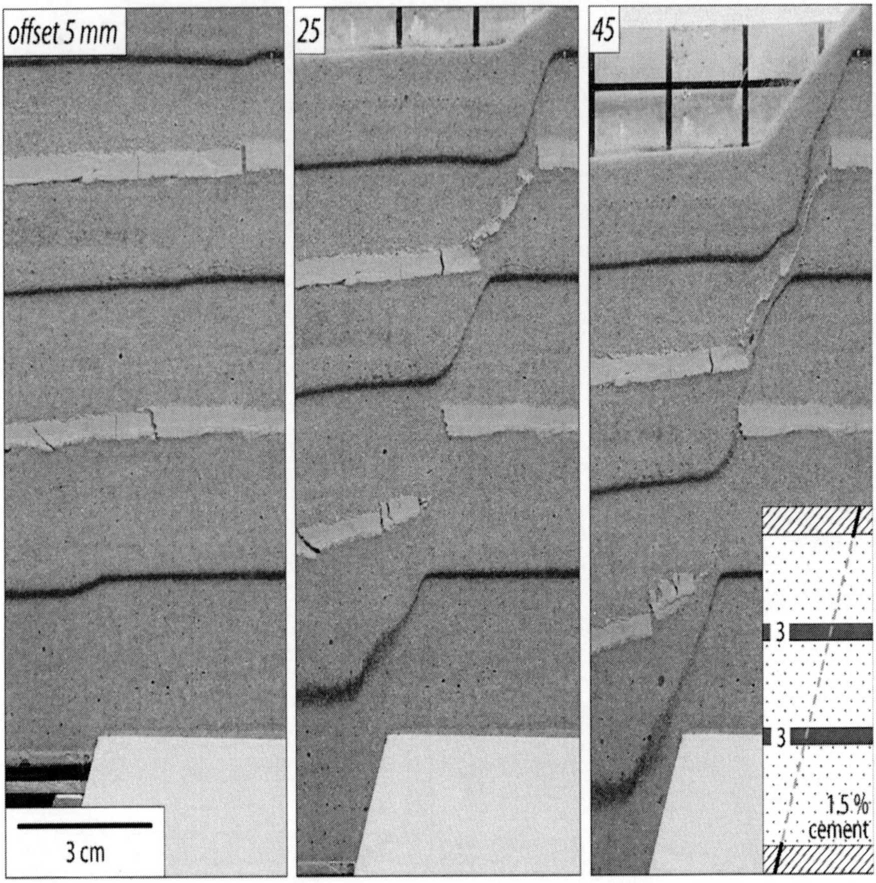

Figure 20. Image sequence showing experiment with two cemented clay layers and top plates Schmatz et al. (2010b). Basement fault offset is 5, 25, and 45 mm. Rigid footwall block (off-white) is stationary, and rigid hanging wall top-plate (silver with black grid) falls into view with displacement. Note initial fragmentation of upper clay layer followed by subsequent shear and abrasion, developing towards more continuous clay gouge. Figure reprinted from Schmatz et al. (2010b) with permission from Elsevier.

Figure 21. Excavated surface of sheared clay layer showing different degrees of mixing of clay with sand from both sides of the clay (light blue areas of fault) and holes in clay gouge (center of dark blue areas). Technique described by Noorsalehi-Garakani et al. (2013). Note sheared lenses in the master shear zone (thicker, pure white sections on fault surface).

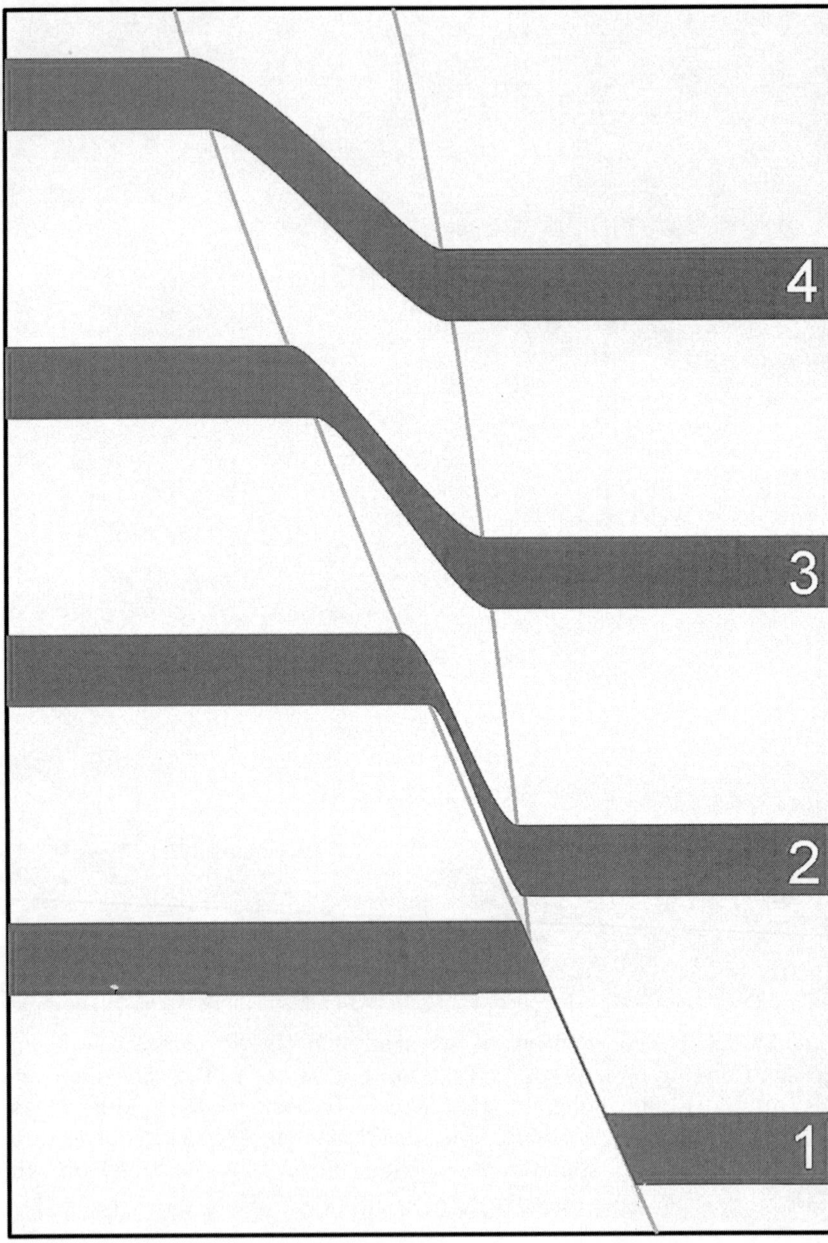

Figure 22. Schematic drawing to illustrate effect of fault zone thickness on clay smear. In fault zone that progressively widens and subjected to homogeneous simple shear for each of four clay beds, clay smear thickness increases with shear zone thickness.

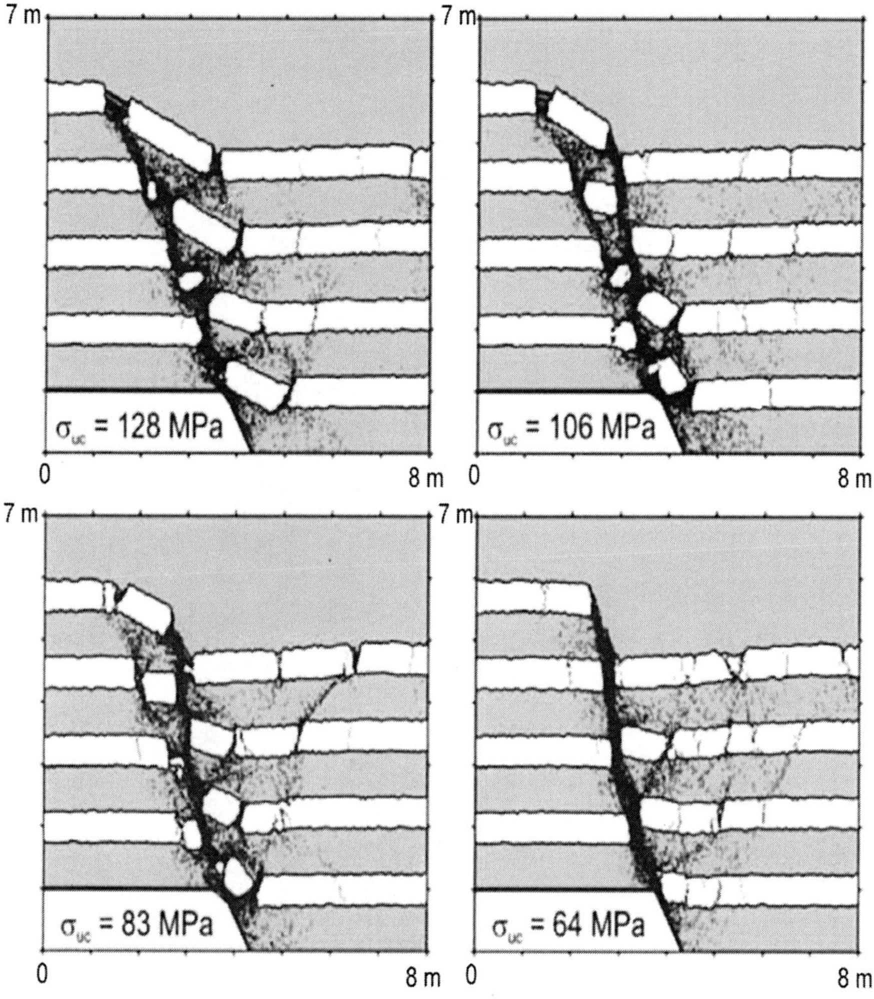

Figure 23. Discrete element models of multilayer faulting above basal fault, with different strength contrasts between layers, showing how complexity and width of fault zone increases with increasing strength contrast, especially when strong layer fails in extension. From Schöpfer et al. (2007b). σ_{uc} is unconfined compressive strength of stronger (white beds) lithology; example for same fault throw and same confining pressure.

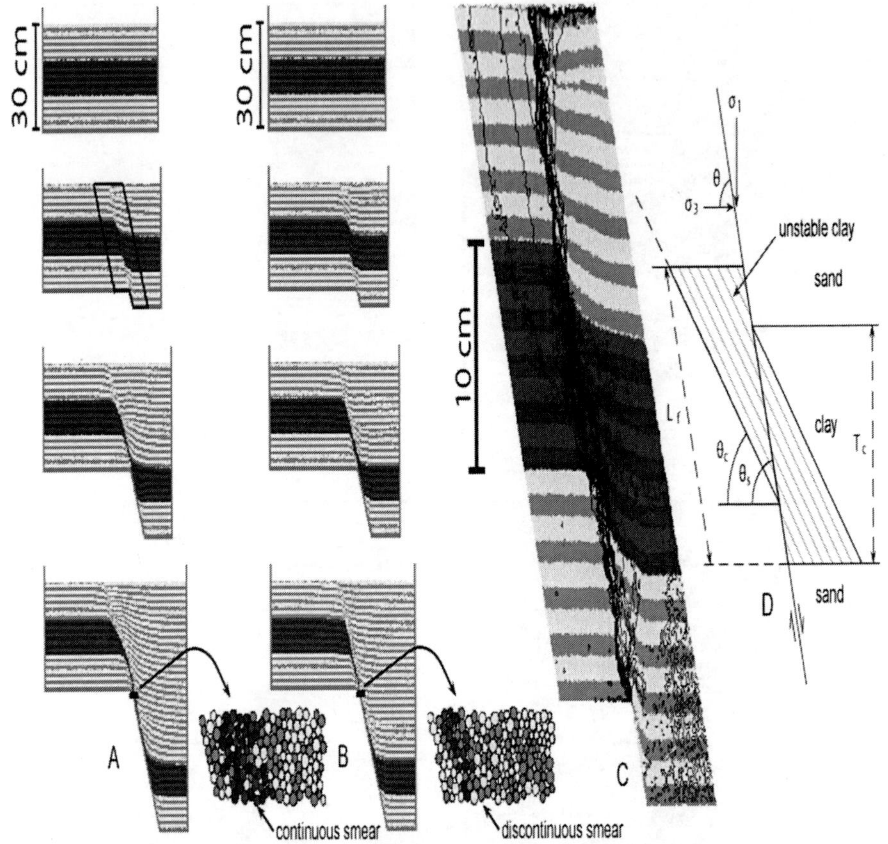

Figure 24. Discrete element models of normal faulting above basement fault (Egholm et al., 2008). B series is model where dark brown layer has same properties as surrounding sand, while in A-series dark brown layer is much weaker (lower friction angle), leading to change in direction of shear zone in soft layer and transport of eroded material into shear zone. Note that in A-series during intermediate displacements, an antithetic fault develops in hanging wall that creates squeezing block geometry.

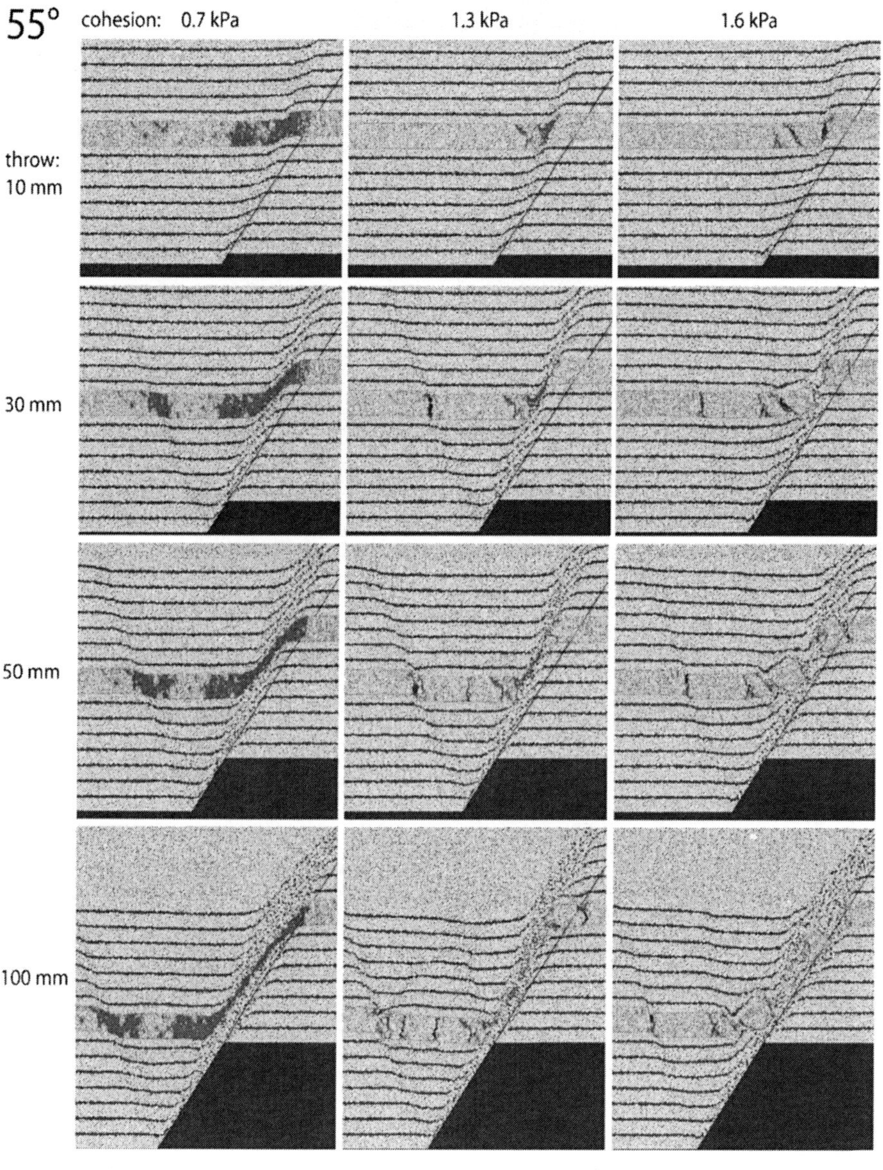

Figure 25. DEM models of normal faulting above basement fault (graben domain) with contrasting strength between clay and sand layer and with different failure modes in clay layer (varying cohesion). Note strong changes in morphology in clay layer for same throw values. Yellow elements are intact; orange elements have failed bonds. From Raith (2012).

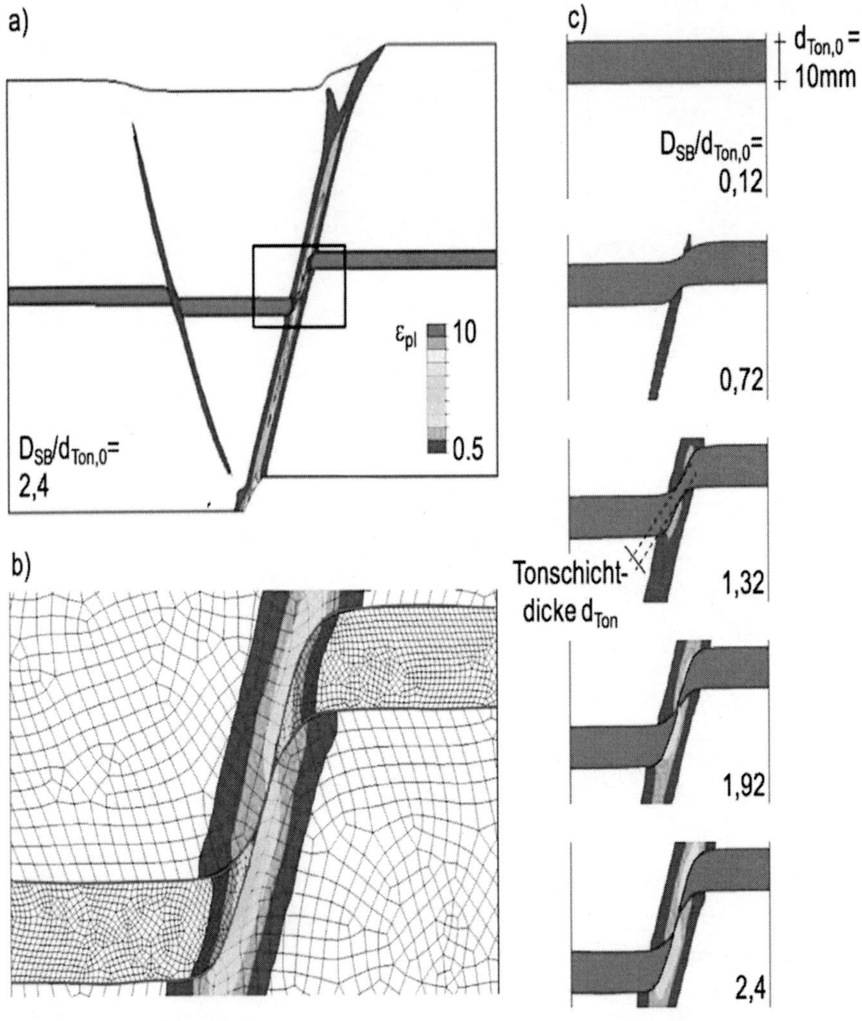

Figure 26. ELFEN model of strong clay layer embedded in sand, sheared in numerical sandbox model above basement fault. Note strong boudinage of clay (Ton) layer. From Kleine-Vennekate (2013).

		Clay				
		<<= Failure in Shear Ductile*		Failure in Extension =>> Brittle*		
Sand	<== Failure in Shear ==> Ductile*	Much softer	Comparable	Stronger	Much stronger	Relative Strength (clay/sand)
		Clay injection	Shear bands coalescence; fault gouge shearing	Fragmentation; abrasion	Fragmentation of clay in sand	Distinguishing Process(es)
		Clay in fault exceeds clay in section	Layered fault gouge with secondary shear bands	Abraded clay fragments in fault gouge	Clay fragments in sand gouge	Diagnostic Fault Structure(s)
	<== Failure in Extension ==> Brittle*	Very much softer	Softer	Comparable	Stronger	Relative Strength (clay/sand)
		Clay injection; clay flows around sand fragments	Shearing of clay; sand fragmentation; sand abrasion	Abrasion/ fragmentation of both sand & clay	Fragmentation	Distinguishing Process(es)
		Sand fragments surrounded by clay	Clay gouge with sand fragments	Sand-clay breccia with fault gouge	Sand-clay breccia; no clay gouge	Diagnostic Fault Structure(s)

NB: Clay in each column has the same absolute strength, and sand in each row has the same absolute strength.
Both classes of sand behavior encompass a range of subclasses not distinguished here.
Fault structure also changes with displacement.

* Ingram and Urai (1999)

Figure 27. Matrix of deformation domains for layered sandstone and shale sequences. Matrix defined in terms of relative strength of clay and sand components and corresponding shear or extension failure modes. Table lists distinguishing processes that occur under different conditions and resulting diagnostic fault structures.

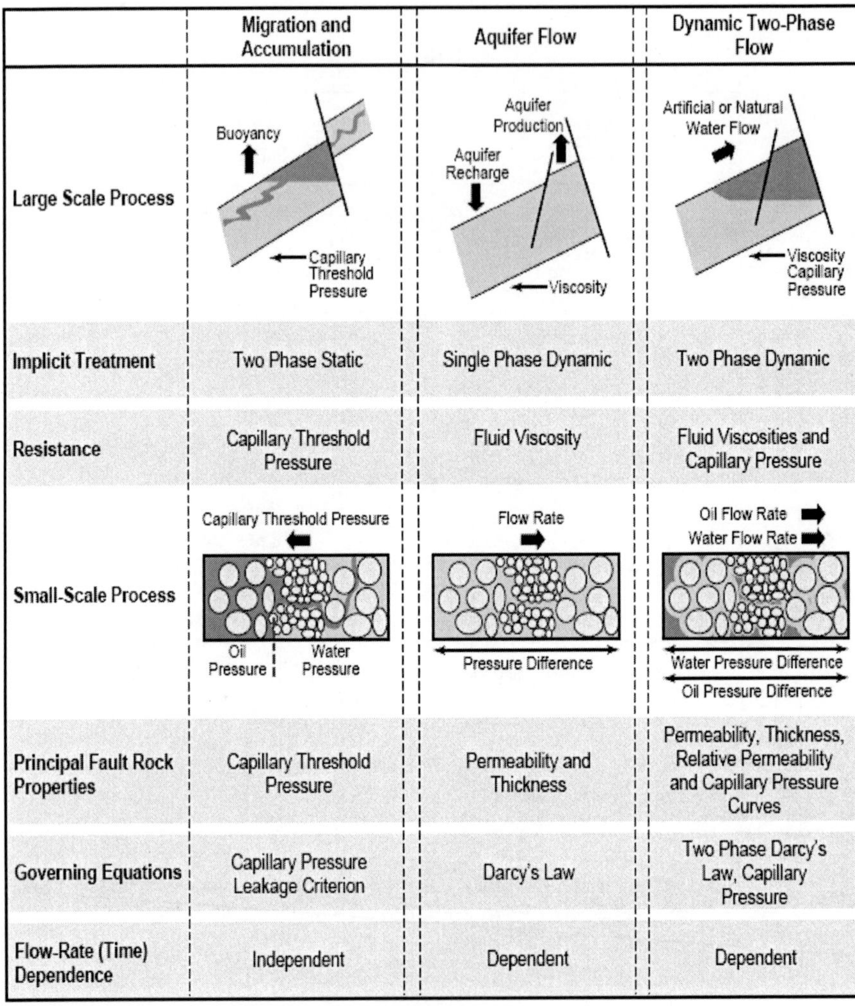

Figure 28. Illustration of various cross-fault flow problems occurring in sub-surface describing physical phenomena that contribute to flow problem and manner in which these phenomena are evaluated. Modified fromManzocchi et al. (2002). Copyright 2002, Geofluids. Reproduced with permission of John Wiley and Sons.

a

b

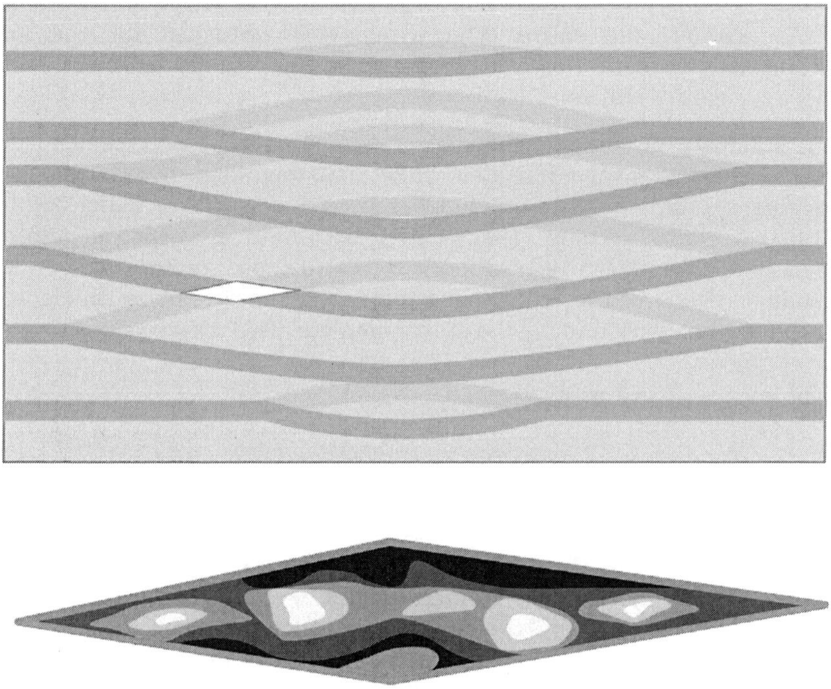

Figure 29. (a) Schematic fault-plane profile or Allan diagram with hanging wall sandstone beds in brown and footwall beds in orange; intervening gray area is shale. Note that only limited portion of fault zone is sandstone–sandstone juxtaposition connections, like beige highlighted one on lower left of profile. (b) Concept of clay smear distribution in juxtaposition window and impact on cross-fault flow. Darkest fringing areas are continuous clay with potential to form capillary seal (no-flow), and lightest area (hole) offers the least resistance to flow (e.g., flow rates = 0.1 rate for sandstone–sandstone connection). As colors become darker, flow rates decrease in order of magnitude steps. Note that in this depiction, although clay smear is present everywhere its contribution to a capillary seal is minimal, and cross-fault flow is restricted to a small area of fault with highest cross-fault fluxes.

Involved reader: those readers interested in learning something about the discipline components reviewed (e.g., experimentalists seeking to find a summary of laboratory experiments in this topic area) are directed to the following additional sections that summarize each component section (Sections 3.1.4, 3.2.4 and 3.3.8).

Expert reader: the entire critical review is intended for the most enthusiastic readers. However, there is the opportunity to select one or more topical area for consideration.

The summary discussion sections in this paper that are recommended to the casual and involved readers reflect the opinions of the authors to express a comprehensive opinion based on the works of many authors. Although these sections are written with few citations to improve readability, we give proper citation credit to those authors in the more detailed review of the individual papers, and those readers should seek proper author attribution in the more complete discussion sections.

1.2. Clay smear definition

Clay smear is a term that was never precisely defined and so has meant different things to different people over time. We struggled with existing definitions to account for structurally coherent but folded lenses of shale like those described by Berg and Avery (1995) in light of the general perception of clay smear as a mono-lithologic, attenuated interval of high shear strain without any contribution of lithologic mixing. Focusing on process models of clay smear development, we found the most general and inclusive definition to be most useful (Table 1). Some will favor a more restricted definition, but we think that perspective hinders an understanding of process. The basis for our definition should be apparent in this review, and we return to this definition in the Discussion section.

Table 1. Table of nomenclature.

Term	Definition	Reference
Abrasion	mechanical wearing, grinding, scraping, or rubbing away (or down) of rock surfaces by friction and; expansion of geomorphic definition to fault processes includes continuous migration of fault zone boundaries into footwall and/or hanging wall with increasing fault offset.	Mod 1
Apparent cohesion	resistance of particles to being pulled apart, due to the surface tension of the moisture film surrounding each particle	2
Boudinage	structure common in strongly deformed sedimentary and metamorphic rocks, in which an original continuous competent layer or bed between two less competent layers has been stretched, thinned and broken at regular intervals into bodies resembling boudins or sausages, elongated parallel to the fold axes.	1
Capillary leak	Condition for which the buoyant pressure at the top of an oil or gas column overcomes the capillary pressure, leading to a condition of oil and gas migration (Ant: Capillary seal; Cf: displacement pressure in AGI Glossary).	Mod 1
Capillary seal	A seal for which the impediment to oil and gas flow is created by capillary forces (Cf: seal).	Mod 1
Clay gouge	A clayey deposit in a fault zone; fault gouge	1
Clay injection	The forcing, under abnormal pressure, of sedimentary material (downward, upward, or laterally) into a pre-existing deposit or rock, either along some plane of weakness or into a crack or fissure (Cf. Sand injection; AGI Glossary).	1
Clay smear	Clay smear forms in normal faults that deform layered sedimentary sequences, typically clastic sequences, and is a type of clay gouge that develops by mechanical processes alone. Clay smear includes the entire fault zone from the most highly deformed clay to stratigraphically coherent fault lenses of shale, thus resulting in a lithologically and structurally heterogeneous fault zone, and typically becomes important once beds are completely offset from themselves. Although clay smear encompasses many fault processes and structures, its main utility is in applications to cross-fault flow problems.	New
Clay smear potential (CSP)	One of a series of definitions of the amount of clay smear in a fault zone derived from the normal offset of a sandstone/shale sequence. CSP is the ratio of the square of a shale bed thickness divided by the fault throw for that bed.	New
Clay smear termination	(1) Umbrella term for processes that cause clay smear to become discontinuous. (2) The discontinuity of clay smear; hole in clay smear.	New
Contractional relay	Synonymous for a contractional jog or overstep that connects two sub-parallel but non-collinear portions of a fault zone and causes local contraction in the wall-rocks as they are displaced	Mod 3

Dilatancy	Increase in bulk volume during deformation, caused by a change from close-packed structure to open-packed structure, accompanied by an increase in pore volume. The latter is accompanied by rotation of grains, microfracturing, and grain boundary slippage; an increase (positive dilation) or a decrease (negative dilation) in volume	1, 2
Drained deformation conditions	A deformation slow enough to allow fluid pressures in the deformed rock or soil body to remain hydrostatic.	New
Fault juxtaposition diagram	A diagram used to predict the juxtapositions of hanging wall and footwall lithologies for a given fault geometry and displacement (Knipe, 1997). It can be used to estimate the sealing potential of a fault (Peacock et al., 2000; Cf: Stratigraphic separation diagram; AGI Glossary).	1, 2
Fault lens	A rock body, deformed or undeformed, that is bordered by two fault strands that link above and below the rock body.	New
Fault-bounded trap	Trap for oil or gas formed by fault juxtaposition of an impermeable bed (e.g., shale) against the reservoir body, or a reservoir that is bounded by a sealing fault.	New
Geocellular reservoir simulation model	A numerical model, often a finite difference model, composed of many grid nodes that contain volumetric information (e.g., porosity and fluid saturations) and connections between neighboring grid nodes that represent flow properties (e.g., permeability or transmissivity). Model is used to calculate flow behavior for defined fluid distribution boundary conditions (e.g., pressure and saturation in a multi-phase model based on an *a priori* definition of flow properties derived from a 3D geologic description of lithotypes and their associated properties.	New
Kinematically preferred	Movements that result in *Kinematic coherence* (defined as the existence of synchronous slip rates and slip distributions that are arranged such that geometric coherence is maintained; AGI Glossary).	Mod 1
Mechanical clay injection potential (MCIP):	This criterion predicts the tendency of lateral clay injection into a pull-apart structure in the fault. $MCIP = \frac{\sigma_1'(1-\sin\phi)}{(1-\cos\phi C)}$ σ_1': maximum principle effective stress, φ: internal friction angle (degrees), C: cohesion.	4
Normally consolidated	Consolidation of sedimentary material in equilibrium with overburden pressure	1
Phyllosilicate framework fault rock	Fault rock developed from the deformation of impure sandstones with phyllosilicate concentrations. The mixture of phyllosilicates and framework silicates generates fault rocks where the porosity and permeability are controlled by the creation of anastomosing networks of microsmears around framework fragments or clasts. The seals arise from the deformation of detrital diagenetic phyllosilicates located between detrital framework grains, or from the deformation of phyllosilicate laminations	5
Polar continua	a type of continuum mechanics (alternately referred to as micropolar continua or Cosserat continua) in which each material point is assigned a microstructure that is equivalent to a rigidly rotating microbody, yielding six degrees of freedom that are divided between translation and microrotation (Cf: textbooks by Gerard A. Maugin).	New
Probabilistic shale smear factor (PSSF):	One of a series of definitions of the amount of clay smear in a fault zone derived from the normal offset of a sandstone/shale sequence. SFF is the ratio of fault throw for an individual shale bed to its thickness.	New
Relative permeability	In multiphase flow in porous media, the relative permeability of a phase is a dimensionless measure of the effective permeability of that phase. It is the ratio of the effective permeability of that phase to the absolute permeability. It can be viewed as an adaptation of Darcy's law to multiphase flow	6
Releasing bend/releasing step	A spatial variation in the orientation of a fault-plane that causes local extension in the wall-rocks as they are displaced around the bend	3
Sand injection	(a) The forcing, under abnormal pressure, of sedimentary material (downward, upward, or laterally) into a pre-existing deposit or rock, either along some plane of weakness or into a crack or fissure; e.g. the transformation of wet sands and silts to a fluid state and their emplacement in adjacent sediments, producing structures such as sandstone dikes or sand volcanoes. See also: intrusion [sed]. (b) A sedimentary structure or rock formed by injection.	1
Seal capacity	the amount of oil or gas that collects when its upward migration is impeded, often expressed as a buoyant pressure derived from the height of oil or gas collected and its corresponding density.	New
Shale gouge ratio (SGR)	One of a series of definitions of the amount of clay smear in a fault zone derived from the normal offset of a sandstone/shale sequence. SGR is the ratio of the sum of all shale intervals times their thickness divided by the fault throw.	New
Shale Smear Factor (SFF)	One of a series of definitions of the amount of clay smear in a fault zone derived from the normal offset of a sandstone/shale sequence. SFF is the ratio of fault throw for an individual shale bed to its thickness.	New
Source bed cut-offs	lines in the hanging wall and footwall marking the boundary between a fault surface and a planar marker (bed, dyke, etc.; Peacock et al., 2000); the line defined at a source bed that contributes clay into the fault zone	3
Squeezing block	A rock body above a layer of ductile clay in the footwall of a developing fault that is bound by two synthetic faults. The specific fault pattern allows a downward movement of the squeezing block which causes a thinning of the ductile clay and results in an injection of the clay into the fault	4
Transmissibility multipliers	Term used in finite difference calculations of subsurface oil and gas flow to reflect various degrees of flow impedance across faults (i.e. factor applied to transmissibility defined between nodes on either side of a fault).	Mod 7

References: New e definition created by authors; Mod # e definition modified from reference cited (below); # e definition used from reference cited.

1. AGI Glossary of Geology.

2. http://www.mindat.org/glossary/apparent_cohesion.

3. Peacock et al. (2000).

4. van der Zee et al. (2003).

5. Knipe (1997).

6. https://en.wikipedia.org/wiki/Relative_permeability.

7. Manzocchi et al. (1999).

A Table of Nomenclature (Table 1) is included because this review touches on many disparate disciplines, some of which may be unfamiliar to some readers. A more complete Nomenclature summary is compiled in Supplementary materials.

2. CLAY SMEAR – A PROCESS TO ADDRESS A PROBLEM

Clay smear emerged as a geologic concept within Shell in the 1970's (F. Lehner, pers. comm. to J.L. Urai, 1997). The concept originally arose from field observations combined with laboratory experiments and a supporting analytical model (Lehner and Pilaar, 1997) with the intent of accounting for trapped gas and oil columns in faulted anticlines. Clay smear processes helped fill in the enigmatic fault zones that Smith (1966) identified as potential seal elements with Weber et al. (1978) the first published reference to this work.

Although Smith (1966) recognized the importance of defining reservoir juxtaposition geometries, it is important to reflect that the subsurface data available for this analysis was restricted to 2D seismic data with limited geophone offsets (with respect to modern data), primitive seismic processing capabilities, but continuous well log information. Because well data were some of the most definitive subsurface data, it is natural to develop an analytical method around them (i.e. fault offset of a stratigraphic section defined by a well). We think that this early decision to apply clay smear to gas and oil accumulations in faulted traps was an essential reason for the development of this area of science in the subsequent decades.

2.1. Clay smear as a capillary seal

Smith (1966) recognized that the properties of a fault zone had the potential to affect oil and gas trapping and subsequent oil and gas production across faults. Weber et al. (1978), however, limited their discussion only to the capillary seal problem. Smith (1980)described how a fault gouge acting as a capillary seal was one of four scenarios encountered in the subsurface that either juxtapose sandstones of the same or different ages and result in a capillary seal or leak (Smith, 1980). The identification of these four potential conditions has remained unchanged since.

Speksnijder (1987) referred to the application of a newly developed computer program for the calculation of clay smear effects in a producing field study, and Bouvier et al. (1989) for the first time used the term Clay Smear Potential (CSP) in a publication. Both of these publications refer to a personal communication from Lehner and Pilaar (1974), thus we attribute the genesis of this idea to Lehner and Pilaar (1974). Fulljames et al. (1997) published the algebraic form of the CSP expression for the first time and described how the observations, analysis, and inferences of Lehner and Pilaar (1997)could be applied to fault trap evaluations as well as cross-fault fault flow properties (including clay smear) for oil and gas production. This limited but nevertheless influential body of literature then formed the basis for all subsequent work on clay smear.

2.2. Alternative clay smear definitions

While the algorithmic description of the concepts described initially by Weber et al. (1978)waited 20 years for the publication of Fulljames et al. (1997), others developed alternative approaches for describing the clay smear process. Based on outcrop observations in Lancashire quarries, Lindsay et al. (1993) derived a dimensionless length parameter, the Shale Smear Factor (SSF). This parameter produces a value that like CSP is larger for greater offset from the source bed (Fig. 3). Perhaps an important realization is that both of these original clay smear definitions, CSP and SSF, were derived based on outcrop observations in fluvial-deltaic sequences with coal interbeds. One may ask: are the fault processes developed in these rock types the same for deformation of sandstones and shales deposited in other environments?

Yielding et al. (1997) reviewed CSP and SSF parameters and introduced two additional empirical equations for describing clay smear in faults (Fig. 3): the Smear Factor and the Shale Gouge Ratio (SGR). The smear factor was defined in such a way as to provide a continuum between CSP and SSF by introducing two exponents into the expression (*m*and *n*). The exponent *m* operates on the shale bed thickness in the numerator of the expression, and the *n*-exponent on the distance (throw) value in the denominator. In principle these exponents could be used to lengthen or shorten the effectiveness of a clay smear from any particular shale source bed. In contrast, the Shale Gouge Ratio (SGR) allows consideration of clay that may exist and be introduced into the fault from shale stringers and thin beds, intervals which might only generate an intermediate V_{shale}(shale volume) value from a gamma-ray well-log that interrogates a volume of rock, allowing a more nuanced view of the stratigraphic section.

In the last decade, however, some publications have appeared with refined algorithms that further attempt to define parameters in the context of the cross-fault flow problem, by incorporating a probabilistic element to show where clay smears may be breached.Childs et al. ('2007) defined a Simple Shear Zone (SZ) method that allowed them to calculate a thickness-weighted harmonic average of the faulted and offset lithologies (akin to a vertical permeability through a stratigraphic section) and a Probabilistic Shale Smear Factor (PSSF) in which holes in the clay smear are considered to exist anywhere along the fault surface (a more complete discussion of this approach comes later).Yielding (2012) further developed this theme by describing a work-process implementation of the PSSF method in the context of SGR modeling.

We think that the current variety of clay smear algorithms is significant in that they implicitly acknowledge a limitation in the existing clay smear model approaches. All models are variations on two geologic parameters: shale bed thickness and normal fault throw. None have any basis in the mechanics of faulting except for CSP, for which the ratio is associated with a general process model and includes the viscosity of the clay (Weber et al., 1978). All models simply state that the more shale in the faulted section, the more clay will be found in a fault zone and have no capacity to account for the rich variety of structures observed in natural faults.

We review the outcrop, laboratory and numerical, and subsurface field studies in this context (Fig. 2), and then combine these evaluations into a summary of the state of knowledge of each of these steps in the clay smear process. We then critically examine the application of this knowledge to problems of sub-surface fluid flow to evaluate its adequacy. Finally, we propose a framework for incorporating a mechanics-based fault process model into the existing geometric models of clay smear development, evolution, and continuity.

3. THREE DECADES OF RESEARCH – EVOLUTION OF THE CLAY SMEAR CONCEPT

3.1. Outcrop studies

Investigations of outcrop examples of normal faults with clay smear are an essential component of any attempt to uncover the processes that create and modify clay smear.Weber et al. (1978) and Lehner and Pilaar (1997) made analyses of normal faults in the Frechen lignite mines an integral part of their early understanding. Since then other authors have approached this problem in different geologic settings, with different observational frameworks, and with different objectives in mind. We think these differences are significant. To consolidate insights across these differences, we constructed an interpretation framework that places observations into three stages of the fault problem that we think are essential for the development and preservation of clay smear, distinguishing processes that:

1. Incorporate shale and mudstone into a fault zone
2. Deform and modify those clay-rich materials in a fault zone
3. Disrupt and terminate a clay smear, ultimately resulting in holes in the clay smear

Although we seek to gain insight about fault process from outcrop studies, that insight comes as an interpretation of the fault products that are observed. In this section we focus on synthesizing the spectrum of fault structures observed in nature and use the final discussion section to relate the fault processes that are explored in laboratory and numerical experiments to the fault products that are observed.

We consider the history of outcrop studies in three stages: (1) an early stage in which the general framework of the problem was outlined and many important issues were identified; (2) an intermediate stage where the problem was considered in different geologic environments; and (3) a modern stage characterized by work that refines and supplements understanding gained previously. We review this literature considering how it has helped advance understanding of the kinematic geometries and evolution of fault zones associated with clay smear, the extent to which general material property deductions influence clay smear, and the resulting relative importance of various fault processes (Fig. 2).

3.1.1. Initial definition of process landscape

Although Weber et al. (1978) referred to the Frechen outcrop observations in their original work, the full results of that work first appeared 20 years later in Lehner and Pilaar (1997). They describe normal faults of the lower Rhine Embayment, deforming Tertiary deltaic deposits in a sandstone-dominated depositional system (3× as much sandstone as shale) and made observations that subsequent authors have returned to while other observations have remained unconsidered.

The key observations made by Lehner and Pilaar (1997) include (Fig. 4):

- A layered fault structure arising from the variable incorporation of offset beds, including the amalgamation of individual clay smears derived from multiple mudstone beds into a composite clay smear (Fig. 5)
- The development of secondary shears that were interpreted as Riedel shears, primarily in the hanging wall and footwall but in some instances defining a strong fabric within the clay smear (Fig. 5)
- The development of folds in the hanging wall and footwall (Fig. 6)
- Structural thinning of mudstone beds in the hanging wall and footwall, sometimes associated with Riedel shears

On the basis of these observations, Lehner and Pilaar (1997) describe a fault zone as a layered shear zone comprising multiple slip surfaces that developed in both space and time. The resulting shear zone varies in thickness depending on the lithologic composition of the faulted interval. Fault segments that overlap often do so in a pull-apart geometry (Fig. 7) creating extensional fault overlaps into which hanging wall and footwall materials may flow (Table 2).

Table 2. Rating summary of outcrop observations: clay smear structures.

Stage	Fault zone structure	Observational quality	Comments
Shale incorporation into fault zone	Structural thinning of HW/FW beds	✪✪✪✪✪	Necessary observation for clay injection process
	Fault lens & relay	✪✪✪✪✪	For fault relays that cause additional shale to become incorporated into fault zone
	HW/FW folds	✪✪✪✪✪	May reflect migration of shear zone boundaries over time
	Fault-propagation fractures & breccias	✪✪✪✪✪	Fractures that develop at a propagating fault tip, perhaps via bending of a stiff bed
	Sand/clay mixing @ shear zone border	✪✪✪✪✪	Grain-scale observations made at clay–sand interfaces
Clay smear evolution in fault zone	Layered fault gouge	✪✪✪✪✪	Preserves offset stratigraphic order, albeit missing some intervals
	Boudins	✪✪✪✪✪	Typically sand lenses encased in clay smear; may be traced back to source sand bed
	Fault lens & relays	✪✪✪✪✪	Fault lenses occur within shear zone where bedding may be undeformed or rotating and shearing into gouge
	Fault segmentation	✪✪✪✪✪	Recognized as multiple slip surfaces developed within the overall shear zone
	Localized vs. distributed shear strain	✪✪✪✪✪	Shear strain concentrated on 1–2 clay smears with less deformed materials between them
Clay smear termination	Riedels and other secondary faults	✪✪✪✪✪	Most effective when developed at high angles to shear zone
	Gouge holes	✪✪✪✪✪	Patch of sand–sand juxtaposition across fault zone
	Gouge thinning	✪✪✪✪✪	Progressive thinning of a clay smear with increasing distance from source bed; taper geometry expressed in CSP definition

Note: Star rating reflects score (red stars) out of maximum potential score (blue stars).

Recognizing the volume problem created when the volume of shale in the fault zone is less than the volume of clay smears that extend over 70 m vertically and 400 m laterally,Lehner and Pilaar (1997) describe a process of clay injection from thinning shale beds adjacent the fault into the pull-apart structures within the fault. They rely upon an interpretation framework of Riedel shears (D-, R-, and R’-shears) that both assist the thinning of shale beds and then modify the clay smear within the fault zones, including the development of a sigmoidal shear pattern observed within the clay smear.

Lehner and Pilaar (1997) observed clay smear developed in every quarry location investigated, leading them to the deduction of clay smear processes that maintain the highest degree of continuity for small to modest offsets from the clay source bed. They offer no observations about how discontinuities in clay smears develop.

Lindsay et al. (1993) offer a complementary viewpoint based on observations from Pennsylvanian fluvial-deltaic sandstones, shales, and coals that are interpreted to have been deformed once all the units were lithified by burial to approximately 2 km. Lindsay et al. (1993) recognize three processes by which clay is incorporated into a fault zone:

1. Abrasion: shale beds faulted past sandstone beds are abraded by the sandstone roughness into a veneer. This appears to occur at scales ranging from the grain-scale to small fractured fragments. The products of the abrasion process include polished and slickensided fault surfaces expressed as veneers in cross-section.
2. Shear: the creation of a simple shear zone includes all of the rocks within the shear zone boundaries; no conditions are defined for the establishment of the shear zone, although drag folding of adjacent beds is described. Shear is interpreted to be responsible for the attenuation of clay smears observed with increasing distance from source beds.
3. Injection: although injection is identified as a process to help create thick clay smears, the corresponding evidence of a thinned hanging wall shale bed is difficult to discern in the field data presented.

In addition to these prominent features, Lindsay et al. (1993) recognized the importance of secondary fault development (i.e. fault relays) for shear zone evolution and understood how this history could affect clay smear processes.

Burhannudinnur and Morley (1997) identified cataclastic sand fragments within a clay smear, which led them to interpret grain-scale mixing of sand and clay. AlthoughBurhannudinnur and Morley (1997) focused more on the geometry of fault networks to assist in seismic interpretation, they nevertheless describe folding of sandstone and shale beds adjacent to fault zones.

Based on these three initial studies, the foundation for clay smear processes was defined for normal faults in sandstone-dominated fluvial-deltaic deposits subjected to varying degrees of lithification.

3.1.2. Broadening clay smear into additional environments

Clay smear developed on the km-scale Moab Fault (Foxford et al., 1998) offers a new scale of observation for mudstones and shales deposited in a continental flood-plain to lacustrine environment. Because fault throws far exceed vertical outcrop dimensions,Foxford et al. (1998) offer no observations about processes that incorporate shales into a fault other than to deduce that the resulting fault zone observations are consistent with tip-line and asperity bifurcation processes occurring during fault propagation.

These same processes are responsible for the creation of fault lenses and an overall layered fault zone structure.

In an important insight, Foxford et al. (1998) state that: 'the number of shaley gouge layers in a fault zone increases with the number of slip zones,' and they describe fault zones that consist of 2–9 slip zones; this is a further elucidation of the layered fault gouge structure. Subsequent work has failed to pursue this insight even though it contradicts the basic premise of all shale gouge calculations, which rely on stratigraphic details of the abundance and thickness of the faulted section while Foxford's statement relates clay smears to the fault zone structure. Moreover, Foxford et al. (1998) are the first to addressYielding et al.'s (1997) claim that SGR is a value testable by field observations. They construct a triangle-diagram with binned values of SGR for various sandstone juxtapositions and compare those values to outcrop observations about the presence or absence of shaley gouge (the more general gouge description is appropriate given the evidence for clay recrystallization processes revealed by fault dating studies by Pevear et al., 1997 and Solum et al., 2005). Foxford et al. (1998) conclude that the 'thickness of the shaley gouge is variable even when juxtaposition and throw are essentially constant,' and no shale gouge is observed when the faulted section contains <20 % shale. They view methods like SGR as a means to account for a multitude of internal slip surfaces that develop in fault zones, creating a: 'layered or anastomosing internal structure' (Foxford et al., 1998).

The first report of clay smear in faults of the Albuquerque Basin (Rio Grande Rift) presented by Heynekamp et al. (1999) describes gravel, sand, and clay incorporated into fault zones from unlithified fluvial system deposits. A motivation for this work was to evaluate fault processes that impact groundwater flow in this area, and thus the fault classification scheme developed by Caine et al. (1996) to consider fluid flow problems was adopted. In this framework, modified to some degree for this unconsolidated sedimentary environment, the fault core is associated with a clay smear and is surrounded by a mixed zone that presumably defines the volume of rock deformed by folding, fault relays, and other secondary faults on either side of the fault core. The damage zone appears to have less hydrologic significance (i.e. improved cross-stratal permeability along the fault dip) in this setting than in the original classification (Caine et al., 1996).

Although the magnitude of fault throws relative to outcrop heights limits observations about how mudstones are incorporated into a fault zone, the observation of paired slip zones bounding the fault core further supports the idea of fault structure as an important element of clay smear processes. More observations for how the fault system evolves include:

- Grain-scale mixing of sand and clay
- Macroscopic mixing of fault lenses in the mixed zone described as 'rootless pods of intact bedding' (Heynekamp et al., 1999)
- Fault thickness, including mixed zone thickness, is influenced by fault geometry
- Greater offset of clay-rich units from source beds than coarser units in the fault zone, resulting in a fault that contains more clay than the adjacent beds

Much of the fault evolution appears to occur in the mixed zone that ranges from undeformed beds where sedimentary structures are preserved to foliated, tectonically mixed materials. This mixing is interpreted to be accomplished by 'attenuation of beds, shear along minor faults and foliation, and mixing from meter-scale blocks to the grain scale' (Heynekamp et al., 1999). Because the fault core/damage zone nomenclature lends itself to describing the hydrologic effect of the final fault geometry, it perhaps limits understanding of how materials evolve as they transition from the protolith to the mixed zone and finally into the fault core. For example, what differences are required for clays to reach the fault core rather than end up as clay veneers that bound lens-shaped slivers of intact coarse-grained sediments in the mixed zone? Heynekamp et al. (1999) observed gaps in the clay veneer (gouge holes) and speculated that these gaps might be transient features.

The work by Heynekamp et al. (1999) is significant in that it is the first example of a multitude of lithologies up to gravel incorporated into a fault zone with the clay component, and it is the first published example to consider clay smear in the context of a spectrum of fault zone elements whose geometry and distribution influence cross-fault flow.

3.1.3. Verification and refinement of fault processes

Subsequent studies include observations that largely fit into the framework defined by the early studies. In some instances the same areas or geologic settings were revisited, which offers the opportunity to independently verify the original observations. In other cases new observations help to further refine our understanding of process, either by bringing a new focus to the same outcrops or by visiting outcrops that preserve processes more completely.

3.1.3.1. Shale incorporation processes

Aydin and Eyal (2002) report detailed field observations around a fault tip and along the dip of a fault to more clearly elucidate the processes that incorporate shales into a fault zone; their results further support the importance of fault segments that overlap in a releasing bend geometry. Doughty (2003) and Faerseth (2006) similarly rely upon overlapping fault segments to entrain shale beds. Eichhubl et al. (2005) offer a more detailed look at fault propagation processes at a sandstone/shale bed interface and interpret incipient distributed shear across a deformation zone in bounding sandstone beds and increasingly localized deformation within the clay smear as a result of granular flow of the clay component. The critical observations in this interpretation include the recognition of sedimentary bedding contacts between sandstone and shale, which indicate incorporation of structurally coherent lenses via a fault relay process. However, it is unclear whether the interpretations of Eichhub! et al. (2005) are limited by assumptions about which fault segments are active in what order and whether multiple slip surfaces are active simultaneously.

Vrolijk et al. (2005b) documented a normal fault system developed over a 2 km strike length cutting stratigraphy that varies considerably over that length. They discovered a fluvial channel stacking pattern that promoted the extensive development of fault relays and lenses that increased the occurrence of folded and tilted coherent shale lenses and a fluvial sandstone-poor interval that results in a simple fault zone with limited fault relay development. Moreover, they documented shale-dominated fault lenses with limited extent along the strike of the normal fault (Fig. 8). Davatzes and Aydin (2005) similarly deduced that the geometry of the fault network was influenced by the geometry of the faulted sedimentary section based on evidence along the Moab Fault, and they concluded that a vertical relay was required to incorporate shales of the Morrison Formation into the fault zone.

One of the single most important insights derived from field studies is that clay smear is associated with a network of fault segments, many of them synthetic to the main fault, that result in the proliferation of fault relays and lenses (Fig. 9). This phenomenon appears at a multitude of scales (e.g., Childs et al., 2009a).

One observation that arises from time to time but escapes significant discussion is the observation that some shale beds are cleanly cut by a fault or contribute far less clay to the fault zone than nearby shale beds (e.g., upper versus lower shale beds of Aydin and Eyal, 2002 and Doughty, 2003; faulted shallow marine deposits of Kristensen et al., 2013). Do shale beds that escape incorporation simply exist in a stratigraphic position for which fault relay formation is suppressed, or do they possess different mechanical properties at the time of faulting (i.e. with respect to other shale beds and surrounding sandstones) that promotes a simple fault zone? It is difficult to answer this question in most outcrop settings, but we note that the incorporation of shale into a fault may depend more on how the fault network develops, compatible with the observations of Foxford et al. (1998), than current clay smear models account for, which treat every shale bed as a potential contributor to the fault zone.

van der Zee et al. (2003) pursued further the idea of clay injection as observed in outcrops in Malaysia, Germany, and Oman to develop kinematic and mechanical constraints for the injection process that they subsequently explored with analytical and numerical models. From these models van der Zee et al. (2003) derived constraints on material properties and stresses necessary for clay injection to occur and constructed a model based on wireline logs to define the potential for which any shale bed in a sequence would inject clay into a fault zone (Mechanical Clay Injection Potential: MCIP).van der Zee and Urai (2005), reporting on the same Miri outcrops, provide a comprehensive description for how shale is incorporated into a fault as a result of an evolving fault relay system (Fig. 10), explore the mechanics of the contribution of deforming fault lenses to clay smear, and established a correlation between SGR and the average values of the highly variable measured clay content of the fault zone.

3.1.3.2. Clay smear evolutionary processes

A nearly universal theme of outcrop studies, either explicitly or implicitly described, is that clay smear exists in a layered fault zone structure (Fig. 5; Lehner and Pilaar, 1997,Foxford et al., 1998 and Heynekamp et al., 1999; Lewis et al., 2002; Aydin and Eyal, 2002, Bense et al., 2003, Doughty, 2003, van der Zee and Urai, 2005, Eichhubl et al., 2005, Vrolijk et al., 2005b, Faerseth, 2006 and Kristensen et al., 2013). Layering in the fault zone preserves the stratigraphic order of the faulted lithologies (Aydin and Eyal, 2002 and Doughty, 2003), even if parts of the stratigraphic section are omitted in some locations (e.g., Davatzes and Aydin, 2005).

Stratigraphic omission appears to be accommodated by boudinage processes (Fig. 11) that may be assisted by small-scale, secondary faults (including Riedel shears). It appears that all lithologies from mudstone to gravels can create smears in a fault zone (e.g., Heynekamp et al., 1999, Bense et al., 2003, Doughty, 2003 and Kristensen et al., 2013), but it is the winnowing of lithologies via a boudinage process with increasing fault throw that allows one lithology to become more abundant in a fault than the other lithologies. It is this variable evolutionary process that caused van der Zee and Urai (2005) to propose a 'preferred smear,' a process that accounts for the attenuation and boudinage of sand layers and thereby allows individual clay smears to amalgamate into a composite smear, enriching the fault zone in clay at the expense of other lithologies.

Numerous authors have described the occurrence of secondary faults adjacent to and within the fault zone, but these observations are made without a kinematic framework that would help compile and evaluate the importance of these small faults in the overall fault structure and in the clay smear evolution, in particular. For example, Aydin and Eyal (2002) conclude that small faults never degrade the integrity of the clay smear in the fault. Although Doughty (2003) interprets small faults in the studied fault zones as Riedel structures, he then neglects to apply the Riedel framework to his fault interpretation in spite of the fact that he attributes thickness changes and clay smear terminations to result from minor normal faults.

Kristensen et al. (2013) investigated the implications of different consolidation states and different stress conditions in a comparative study of three outcrop locations in Denmark. In the shallowest setting where recent (post-glacial) shallow marine deposits are faulted, the resulting fault zone includes smears of sand, silt and clay lithologies. Some sand grains are mixed into the clay smear, and some clay beds are cleanly cut by the fault without appearing to contribute at all to the fault zone. In the deepest Jurassic outcrop the presence of cataclasis is a distinguishing characteristic. Kristensen et al. (2013)document progressive mixing of sand and clay components with increasing fault offset, although thin sandstone layers developed between thicker shale layers are absent in the fault. The intermediate depth example also defines a layered fault zone with little mixing of grains, and both brittle and ductile deformation of shale beds is observed. The implications of this work are that fault processes evolve as a function of the changing consolidation state of the faulted materials and the stresses available at the time of faulting.

3.1.3.3. Clay smear in over-consolidated rocks

Although clay smear is usually associated with soft, ductile shales, there is evidence that clay smear-like structures develop when fragments of stronger, overconsolidated shale are incorporated into fault zones. Holland et al. (2006) describe the evolution of highly over-consolidated shales into clay smear via fracturing, faulting, and abrasion processes. They document the creation of a high porosity, low permeability clay smear from an initially low porosity, low permeability shale, and because that transformation takes place without any significant chemical or mineralogical changes, the resulting clay gouge may look indistinguishable from a clay smear (e.g., sampled in a core); a reworked shale is a soft clay, while a reworked sandstone is a cataclasite although both are produced by purely mechanical processes. Although the examples shown by Holland et al. (2006) represent an end-member behavior, they offer a good example of how clay smear may arise in a fault zone from brittle processes.

3.1.3.4. Clay smear applied to hydrologic problems

The work by Heynekamp et al. (1999) illustrates how fault zone structures are defined for the purpose of investigating the effect of clay smear on cross-fault flow, but the focus on defining the final geometry and flow properties of various fault elements sometimes limits the insights for how that final geometry was achieved. Notwithstanding this limitation, these studies provide some of the most compelling observations for terminating clay smears.

For example, Bense et al. (2003) describe how: 'gravel pebbles trapped in the fault plane cause discontinuities in the clay smear.' The pebbles create asperities that can disrupt clay smear continuity when the pebble size approaches the thickness of the clay smear.Caine and Minor (2009) describe laterally persistent clay-rich fault cores (clay smears) in a sandstone-dominated setting but identify a few rare exposures where a clay-rich core is absent and hanging wall and footwall sandstones are juxtaposed. They recognize that a key factor is the: 'identification of the size and location of relatively high-permeability "hydraulic holes" in ...faults' (Bense et al., 2003) that compartmentalize an aquifer and implicitly acknowledge the difficulty in outcrop studies for providing this characterization.

Loveless et al. (2011) describe many fault zone elements associated with clay smear processes: a fault zone consisting of multiple slip surfaces; drag of thinned lithologies in the fault zone; incorporation of coherent fault lenses into the fault zone; grain-scale mixing of different lithologies. All of this occurs in a stratigraphic section without shale, and thus there is no fault core containing clay smear. Although this outcome seems obvious, the ubiquitous development of similar fault structures shows that clay smear develops in the context of a more universal fault structure evolution.

3.1.3.5. Processes that terminate clay smears

Although this facet of the problem is the most important for determining cross-fault flow, there are few outcrop observations that address how this actually occurs. Firstly, holes are hard to find given the inherently 2D nature of outcrop exposures. Secondly, when they are described, they occur related to thinning and attenuation of the clay in the fault zone, but it is hard to say whether thinning is a necessary or just helpful condition. It is clear, however, that attenuation and thinning is by no means a sufficient condition for holes to form. There is as yet no explanation for why a hole in a clay smear forms at any particular point. Doughty (2003) pointed to small faults as the cause of observed holes, and Bense et al. (2003) attributed the holes they saw to gravel roughness elements in the fault zone. van der Zee and Urai (2005) claimed there were no holes observed in dip sections of faults investigated in Miri, Malaysia, but when they found holes in the clay smear in fault strike sections exposed on benches, they were unable to rule out the possibility that a parent shale bed was absent at that location and thus the cause of the hole.

Childs et al. (2007) studied clay smear processes in deep-water turbidite channel and fan deposits cut by normal faults in coastal exposures of New Zealand. The observations made at these outcrops appear to contradict generalizations made from previous studies in settings that are more dominated by sandstones, including the following:

- 'Clay smears maintain continuity for high ratios of fault throw to clay source bed thickness (c. 8; *approximately 8*), but are highly variable in thickness, and gaps occur at any point between the clay source bed cut-offs at higher ratios'
- 'data demonstrate that there is no clear relationship between distance from the source bed and smear thickness, and that thickness immediately adjacent to the source bed may be as low as those distant from the source bed'

The result of these observations is the definition of a Probabilistic Shale Smear Factor (PSSF) that attempts to account for heterogeneous fault zone structure by allowing the finite probability of a hole developing somewhere on a clay smear according to a specified probability density function. The rationalization for this approach is the inference that: 'the locations of smear breaching are controlled by strain distributions, and slip surfaces, within the fault zone' (Childs et al., 2007), a statement that is compatible with Foxford et al. (1998), although the precise structural mechanisms that cause these holes remain undefined. What is remarkable about this study, though, is that holes may form before there is any significant thinning by simple shear processes in the fault zone, the first realization of this fact since the original conceptualization of a clay smear as a taper from a source bed.

3.1.4. Summary insights from outcrop observations

Field observations describe fault structures resulting from processes that control the presence, morphology, distribution, and continuity of clay smears (Table 2). However, when broken down into the essential elements necessary to address the clay smear problem, it becomes clear that there is substantial room for improvement in all dimensions of the problem. A number of studies have shown that the path to a deep understanding includes outcrops that contain the source beds on both sides of the fault, but this in turn makes it very difficult to find outcrops that provide this exposure of a fault with as much as 25 m throw.

There has been reasonable progress toward developing an understanding of the processes that incorporate shales into faults and the processes that further deform those clays in the fault (Table 2), even though some elements like folding appear poorly defined by field observations, especially in light of potential folding associated with propagating fault tips. Perhaps a large part of this state is the difficulty in deducing kinematic histories of all the minor elements in a fault zone; for example, interpretation of folding and minor faulting interactions leading to sandstone boudinage (Fig. 12). With greater fault offset, the inference of these interactions becomes much more challenging. One important insight that does emerge from outcrop studies is that the kinematic fault network evolution, in the sense of Childs et al. (1997) and many other authors, is essential for defining how fault zones evolve. For example, Cilona et al. (2015) interpret clay smear in a fault with primary strike-slip offset and minor dip-slip displacement; is dip-slip of gently dipping beds necessary for shale incorporation in the fault with subsequent evolution dominated by the strike-slip component, and if so, did dip-slip occur before strike-slip displacement? The role of kinematics will be further developed in the discussion section.

Most remarkable, however, is the paucity of helpful observations for processes that terminate clay smears. This aspect of the problem has been recognized as critical for predicting the flow properties across faults since Weber et al. (1978), yet progress is limited. Are outcrops poorly suited for investigating this aspect of the problem? It would appear that non-destructive 2D outcrop observations hinder investigation of structures associated with holes because there are always uncertainties that arise from the geology just beyond the outcrop surface; even studies with minor amounts of excavation (e.g.,Kristensen et al., 2013) illustrate the benefits of seeking the three-dimensional geometry and continuity of fault zone components.

A composite view of the elements present in a fault zone with clay smear is becoming clear (Table 2), but the distribution of these elements in any individual fault zone is uncertain. It is also difficult or impossible to construct a detailed 3D block model that is retrodeformable, a fact which probably reflects an incomplete understanding of the kinematic evolution of the fault zone, especially the role of the secondary fault structures. Because insights into the material properties and stress conditions resulting in different structures are limited, it may be difficult to achieve much progress with outcrop observations alone. Moreover, the necessary observations for identifying the few most important processes that control clay smear evolution remain elusive.

The goal of outcrop studies is to define fault structures and associated lithologic components in a way that allows analyses comparable to subsurface studies and provides the structural and geometric characteristics necessary for comparison with laboratory and numerical models. We propose that the following observations are critical to achieving this goal:

1. Multiple transects with clay fraction determinations along the dip and strike of the fault zone (e.g., van der Zee and Urai, 2005)
2. Geometric aspects of fault (spatial components), including:
    a. Thickness of each of the structural and lithologic components
    b. Fault throw (in lieu of displacement vector)
    c. Stratigraphic column of the faulted section, including Vshale characterization
3. N:G of section (proportion of lithologies on either side of Vshale cutoff)
    a. ii. Average bed thickness for each lithology (for a defined Vshale threshold)
    b. Number of shear zones in each fault transect
4. Comprehensive listing of diagnostic fault structures (developed in following sections;Fig. 27)
5. Detailed fault description of all fault components, including deformation in sand (stone); from one edge of fault zone to the other
6. Along-dip lengths (and continuity) of all structural and lithologic components (i.e. basis for subsurface fault zone mapping)
7. To the extent possible, 3D definition of fault systems (i.e. along fault strike)
8. All scales of observations from microscopic to largest scale possible in each outcrop (which must be defined)
9. Collect data in such a way that observations and measurements are reproducible

Outcrops will always be limited in the dimensions of observation and thus over-estimate the continuity of structural components. This limitation needs to be embraced in such a way that all clues in the outcrop that indicate further dimensions are observed and recorded.

3.2. Laboratory and numerical models of clay smear

Laboratory and numerical experiments are undertaken in order to elucidate physical processes that result in faults with clay smear. Because it is possible to follow the incremental strain history and to measure stresses and material properties in experiments, they form a crucial part of our understanding of clay smear processes. In this section we review the available literature on laboratory and numerical experiments, summarize what we think has been learned from these studies, and discuss where we think future areas of research may contribute to further advances in our understanding. Often the results of laboratory and numerical experiments are used to compare the effectiveness of clay smear algorithms in the subsurface – later in this section we discuss the many pitfalls of this approach.

Outcrop studies reflect three important stages in the development and maintenance of a clay smear: (a) an initial stage responsible for incorporating the necessary clay into a fault zone; (b) an intermediate stage where clay smear evolves as the fault accumulates large shear strain and initial holes may be closed; and (c) a final stage where clay smear is disrupted and holes form. As was described in the Outcrop Studies section, this generic history of the continuity of a layer during fault evolution exists regardless of whether clay is involved in the faulted section. In our review of the experimental and numerical studies we follow this same scheme.

The search for the answer to the question – how does a clay layer in sand evolve into a more or less continuous clay smear in a fault zone? – should start with a thorough understanding of the relevant processes in homogeneous materials (for example by replacing the clay layer with sand, just colored differently). Fault evolution in pure sand is by no means understood, but what is known provides a useful baseline to study how the difference between clay and sand affects the process of clay smear evolution. Interestingly, experimental and numerical studies use similar model geometries and suffer similar problems with simulating localization and fault evolution processes.

If these processes are understood in the laboratory and numerical simulations, the next non-trivial task is to upscale these results to faults in nature. Does any individual experiment relate to nature at a scale of 1 m or 100 m in size? When does the grain size chosen in the laboratory or grid size chosen in the computer influence the observed fault processes in a way that biases the understanding of faulting when results are up scaled to nature? The final, most difficult question asked by subsurface flow applications is: when is the clay smear discontinuous in the subsurface? Keeping this in mind, experimental and theoretical modeling of clay smear can be seen as a truly challenging task. On the other hand, even if the results are incomplete, these studies shed light on processes or define structural domains which are otherwise difficult to identify from final, total finite strain examples represented in outcrops, and if one can demonstrate that the results are applicable or can be scaled to natural prototypes, they then prove useful. Our final task in this section is thus to summarize how the available laboratory and numerical experiments are appropriately used in nature.

3.2.1. Experimental study of clay smear processes

Experimental research on clay smear has been done in a range of configurations and corresponding boundary conditions. The short review below attempts a comparison of these different experiments, trying to identify structures and processes that are independent of the experimental technique and those that occur both in numerical simulations and in laboratory experiments.

A good starting point for this discussion is shear band formation in homogeneous granular material, which has been studied extensively since the late 1950's (Fig 13). It is well known that in triaxial tests (Vardoulakis and Sulem, 1995) development of a single well-defined shear band is complicated by the symmetry of loading; initially a complex failure pattern evolves of multiple incipient shear bands. Many studies of shear band formation in sands have been conducted using plane strain compression testing. Here, there are many uncertainties regarding when shear bands initiate, and at what stage of the loading history bifurcation and localization occur. 3D image correlation techniques to measure the full velocity field in the sample (Hall et al., 2010) are required to address these questions. In direct shear experiments, curved shear bands form a lens before a through-going shear zone is established. Shear bands also develop in ring shear experiments. Although the shear band is prevented from developing at any orientation in a ring shear apparatus, very large displacements are possible. In sandbox experiments, a multitude of shear band structures develop depending on boundary conditions.Balthasar et al. (2006) compared a number of experimental configurations and discussed existing constitutive models for shearing clays, concluding that there are no calibrated and validated models for the rheology of sheared clays at present.

3.2.1.1. Ring shear experiments

The classic papers on experimental study of shearing in granular aggregates and the development of clay smear are Mandl et al. (1977) and Weber et al. (1978). A ring shear apparatus was used to study the development of shear zones, together with in-situ stress measurements, microstructural study of the deformed materials and permeability measurements. The experiments were analyzed in great detail and clearly identified the limitations of such experiments in the study of fault zones in nature. Shear stress was applied to the samples by rough, permeable, ring-shaped plates at the top and bottom of the shear zone. A transparent annulus was illuminated from inside so that samples saturated with pore fluid of the same refractive index remained transparent, and the development of shear zones, clay smears and slip planes could be followed in real time. Materials and normal stress were chosen to either allow or suppress dilatancy and to cause grain crushing in the shear zone at high stress. Principle stress orientations were measured inside the samples using photoelastic cylinders. Deformed samples showed slickensides on the master slip plane and secondary slip planes in Riedel orientations. The photoelastic stress measurements show that in the beginning of the shearing process the direction of maximum compressive stress rotated from vertical into a position at an angle of 45° to the horizontal shear direction, and the authors concluded that: 'the shear band produced between the rigid platens of the apparatus is bounded by planes of maximum shear stress rather than by Coulomb-type slip planes.' At the same time the observations of oblique sets of minor shears indicate the tendency of the material to deform in accordance with Coulomb's slip concept, which obviously is suppressed by the specific type of kinematic boundary constraints (Mandl et al., 1977). Thus the trade-offs between the inherent nature of the fault structure and the structure imposed by the kinematic constraints of this particular experimental design, selected for the high shear strains allowed, became established.

For the clay smear experiments (Fig. 14), the apparatus was filled with alternating layers of sand and remolded clay (Weber et al., 1978). Results show the clay sheared to form a continuous, multi-layered clay smear along the shear plane. An important observation was the formation of wedge-shaped sand intrusions in the clay (pointing into the shear direction). In movies of these experiments made through the transparent outer ring, these sand wedges were observed to move into the clay layer, producing a local thickening (injection) of clay in the shear zone as a result (G. Mandl, pers. comm. 1997). The clay smear was mixed with sand grains and had a low permeability to flow across the shear zone.

Sperrevik et al. (2000) used a ring shear apparatus at effective stresses corresponding to a depth of about 50 m to shear sand-clay sequences. They used different clay types, with undrained shear strength between 50 and 350 kPa, and water content between 19 and 50%. Removing the upper sand exposed the clay smear, allowing measurement of its continuity (Fig. 15). The development of clay smear was discussed as a function of the competence contrast between the clay and sand. Many of their results are comparable to those of Weber et al. (1978). Clay, which was less competent than sand, was interpreted to be ductile, and formed clay smears along the shear zone. The compaction of the sand during deformation led to work hardening and increased its competence contrast with the clay, which smeared along the shear zone in a ductile manner. In experiments where the sand compacted during deformation, a clay wedge formed by drag of the clay into the shear zone. The thin (1.5 mm) clay smear was continuous, in some cases along the whole shear zone. With increasing distance to the source clay, increasing amounts of clay mixed with sand were found, together with discontinuities in the clay smear. In some cases the clay formed undulatory bumps over Riedel shears in the sand, but in other cases it lost its continuity in these structures. Multiple initial clay segments led to a composite, layered clay smear, which was sometimes discontinuous in 3D with a thin sand wedge connecting the sand on both sides of the clay.

When experimentally deformed clay was more competent than sand, it behaved in a brittle manner and formed isolated angular fragments. Here, the shear zones dilated and the sand strain-softened and became less competent than the clay. Sperrevik et al. (2000) propose that the transition is a function of the stress conditions, initial porosity of the sand, and the mechanical properties of the clay. Unfortunately the full mechanical properties (e.g., dilatancy transition) of the individual clays were not determined so that it is difficult to establish the exact deformation mode of the clay. Based on rock mechanics considerations, the absolute value of the clay's brittleness must also play a role, not only its contrast to the sand. An additional problem that Sperrevik et al. (2000) could have discussed is the initial state of stress (e.g., as can be visualized using photoelastic techniques) because it is difficult to pack and load a sample such that initial stress is homogeneous.

Clausen and Gabrielsen (2002) and Clausen et al. (2003) built on the work of Sperrevik et al. (2000) using the same apparatus and sand with a wider range of clays (undrained shear strength between 20 and 700 kPa) in drained shear experiments which have shown good reproducibility. Detailed measurements of the shear stress and volumetric strain in the samples are accompanied by thin section analyses of deformed samples. The authors classified three types of structures: absence of a clay smear with only clay fragments at low stress, a mixture of clay and sand or patchy clay in a sand matrix at intermediate normal stress and a semi-continuous clay smear at high normal stress. The normal stress required for the transition between these continuity domains depends on the clay type and is higher for stiffer clays, as expected. Clay wedges or lenses were observed close to the source layer in many experiments, interpreted to be associated with initial fracturing of the clay with the assistance of Riedel shears. Clay smear continuity and area of shear plane covered with clay became larger when the normal stress was increased. The water content of the clay decreased with increasing strain in compacting samples. Thin sections of deformed samples show that both sheared clay and mixed sand-clay are present in the clay smears.

Cuisiat and Skurtveit (2010) presented results from a high-stress ring shear apparatus using uncemented, normally consolidated sand and clay at stresses corresponding to burial depths up to 1500 m. The samples all reached a steady shear stress during the experiments and compacted progressively. It is unclear if compaction is due to grain rearrangement in the different phases or mixing of sand and clay (cf. Schmatz et al., 2010a and Schmatz et al., 2010b). Sectioning of the samples after the experiments, corresponding thin sections, and permeability measurements across the shear zone were used to investigate clay smear continuity, thickness and evolution of permeability (Cuisiat and Skurtveit, 2010). Granular flow, grain mixing and cataclasis all contribute to the resulting clay smear structure as a function of effective mean stress. Under conditions of granular flow, clay smear led to strong permeability reduction across the shear zone. At greater effective stress, permeability reduction by grain crushing was of a similar order as by clay smearing. Cuisiat and Skurtveit (2010) interpreted that in addition to clay smearing, drag and injection of clay along the fault plane also occurs, together with mixing of clay and sand in the fault core, although the evidence for this seems inconclusive. The thickness of clay smear was higher for thicker clay source layers, while reducing the clay layer thickness to one half of the reference layer produced a thin and discontinuous clay smear. Shearing of multiple clay layers produced a layered, composite clay smear 2–3 times thicker than that for a single clay layer. The authors concluded that the experiments were consistent with a first order correlation between SGR and seal capacity.

Sadrekarimi and Olson (2010) studied sand samples with colored marker layers in a ring shear apparatus that had a transparent outer ring to allow direct observation. The samples deformed homogeneously before peak stress, followed by localization and slow widening of the shear band until a stable thickness of one grain diameter was reached.

In summary, the main advantage of ring shear experiments is the very large shear strains allowed by the apparatus. Intensive shearing in the thin shear zone imposed by this design occurs by granular flow and mixing of sand and clay, by abrasion of brittle clay fragments, or by cataclastic flow at high normal stress. The full spectrum of clay smear continuity from discontinuous clay fragments to continuous clay smear was found in the shear zone, depending on effective stress, and the transition from discontinuous to continuous occurs at higher stress for stronger clays. The key parameter for overconsolidated clay in cohesionless sand is the effective stress, which suppresses brittle failure of the clay and enhances shearing. Cataclastic flow produces similar geometries. A disadvantage of ring shear experiments is the strong kinematic constraints imposed by the fixed geometry of the apparatus. Localization patterns are prescribed by the rotating outer rings so that the sample can only develop the thin shear zone in the plane of the forcing rings.

3.2.1.2. Direct shear experiments

Karakouzian and Hudyma (2002) present a novel apparatus for investigation of clay smears. Samples consist of alternating layers of sand and clay, encased in a transparent tube, and an axial plunger with a half-circle cross section which pushed half of the sample past the stationary part. Deformed samples show deformation bands with sheared clay and the development of lenses, similar to the structures that develop in a direct shear configuration (Thornton and Zhang, 2003). The authors conclude that direct (real time) observation is a useful addition to more sophisticated deformation experiments.

Urai et al. (2003) used artificially prepared layered samples with sand, kaolinite, illite and smectite. Samples were deformed to shear offsets up to twice the layer thickness in a geotechnical direct shear apparatus. After saturation and initial compaction the sample was inserted in the shear apparatus and deformed under drained conditions. Besides recording the stress-strain behavior of the aggregates, a detailed study of the internal structure of the fault zones was done by serial sectioning of the deformed samples.

Structures developed in end-member samples were quite different, with relatively wide deformation bands in the sand and much sharper deformation bands in the clays. Shear bands were initiated at the edges of the moving sample chamber and propagated towards the center of the sample. The primary effect of inserting several layers of clay was to increase the width of the deformation bands by creating an additional degree of freedom of the system due to layer-parallel shear in the clay layers. Changing the type of clay layers caused less dramatic changes in the final structure. In general, continuity of the sand layers across the shear zone was maintained in experiments with thin multilayers. However, in experiments containing one layer of clay sandwiched between much thicker layers of sand, Urai et al. (2003) observed loss of continuity of the clay layer.

The experimental study of clay smear has made surprisingly few links to the extensive body of literature on shear deformation and the evolution of frictional behavior in clay gouge (e.g., Rathbun and Marone, 2010). In this study, water-saturated layers of a granular material analog to a fault gouge were deformed in the double-direct shear configuration. Strain markers helped define shear localization as a function of dilation in response to perturbations in shear stress and rate/state friction response to shear velocity perturbations. Although the different layers only differed in color and had the same mechanical properties, the deformed layers had a complex attenuated shape. These experiments show the complexity of the shearing process and the wide range of parameters that affect the final geometries of fault zones.

Giger et al. (2011) present a novel design of the direct shear experiment to deform large (0.3 × 0.3 × 0.6 m) rock samples under high pressure up to 36 MPa, fluid-saturated and with the possibility for fluid flow measurement across the shear zone. The most important innovation in design was the relaxation of the sharp displacement boundary condition at the edge of the sample. They use a high viscosity fluid to seal the sample around a 1 cm wide gap in the contact between the sample and the loading plates so that the shear zone localizes more broadly than in conventional direct shear tests. The philosophy behind this adaptation is that: 'the increasing complexity and inaccuracy of the mechanical results ... were deliberately accepted ... in exchange for producing a more realistic fault zone' (Giger et al., 2011). These boundary conditions determine the width of the shear zone, which does increase towards the middle of the sample, but far less so than in conventional direct shear tests. There is a small rotation of the sample's lower and upper halves with respect to each other, and it is not completely clear how this influences the displacement field in the shear zone and the formation of Riedel shears.

Ciftçi et al., 2012 and Giger et al., 2013, and Çiftçi et al. (2013) used this apparatus in a series of experiments to measure transport properties and stress-strain behavior as a function of clay content and strength contrast between the sand and clay. Artificial and natural rock samples were studied by CT-scans to image the geometry of the sheared clay layer (Fig. 16), which was found to depend on stress and on material properties; brittle clay forms segmented smears while ductile clays form more continuous smears. No tapering of the smear away from the source layer was observed. Increasing normal stress increased shear zone width slightly (width is mainly controlled by the gap between the loading grips) and correspondingly there was more clay in the shear zone. The authors interpret their results to be in reasonable agreement with the SGR model. The main processes in the generation of clay smears are interpreted to be brittle processes of 'slicing' and wear rather than 'ductile drag or plastic flow' (Çiftçi et al., 2013). Highly overconsolidated and cemented clay deforms by dilatant fracturing at low displacements. Clay, with Unconfined Compressive Strength (UCS) equal to or greater than the UCS of the matrix sandstone, forms continuous clay smears. Morphology of the clay smears thus vary widely, depending on stress and brittleness of the clay. However, the authors point out that initial segmentation of the clay layer sometimes evolves into a continuous clay smear because the clay fragments in the shear zone are reworked to clay in critical state. The smearing in strong clays can start with brittle failure producing fragments of clay in the shear zone which are then abraded during progressive shearing, as described byHolland et al. (2006) and Schmatz et al. (2010b), and followed by the formation of clay smears along Riedel shears (Fig. 17). With softer clays, a more uniform clay smear is formed, with lower average thickness of the clay smear.

In summary, the advantage of direct shear experiments over the ring shear apparatus is the benefit of larger samples, which yields results that are closer to the heterogeneous structures observed in nature. The geometry of localization is complicated in direct shear devices, starting from the lines of discontinuity around the central plane in the sample propagating inwards, forming a lens-shaped zone of deformation in the samples. This lens has some resemblance to lenses observed in outcrops, and in the center of experiment the sample has more freedom to develop a complex localization pattern than in a ringshear device. Interestingly, the relaxation of the sharp discontinuities at the sample boundary by Çiftçi et al. (2013) produces a much more diffuse lens in the sample.

3.2.1.3. Triaxial shear experiments

In triaxial experiments the suppression of strain localization by rotational symmetry in the sample is usually solved by using pre-cut cylinders that are subsequently sheared (e.g.Savage et al., 1996). Takahashi (2003) used a variation on this geometry and presents a detailed study of the shearing of a siltstone layer embedded in sandstone, deformed by sliding along pre-cut shear planes in the sand. Thickness of the sheared layer was measured together with displacement, microstructure of the deformed samples was studied by optical and electron microscopy, and fluid flow across the sheared layer was monitored during the test.

The sandstone deformed by cataclastic shearing with the shear zone widening during deformation, and the sheared siltstone varied from thick and continuous to thin and discontinuous. Takahashi (2003) proposes that the sealing of the fault was mainly a function of effective stress. Permeability measurements indicate three permeability regimes that arise out of the deformation. In regime 1, permeability is rapidly reduced by 1–1.5 orders of magnitude by compaction until the sample reaches its yield stress, at which point inelastic deformation starts. In regime 2, permeability decreases, while it gradually increases in regime 3. The transition between these regimes depends on effective normal stress; higher stresses are proposed to enhance abrasion of the siltstone and incorporation of more smear into the shear zone.

Permeability increase in regime 3 is due to thinning of the smear by sand grain erosion. Post-mortem observations suggest that the amount of cataclasite and gouge derived from the sandstone increases after loss of continuity in the siltstone smear and that cataclastic sand gouge keeps the permeability lower than the initial permeability. The thickness of the zone of sheared siltstone plays an important role in the hydraulic properties of the sample. We note that it is not clear if the two pre-cut planes in the sandstone were slightly misaligned and thus non-coplanar after assembly with the siltstone layer sandwiched in between (cf. the top block in the sandbox experiments ofSchmatz et al., 2010b).

Crawford et al., 2002 and Crawford et al., 2008 report experiments on mixtures of fine-grained quartz and kaolinite gouge to establish their strength and fluid flow properties. In general, permeability decreases with increasing kaolinite content. Porosity shows a minimum that evolves with increasing effective pressure as predicted by an ideal packing model. Sheared gouge samples show a reduction in frictional strength with increasing clay fraction. Permeability variations as a function of clay content and shear deformation are explained reasonably well with simple models based on theory of mixtures (c.f. comparison of laboratory data with models in Crawford et al., 2002). Clay content had the largest effect on permeability, and shear deformation affects permeability of quartz-rich gouges more than clay-rich gouges. Shearing mechanism, as defined by Lupini et al. (1981) is in three modes: a turbulent mode, a transitional mode, and a sliding mode, depending on dominant particle shape and the coefficient of interparticle friction.Crawford et al., 2002 and Crawford et al., 2008 conclude that clay mineralogy is as significant as clay content in controlling fault rock permeability, and strain is of secondary importance for permeability evolution.

In summary, triaxial experiments show results similar to the ring shear and direct shear experiments: different failure modes of the clay layer lead to either macroscopically ductile shearing or initial fragmentation followed by reworking of the fragments during shearing in the sand. Probably the strong boundary conditions in these tests prevent deformation modes other than shearing in the shear zone pre-defined by saw cuts, although increasing the thickness of the zone of shearing has also been shown to lead to thicker clay smear. Granular mixing is recognized as an important process in many experiments, and the measurements of shearing in artificial gouges provide important clues on this process. However, the rates of this process and the properties of the mixture are far from understood.

3.2.1.4. Sandbox experiments

Sandbox experiments, (dry or water saturated) of clay smear evolution are done at the lowest effective pressure of all experiments discussed in this review, with well characterized nonlinear rheology of model materials. In spite of these limitations, they offer the most realistic set of geometric boundary conditions because of the significant body forces and the large degree of freedom of the models to develop fault geometries like relays and lenses, which are common in nature.

van der Zee et al. (2003) present analog models of dry sand above a basement fault dipping at 45°, with a layer of relatively very soft (yield strength 20 Pa) oil-water emulsion with an elastoplastic rheology to model clay. Movement on the basement fault creates normal faulting in the multi-layered sedimentary overburden with small apparent cohesion (Maksimovic, 1989) and a friction angle of around 35°. In experiments with a thin overburden of 1.5 cm, the fault zone develops with the soft layer derived from the hanging wall sheared along the fault, creating a continuous and asymmetric smear that becomes discontinuous at high strain. With a thick overburden of 9.5 cm, a locally thick fault zone of soft clay analog was created by injection into the fault zone from the layer in the footwall (Fig. 18). The basement fault formed a releasing step in the soft layer, and a second fault was initiated to form the squeezing block, causing the soft layer to become injected into the fault. The differences between the thin and thick overburden cases are explained by considering the contrast in strength between the sand and the emulsion at the given overburden stress. For a thick overburden the strength contrast between sand and soft emulsion is much higher. At a certain overburden thickness the squeezing block is spontaneously initiated in the footwall, leading to injection of the soft layer, but the mechanism needs more study.

Schmatz et al., 2010a and Schmatz et al., 2010b deformed water-saturated, layered sand–clay models above a rigid basement cut by a fault. The models had a free top surface or a faulted top block, resembling weaker layers (sand and clay) sandwiched between two strong and stiff layers cut by two co-planar faults. Models were deformed between two lubricated glass plates to allow real time observation and measurement of the velocity field by Particle Image Velocimetry (PIV). The thick clay layers were undrained, while pore pressures remained hydrostatic in the sand. Material properties were characterized by a series of standard geotechnical measurements. The experiments were all in the 'precursor structural domain' defined by Nollet et al. (2012) in which the initial deformation is accommodated on a steep antithetic fault to the basement fault. After the first increments of deformation, the fault migrates toward the kinematically preferred plane coplanar with the basement fault (Fig. 19). Thus, different segments of the final shear zone have moved at different times, and the clay smear increases the shear zone's thickness. The models are sufficiently large to form releasing or restraining relays in the clay layer, which have a major control on fault-zone structure. In general, high strength contrast between sand and clay leads to a more complex fault zone. Because sand undergoes boudinage while weak clay layers continue to shear and become amalgamated, clay is found in greater abundance in the fault zone than in the faulted stratigraphy. Fragments of brittle clay (Fig. 20) may be reworked with ongoing deformation following an initial loss of continuity during an earlier stage of deformation. Thus, the continuity of a clay smear within a layered sequence can increase after an initial decrease. Thin, weak clays deform continuously over large displacements, and the volume of clay smear increases as sand mixes into clay at the margins of the shear zone.

An unexpected outcome of the Schmatz et al., 2010a and Schmatz et al., 2010bexperiments arises from the evolution of clay thickness in the shear zone with increased fault offset. That thickness should decrease by simple shear, yet thickness varies little with offset. This led to the observation of sand grain incorporation at the margins of the sheared clay, effectively inflating the clay volume by a process of sand-clay mixing. Based on these observations, we think mixing is an integral part of clay smear processes.

Noorsalehi-Garakani et al. (2013) present observations of clay smear continuity in water-saturated sandbox experiments similar to those of Schmatz et al., 2010a and Schmatz et al., 2010b, but in this follow-up study the sheared clay layers were excavated after deformation (Fig. 21). Standardized model materials were characterized in detail. The sheared soft clay layer above a 70° dipping basement fault reveals a complex, natural-looking fault zone architecture with relay ramps, breached relays and fault lenses. The clay smear shows clear variations in composition and thickness and becomes locally discontinuous, forming holes at throw-thickness ratios >7. In addition to fault segment migration into the footwall and/or hanging wall and tectonic telescoping in the relays, the thin, continuous parts of the clay smear is formed by mechanical mixing of sand and clay. Deformation in the hanging wall was stronger and more complex than in the footwall.

A complementary perspective on the clay smear problem is provided by van Gent et al. (2010) and Kettermann and Urai (2015) who deformed thick, brittle hemihydrate layers with thin, weaker sand between them. These experiments may provide clues for faults where the sand is cemented and the clay is weak. In these experiments, the fragmentation of the brittle layers made the fault zone wider and more complex (cf.Schöpfer et al., 2007a and Schöpfer et al., 2007b) with the weak sand flowing around the blocks. An interesting observation is the gravity-driven downward flow of sand into open fractures, a process which may be relevant in faults in some carbonate-clay sequences.

In summary, the main advantage of sandbox experiments over other experimental designs is that the boundary conditions allow a much more realistic development of a network of fault segments that evolve into a thoroughgoing fault zone by coalescence of segments and by migration of the zone of shearing. As observed in all other experimental setups, a key parameter in sandbox experiments is the failure mode of the clay layer; if the compressive strength of the material is much stronger than the mean effective stress, the clay layer will fail in extension and fragment, with reworking of the fragments during further shearing. A disadvantage of sandbox experiments is the low effective stress, which prevents simulation of cataclastic flow.

3.2.1.5. Discussion of experimental studies

Since the pioneering work of Mandl, Lehner and coworkers, forty years of experiments on clay smear processes have produced many interesting results. There have been important developments in technology and measurement techniques, such as image correlation, 3D modeling, etc. It has also become clear that full analysis of many different parameters, such as fluid flow, velocity field, stress field, in real time and 3D, is required to learn more about the complex, nonlinear fault processes. It is also clear that to understand clay smear, one needs to understand the whole fault zone, including the structures in the adjacent sand as well as those in the clay.

All experiments produce shearing in a sample with a shear zone at high angle to layers embedded in sand. However, there are large differences in boundary conditions, leading to very different geometries of localization. The challenge in experiment design is to produce a large displacement in a small sample (Table 3). Thus experimental geometries like the plane strain biaxial test, which would allow the localization of deformation to be mainly determined by the mechanical properties of the sample, have been underexplored. The boundary conditions chosen to reach high strains place different restrictions on how the zones of shearing are localized. The strongest restrictions are probably present in the ring shear and pre-cut triaxial experiments. In both of these systems, a thin (of the order of several grain sizes thick) zone of shearing is forced at both edges of the relatively small sample, from where the shear zones propagate towards the center.

The direct shear experiments and sandbox experiments have samples almost an order of magnitude larger than in other experimental designs, and the structure of the shear zone is correspondingly more complex and more similar to fault zones in nature. In these larger set-ups, the shear zones develop more segmented faults and fault lenses, relays and Riedel shears. The sandbox experiments have the least restrictive boundary conditions because the shear zones are allowed to migrate laterally with an unconstrained width. It is interesting to note that granular flow and cataclastic flow of clay seem to produce similar geometries as long as the boundary conditions are comparable.

Table 3. Comparison of main parameters in clay smear experiments.[a]

	Sandbox	Ringshear	Direct shear	Triaxial	Comments
Effective mean stress	☆☆☆☆	☆☆☆☆	☆☆☆☆	☆☆☆☆	Triaxial cell can reach highest stress
Shear offset	☆☆☆☆	☆☆☆☆	☆☆☆☆	☆☆☆☆	Up to 40 cm in ringshear
Fault zone area	☆☆☆☆	☆☆☆☆	☆☆☆☆	☆☆☆☆	Up to 40 × 40 cm
3D geometry of fault zone	☆☆☆☆	☆☆☆☆	☆☆☆☆	☆☆☆☆	
3D geometry of clay smear	☆☆☆☆	☆☆☆☆	☆☆☆☆	☆☆☆☆	
Freedom to localize	☆☆☆☆	☆☆☆☆	☆☆☆☆	☆☆☆☆	Improved in [illegible] et al. (2013)
Dilatancy measurement	☆☆☆☆	☆☆☆☆	☆☆☆☆	☆☆☆☆	
Stress measurement	☆☆☆☆	☆☆☆☆	☆☆☆☆	☆☆☆☆	
Velocity field	☆☆☆☆	☆☆☆☆	☆☆☆☆	☆☆☆☆	Using DIC
Microstructural investigation	☆☆☆☆	☆☆☆☆	☆☆☆☆	☆☆☆☆	Direct observation and post mortem microscopy
Fluid flow measurement	☆☆☆☆	☆☆☆☆	☆☆☆☆	☆☆☆☆	
Body forces	☆☆☆☆				

Note: Star rating reflects score (red stars) out of maximum potential score (blue stars).

[a]Higher rating represents a parameter closer to subsurface prototype or better access to measurement.

In all types of experiments, shearing in the thin high strain shear zone occurs by granular flow and mixing of sand and clay, and by abrasion of brittle clay fragments if present. Cataclasis of sand occurs at higher stress and produces similar geometries. Thin multilayer clay smear forms if several clay source layers are present, and in those instances where mixing and abrasion are sufficiently active, the sand-clay layered gouge will transform into a more homogeneous sand-clay mixture (the volume change caused by this process is an additional driving force). The evolution of the internal structure of this thin shear zone is poorly known as little of the process is resolved in today's experiments, by for example electron microscopy.

All experiment types point to the importance of the failure mode in the clay layer – if the relative values of mean effective stress and strength are such that extensional fracturing or shear fracturing result in clay fragmentation, continuous clay smears may only form after abrasion of the fragments in the shear zone. In other words, after an initial phase of fault permeability that is little reduced from the juxtaposed sandstone permeability, permeability of the fault decreases as clay fragments become abraded and distributed along the shear zone.

Lateral injection of clay into the fault is shown to occur when clay is much weaker than sand. This process can locally produce large increases in the amount of clay in a fault, but many questions about the process remain. Even less studied is the effect of brittle fragmentation of sand layers with soft clay redistributed around the sandstone fragments.

To a first approximation, deformation in thin shear zones is heterogeneous simple shear. A number of workers have calculated the deviation from this simple shear as a measure of strain heterogeneity (Schmatz et al., 2010b). It is also surprising that in most experimental studies the thickness of shear zones is overlooked as an important parameter (Fig. 22), although simple geometrical arguments illustrate how this parameter affects the shear strain and the fault zone geometries that result from that shear strain.

In faulting at upper crustal temperature and pressure, there is extensive evidence for solution transfer processes and frictional healing (e.g. Gratier, 2011). The role of these processes is unexplored in the clay smear community. Also, the extensive literature on clay gouge dynamics has yet to find its way into the clay smear literature (e.g., Byerlee, 1978, Haines et al., 2013 and Vrolijk and van der Pluijm, 1999), although we recognize that these processes represent an extension of the clay smear problem beyond its classical definition.

3.2.2. Analytical models of clay smear

The first mechanical model of clay smear was developed almost 40 years ago (Lehner and Pilaar, 1997 and Weber et al., 1978). Based on detailed field observations, these authors argued that the mechanism of clay smear emplacement must involve both shearing in the fault zone and continuous supply from the source bed. Lehner and Pilaar (1997) proposed that 'traversing a shale source bed, a normal fault has to be offset in the direction of the downthrown block in order to make room for the emplacement of a clay smear. Evidently, this mechanism can also provide an effective way of unloading the source bed along the rupture faces, thus facilitating the extrusion of plastic clay material. The extruded material will fill the gap created by the pull-apart mechanism. It will thereby enter the shear zone proper, where it will be subjected to the slow shearing imparted by the relative displacement of the fault blocks, to be smeared out along the fault.'

To model the extrusion process, the authors assumed that the clay layer has an effective viscosity, which implies that under otherwise identical conditions, one thicker clay layer extrudes more clay than two thinner layers with the same total thickness, expressed by the often cited $w \sim h^2$ relationship (Lehner and Pilaar, 1997). In many ways, this paper was far ahead of its time, combining detailed field observations with a kinematic-mechanical model, clearly distinguishing between the initial localization process and the shearing in the fault zone.

van der Zee et al. (2003) investigated the model of Lehner and Pilaar (1997) further, this time using a time-independent Mohr Coulomb rheology for the clay. Based on field observations van der Zee et al. (2003) propose that clay injection is associated with footwall collapse and the formation of a 'squeezing block' above the clay layer. For this model it was shown that no injection occurs when the horizontal stress in the releasing segment reaches zero if the differential stress fails to achieve a state of failure in the hard clay bed. No explanation is given for the origin of the squeezing block. They propose the Mechanical Clay Injection Potential (MCIP) condition for the onset of injection, defined as follows:

$$C = \frac{\sigma_1' \times (1 - \sin \phi)}{2\cos \phi} \quad (1)$$

where C = cohesion (MPa), σ_1' = maximum principle effective stress (MPa), and ϕ = angle of internal friction (degrees). This expression states that above a certain cohesion value, a clay layer will be prevented from extruding into the fault zone. Finite element models (with pre-existing contact elements) are in good agreement with the equation above. This equation also describes the transition in failure mode from extension to shear failure found to be important in many experiments. An interesting conclusion described by van der Zee et al. (2003) is that if the lateral dimension of the squeezing block scales with clay layer thickness, the $w \sim h^2$ relationship proposed byLehner and Pilaar (1997) also follows. The formation of the squeezing block is seen as another example of the increasing fault zone complexity with competence contrast, as discussed by Schöpfer et al., 2007a and Schöpfer et al., 2007b.

Welch et al. (2009) adopted the idea of the trishear kinematic model of fault propagation folding and developed a 2D, kinematic, quadshear formulation to model fault-related folding in a ductile interval sandwiched between brittle layers. The models offer interesting ideas of the different possible kinematic styles formed as a function of the location of the faults on both sides of the layer and of the geometry of the localized zone in the quadrilateral lens of deformation between the two faults. However, the model offers no mechanical understanding of why the localization stops with progressive deformation, nor a way to predict when a fault's deformation will delocalize into the shear zone.

3.2.3. Numerical simulation

The simulation of localization in shear zones and extension fractures in general is as challenging a numerical problem as is the experimental investigation of large shear deformation in geomaterials. Besides the continuum mechanical approach, improved by extension to polar continua and adaptive remeshing, numerical methods with discrete particles are used to explore these problems. In general, there is agreement that numerical models in the state of the art are unable to accurately reproduce the details of patterns of localization, and the same holds for patterns of shearing and evolution of Riedel shears in a fault zone. However, numerical models are nevertheless helpful for gaining insights into the complex evolution of stresses in a layered section with contrasting mechanical properties, an area of investigation where analytical models offer little benefit.

3.2.3.1. Discrete element modeling

Numerical modeling in geomechanics with discrete particles is based on Cundall and Strack (1979). Material properties are defined at the particle level, and macroscopic properties result from interactions between the particles. Cundall et al. (1982) show that the method yields good correspondence with theoretical models of bifurcation (Rudnicki and Rice, 1975 and Vardoulakis, 1989) and with experiments on sand (Thornton, 2000 and Zhou and Chi, 2002). Tensorial variables such as stresses and strains are also calculated (Cundall et al., 1982, Luding and Hermann, 2001, Hardy and Finch, 2005,Hardy and Finch, 2006, Finch et al., 2004, Zhao et al., 2007, Egholm, 2007 and Egholm et al., 2007). Deformation to very large strains in fault zones has been modeled using discrete elements, including dynamic rupture processes by Wang et al., 2006 and Abe et al., 2006, and Latham et al. (2006). Compared with continuum methods (Section3.2.3.2), the discrete methods hold significant advantages when it comes to modeling very large deformation in granular flows. Local dilation, localization of strain and fragmentation is easily simulated. The disadvantage is that the complex constitutive models required to accurately model real geomaterials are implemented only with difficulty. The stress-based discrete element method (SDEM: Egholm, 2007) solves some of these problems. By reproducing the 'Benchmark' sandbox models they show that the method generates shear zones at angles in agreement with general observations. Additional limitations for clay smear modeling include the number of particles and thus resolution of the model, as well as the shape of particles which are round rather than typical clay flake forms. Recent developments such as the SPOT method (Rycroft et al., 2010) promise large increases in computation speed.

3.2.3.2. Continuum modeling

Herle and Feda (2001) give a fundamental overview of the different finite element simulation methods and discuss problems related to shear zone growth, such as singularities and strain gradients at the tips of shear zones and mesh dependency in classical non-polar continua. For softening materials the problem is ill-defined. The effect of different material properties on localization is poorly understood (e.g., Herle and Feda, 2002, Hügel, 1995 and Belytschko et al., 1994). A method to keep the shear zone thickness independent of discretization is proposed by Cosserat and Cosserat (1909), who introduce a characteristic length. Apart from additional degrees of freedom, which are defined at the level of elements, material characterization also needs to be extended (Mühlhaus, 1986) to properly address the localization problem. The calculated shear zone thickness is thus adjusted to correspond to experiments. Another approximation to softening is achieved by hypoplasticity (e.g., Kolymbas, 1988, Niemunis, 2003 and Nübel, 2002; and Gudehus and Nubel, 2004). A recent approach to address localization is ELFEN by Rockfield (Crook et al., 2006) which uses Lagrangian methods and adaptive meshing, together with a constitutive model based on critical state concepts and global energy dissipation regularized by fracture energy. These models also compare well with the Benchmark tests (Buiter et al., 2006 and Santimano et al., 2015).

3.2.3.3. Discrete element models of clay smear

Schöpfer et al., 2006, Schöpfer et al., 2007a and Schöpfer et al., 2007b present a major study of the localization of deformation in mechanically stratified multilayer sequences above a basement fault, using the PFC3D Discrete Element code. They show that deformation first localizes in strong, brittle, layers with dip relays in weak, ductile, layers. Faults under sufficiently low pressure-high strength conditions form by initial Mode I fractures in the brittle layers, together with low amplitude monoclinal folding. The fractures are linked by shallow-dipping faults in the weak layers. The models are similar to geometries observed in natural faults with brittle layers, including large changes in fault dip, fault branches, normal drag, and progressive linkage of fault segments. Fault zone width increases with increasing displacement. Faults in models with high strength contrast and low confining pressure are strongly segmented because of different modes of failure in the different layers (Fig. 23). At higher pressure and lower strength the models show a transition from extension to hybrid failure to shear fracture and an associated decrease in fault zone width and complexity.

Egholm et al. (2008) considered the case where an upward propagating fault in sand above rigid basement intersects a weak clay layer (with lower friction angle), refracting to a shallower dip in the clay and continuing at the initial angle in the sand above. This creates a contractional relay in the clay and locally increases the mean stress in this zone, promoting flow of the clay into the fault zone (Fig. 24), similar to the squeezing block of van der Zee et al. (2003). In this model, the volume of clay available for injection also scales with the thickness of the clay layer. SDEM models with the dimensions of sandbox models were run above a rigid basement fault dipping 65°, and with large rheological contrast between sand and clay (35° and 5° friction angle and a low cohesion). Comparison of the results produced with and without a clay layer show that in the case of a weak clay layer, shearing in the sand above the clay migrated into the footwall, progressively eroding the clay layer and transporting it into the fault zone. Initially, deformation in the lower sand localizes in a steep precursor fault, which migrates into the footwall. The thickness of the sheared clay in the fault zone increases with decreasing clay friction angle. The Egholm et al. (2008) results are thus at variance with the model of Lehner and Pilaar (1997) who proposed that the zone of localization steepens in the clay layer. In both cases, however, the deviation from a planar fault effectively widens the fault zone and increases its complexity. Egholm et al. (2008)propose that if clay beds are cohesive, steep tensile joints could produce the pull-apart ofLehner and Pilaar (1997). In this case, however, the model of van der Zee et al. (2003)predicts that the clay will not fail and should support an open fracture. Consequently, van der Zee et al. (2003) expect that the clay will fragment in this case and flow into the fault zone where it is eroded and smeared along the fault, as in the experiments of Sperrevik et al. (2000). Clearly more work is needed to explore how fault propagation creates a segmented fault network that affects the incorporation of clay into a fault zone and its early deformation.

Raith (2012) investigated models similar to Egholm et al. (2008) and Schöpfer et al., 2006, Schöpfer et al., 2007a and Schöpfer et al., 2007b, and studied the effect of cohesion of a simulated cemented clay layer and the angle of the basement fault. However, in these models the clay was stronger than the sand. Results from these numerical simulations reproduced the graben and precursor domains of Nollet et al. (2012). In both domains, the increase in cohesion of the cemented layer produces large differences in the structural evolution. When clay cohesion is sufficiently high, tensile stresses in the cemented clay layer lead to fragmentation, discontinuous clay and a complex fault zone (Fig. 25). When cohesion in the clay layer is low, it softens and is distributed along the fault zone.

TerHeege et al. (2013) used a 2D DEM code to model direct shear experiments in large samples of Çiftçi et al. (2013) and Giger et al. (2013) discussed above. Numerical results were discussed in terms of four processes: drag of the clay layer, with tapering towards the fault, 'slicing' of the clay layer along slip planes, abrasion of the clay fragments and mixing of clay with rock fragments, and lateral flow of clay associated with dilation in the fault zone. Micromechanical parameters were calibrated by simulating 2D biaxial tests. The choice of the 2D direct shear setup was justified by arguing that (1) the clay smear is only passively loaded by the applied normal stress so that deformation of the source layer and associated changes of mechanical properties prior to the onset of displacement is limited; (2) entrainment of clay in the fault zone can be better studied in terms of end-member mechanisms, such as slicing, drag, wear and flow; (3)

geometrical effects influencing the volume of clay entrained in the fault zone are minimized. Results of the simulations were similar to the experiments, but with shear zones less sharply localized. The authors used critical stress ratios (proximity to mechanical failure, see Eq.(1)) to define a criterion for clay smear failure and the construction of a smear breach diagram to constrain the sealing capacity of faults containing clay smears, which was higher for higher clay content of the source beds and for higher normal stress. One important point about this paper is that for realistic description of the mixing of sand and clay during shearing, a 3D model is essential, because the flow of clay into the pore space in the sand occurs through the pore throats, even when two sand grains touch (in 2D grain-scale mixing is prevented by the geometry of the experiment).

3.2.3.4. Continuum models of clay smear

Simulations of normal faults with clay smear caused by hypoplasticity (Gudehus and Karcher, 2007) present examples of deformation of sand-like and clay-like materials including creep terms for clay. First results show shear band patterns, which are compared to model experiments and natural fault patterns. The model uses a normal fault with a clay layer in sand above a basement fault. A simulation of shearing of the sand-clay layers is presented, but unfortunately the information provided about the models (Gudehus and Karcher, 2007) prevents critical evaluation of the results. An interesting first result is that with increasing offset the clay smear thickness stabilizes at ca. 15% of the source layer thickness, whereas the fault zone gets wider. The authors propose that local remeshing with polar quantities could resolve finer details of the fault zone structure.

Nollet et al. (2012) used ELFEN to study the effect of basement fault dip on the evolution of shear band patterns in sand as observed in sandbox experiments. Results reproduced the range of structural styles observed: a basement fault dip of 60° and lower results in a graben structure with antithetic shear bands, while a basement fault dip of 70° and steeper initiates a precursor shear band followed by a synthetic shear band close to coplanar to the basement fault. The authors propose two structural domains in which the first-order structural style and deformation patterns are only weakly dependent on the details of the rheology of the model materials and are also observed in sandbox experiments for the same set of boundary conditions but with different material properties. Note that the internal structure of the shear zones is only generally resolved by these models; the shear zones essentially deform by simple shear, and the layered display is an attractive demonstration for what one may expect in a homogeneous material.

Kleine-Vennekate et al. (2014) and Kleine-Vennekate (2013) used ELFEN to model the evolution of localization. The first numerical model series used a homogeneous model extended laterally until a shear band initiated from a soft element (i.e. a single node within the model with artificially weak mechanical parameters assigned to insure that failure occurs first at this node). Results show a shear-band dip arising as a function of friction angle and a weak function of dilation angle. The shear band becomes one element wide during fully developed shearing, while the fracture energy is kept constant by regularization. The ratio of maximum displacement to shear band length is much larger than for faults in nature, perhaps because the ELFEN model allows no work hardening phenomena. Two non-coplanar shear bands coalesce in a relay, while closely spaced shear bands develop a lens structure. In mechanically layered models the shear band can change dip (as expected from friction angle and dilation angle) when the shear band propagates continuously. In models where localization is discontinuous and the shear band jumps across layers, coplanar shear zones are formed on either side of the clay layer, elegantly explaining why in many sand-clay sequences in nature faults maintain a single orientation across layer boundaries. Such models have the potential to capture the initiation of the clay smear process as purely as possible, but unfortunately numerical problems encountered when running models to very large displacement have so far prevented this result. For modeling of sand-clay experiments above a basement fault in a sandbox, material properties were carefully calibrated based on tests of the model materials, and simulations of experiments without a clay layer also showed close correspondence with measurements. In models with a clay layer, different clay strengths were investigated. Although the continuum formulation fails to simulate the breakup of the clay layer, models with higher strength clay show a much stronger thinning with increasing displacement (Fig. 26), especially when the clay layer is so strong that tensile stresses are generated.

3.2.3.5. Discussion of numerical simulations of clay smear

Numerical modeling of clay smear processes has been less intensively explored than laboratory experiments. Numerical studies have the advantage of detailed and systematic parameter studies. Interestingly, some of the same problems of scale (e.g., shear zone thickness and development of secondary Riedel shears in the shear zone), resolution, and boundary conditions which are present in experiments are present in numerical models as well. Thus, in many ways experimental and numerical modeling of clay smear are complementary, and it would be very helpful in the future to develop benchmark models which are investigated by both methods to allow comparison and cross-fertilization.

In numerical models it is less clear what aspects of clay are included in the description of the material – this is true for both FEM and DEM models. For example, in a DEM model one may include a reasonable description of macroscopic mechanical properties, while the particle size of the Discrete Elements hinders the effects of grain scale mixing. Or, in a FE model clay softening and localization may be included in the formulation without a correct description of the thin shear zones that develop in clay or of the slicing and fragmentation often observed in experiments. As in experiments, benchmarks to compare the results of different codes would be very helpful to evaluate the results of different calculations.

One problem that clearly requires more numerical work is the evolution of Riedel shears, which are yet to be produced in the numerical models discussed above. Mixing of sand and clay is modeled poorly with existing DEM models; a much larger difference in grain size between sand and clay and construction of 3D models are needed to allow clay particles to migrate through sand pore-throats. Failure mode in one of the layers has a major control on the results of numerical simulations, but very little systematic work has been done on the rates of reworking of brittle fragments during shearing in the fault core (i.e. representing an evolution in the failure mode during progressive shear deformation).

3.2.4. Discussion and summary: laboratory and numerical models of clay smear

We set out in this section to summarize the literature about fault processes that create and disrupt clay smear. There are a multitude of laboratory and numerical approaches applied to the problem, and each approach enjoys advantages and suffers limitations imposed by either a laboratory or numerical experimental design. No single approach describes the entire mechanical and kinematic evolution of a fault zone, but rather that understanding lies at the intersection of all these approaches where the physical insight contributed by each method is free of design limitations. Unfortunately there is uncertainty about the limitations of all of these methods.

An important process revealed by the existing laboratory experiments is heterogeneous shear to large strains. This plays an important role in developing layered gouge in sandbox experiments. Grain-scale mixing (grain-scale abrasion) is important in many experiments, as shown by microscopy. More coherent fragments of clay are also observed to disaggregate during shearing under sufficiently high mean effective stress, reflecting a strain-softening evolution.

Sandbox experiments also provide information about the initial localization processes, reflecting an early history of fault segment evolution as fault orientation adapts to the changing stress state and misaligned fault segments coalesce and evolve into a geometrically simpler fault zone, including fault relays and fault lenses. Secondary faults, such as Riedel shears, are less common in laboratory experiments and numerical simulations than they are in outcrops. Secondary fault structures may also contribute to the rupture of a clay smear and the development of holes, but experiments also reveal the development of holes early in the fault development if brittle clay fragments are fractured and plucked from the source bed. This idea of initial clay fragmentation and evolution into a clay smear via strain-softening processes is a more recent insight that has presumably influenced the development of probabilistic clay smear approaches as applied to subsurface predictions (e.g., Childs et al., 2007 and Yielding, 2012).

3.2.4.1. Proposed mechanical framework

We propose a matrix of deformation regimes that predict where clay smear may form and the processes that contribute to clay smear in each of those regimes (Fig. 27). The matrix is developed in terms of the relative properties of sand and clay; absolute properties are normalized to the stress state at the time of fault deformation. For example, an uncemented clay layer may fail in shear at low effective stress, while a stronger, stiffer, and more cohesive clay may fail in a similar deformation mode if effective stresses are high. In this way both the relative and absolute properties of sand and clay are considered in the context of the stress state of fault deformation.

Within this matrix (Fig. 27) we recognize that much of the existing work on clay smear is concentrated in the upper left-hand corner of the diagram where both clay and sand fail in shear and where clay is softer (weaker) than or comparable to the associated sand. If effective stress is relatively high and stresses in the clay remain compressive, the sheared clay is initially continuous with its thickness variable over the different segments of the shear zone. With progressive shearing a layered gouge will develop, potentially assisted by lateral fault migration, telescoping and an increasing amount of mixing. Ultimately the clay smear may become discontinuous by attenuation and additional processes like Riedel shears that produce a hole. If clay is very weak, lateral injection may locally increase the amount of clay in the fault.

However, the other parts of this matrix are also relevant in nature. For example, the lower left-hand corner may reflect a calcareous sandstone with some early calcite cement resulting in a high cohesion for the sandstone. The upper right-hand corner may reflect conditions associated with sand injections in which sand failure in shear is dominated by fluidization processes while clay with minimal but finite cohesion fails in extension at the low effective stress levels that arise with sand fluidization and mobilization. If the ratio of effective stress and strength are such that extensional fracturing leads to clay fragmentation, a fault zone with clay fragments embedded in sheared sand is formed. Continuous clay smears may form only after shearing and abrasion of the fragments, although the rates of erosion, abrasion and reworking of fragments to produce continuous gouge are very poorly known. The lower right-hand corner represents the case of faulting over-consolidated rocks that are sufficiently cohesive to fail in extension under the relevant effective stress.

We view this matrix as an initial framework of processes because we recognize that current understanding is incomplete. We expect future research will modify this matrix while in the meantime it provides a basis to motivate that research. Moreover, this framework can also serve as an initial basis for an evaluation of what conditions, expressed as failure mode, are best suited for different clay smear algorithm approaches.

3.2.4.2. Outstanding technical issues

There are four general areas where the current laboratory and numerical experimental approaches face limitations:

1. Erosion and reworking of fragments to produce continuous gouge. Reworking of claystone at sufficient mean stress produces clay at critical state, while reworking sandstone produces a cataclasite. The higher the initial cohesion, the greater the softening in the fault zone. Although the mechanical and transport properties of quartz/clay mixtures is known to some extent (Lupini et al., 1981, Crawford et al., 2008 and Tembe et al., 2010), the mechanics of the abrasion process are ill-defined, yet this is what determines the mechanical properties of faults with evolving clay-rich gouge. Faults with patchy and layered shear strength are probably the rule.

2. Scaling experimental results to nature: Careful reading of the literature on experimental and numerical modeling of clay smear often brings up the questions about the scale and level of detail one aims to understand in a study. Keeping in mind the enormous literature on fault zone evolution and our still rather incomplete understanding of this important domain, this question is important but often not well answered in the different studies. Considering a natural example of a normal fault zone with perhaps 20 m throw in a sequence of claystone and sandstone, it remains unclear which aspects of this fault zone are lacking in the present set of models. Moreover, fault zone thicknesses vary in nature and laboratory experiments, but this facet of the problem is rarely measured or reported.

3. Incomplete definition of fault processes: For example, rates of mixing of sand and clay are poorly modeled with existing DEM simulations because a much larger difference in grain size between sand and clay is

needed to allow clay to go through sand pore throats. Moreover, the effects of diagenesis, pressure solution, and other deformation mechanisms in clay play an unknown role in the evolution of clay smear, and little of this has been systematically investigated.

4. Fault geometric evolution: Before much progress can be made in experiments and numerical simulations of clay smear, a better understanding of the effect of model geometries and boundary conditions on the evolution of fault zone architecture is required. We need well-defined benchmarks to compare experiments and simulations. If a numerical model can predict the evolution of structures in sandbox experiments then perhaps upscaling to nature is possible.

Experiments and numerical modeling are complementary, and the deliberate pursuit of both in concert offers the opportunity to achieve greater confidence in the developing understanding of physical fault processes and the means to extend results beyond which the laboratory or computer easily allows. In addition, it is clear that the ability to compare mechanically layered models with homogeneous systems in order to understand the difference caused by the clay layer could yield substantial insights by better clarifying when the presence of a clay layer matters. Lastly, a mechanism for incorporating an improved physical understanding of fault processes needs to be implemented for predicting the effects of clay smear development for subsurface flow problems. Comparing existing algorithms against experimental results has limited value if the comparison is done without careful scaling and outside of the context of geometric and geomechanical boundary conditions.

3.3. Field studies

Field studies include the subsurface application of clay smear models to oil, gas, and water flow. Clay smear processes are among the most important mechanisms to alter cross-fault fluid flow because they introduce the lowest permeability component of a stratigraphic section into a fault zone that cuts that stratigraphy at high angles.

Field studies contribute to our understanding of clay smear products because fluid flow interrogates the entire fault zone, seeking out the most permeable part of a fault where clay smear is least developed or continuous. However, field studies are limited in that the clay smear process is generally inferred rather than observed, and other processes may yield similar results. Moreover, most of the reported subsurface field studies fail to document many of the critical observations necessary to reproduce the presented interpretation (a criticism levied by those attempting to perform clay smear calibration studies; e.g., Bretan et al., 2003). While knowledge has been gained by the field studies reviewed here, there is the potential to learn far more from well-designed, comprehensive subsurface studies as two recent examples illustrate.

3.3.1. Definition of cross-fault flow

A common usage in the literature and the petroleum industry is to describe faults as either 'sealing' or 'leaking.' Seals are well defined when capillary properties in fine-grained rocks create no-flow boundaries for non-wetting oil and gas fluids (e.g., Urai et al., 2008) and lead to subsurface oil and gas accumulations (Fig. 28). Failure to hinder oil and gas migration is caused by 'leaking' faults that fail to trap 'significant' oil or gas accumulations.

Unfortunately the seal/leak terminology is also loosely applied to cross-fault flow problems when wells create fluid potential differences that lead to fluid flow (Fig. 28). In this instance a fault becomes sealing when flow rates at a well are deemed 'insignificant,' a threshold that may be problem specific. As a practical matter if often turns out that substantial differences in well flow rates are achieved with a limited range of fault flow properties, but it is nevertheless more informative to consider the problem of cross-fault flux in terms of a continuous spectrum of flow properties rather than the binary leak/seal properties useful for the capillary seal problem (Fig. 28).

Most workers (e.g., Yielding, 2012) agree that the geometry of juxtaposed permeable intervals is the primary control on cross-fault flow as Smith (1980) originally proposed, but where debate remains is in reconciling the apparent ubiquitous presence of clay smear in some fault zones with the inherent uncertainty in juxtaposition geometries. For the case of capillary seals in fault zones, two approaches developed: (1) a probabilistic method that simulates many fault juxtaposition geometries in a complete trap system of leaks and spills (James et al., 2004); or (2) a method using one of the clay smear algorithms (Fig. 3) 'calibrated' for local oil and gas accumulations. The result is a discrepancy in the area of a fault juxtaposition window that may be covered by a capillary clay smear seal (Fig. 29). We think that juxtaposition uncertainty has been under-appreciated in the literature and thus we infer only a small contribution of capillary seal from clay smear (Fig. 29). This distinction is important because it bears on the consideration of 'calibration' studies, although Dee et al. (2007) have shown that it is possible to tune analyses to yield similar results by both approaches.

Fig. 29 also illustrates how a juxtaposition area beyond the fraction covered by a capillary clay smear gouge, clay smear has a variable and irregular distribution of flow properties. It is within this conceptual framework of uneven effects of clay smear developed in a fault zone that we approach the Field Studies section.

3.3.2. Field studies outline

Field studies incorporate the full spectrum of flow from static, no-flow conditions where clay smear acts as a capillary seal, to full 3D flow studies where clay smear restricts cross-fault flow of oil, gas, and water fluids that is focused along stratigraphic beds (Fig. 28). For capillary seal studies the inference of clay smear products is derived from a stratigraphic model of the distribution of sandstone and shale lithologies and a structural model that defines the position and offset of normal faults that displace the stratigraphic section (Table 4). Measured subsurface data used to infer the effectiveness of the clay-smear capillary-seal include oil and gas column heights, wireline saturations of oil and gas from well logs that define the spatial distribution of fluid types in the subsurface, and wireline pressure data used to define fluid pressure with depth in respective fluid types (Table 4). Sometimes core materials are available with faults in them, and the flow properties of the sampled faults are analyzed in the laboratory (Table 4).

Table 4. Summary of subsurface field studies and data types applied.

Reference	1D strat model[a]	3D strat model[b]	3D seismic data[c]	Oil & gas column heights[d]	Wireline saturations[e]	Wireline pressures[f]	Fault core & characterization[g]	Single well tests[h]	Well interference tests[i]	Downhole pressure history[j]	Well rate history[k]	Mass balance calculation[l]
Bentley and Barry, 1991*	Y	y	Y		Y	?	?	?	?		y	Y
Berg and Avery, 1995	Y			Y	Y		Y					
Bouvier et al., 1989	Y		Y		Y							
Childs et al., 2002	Y		Y		Y	Y						y
Childs et al., 2009b	Y	Y	Y		y	y						
Davies et al., 2003	Y		Y	Y	Y					Y		
Ottesen Ellevset et al., 1998	Y		Y	Y	Y		Y					
Fristad et al., 1997	Y		Y		Y	Y	Y					
Gibson, 1994	Y		Y	Y	Y		Y					
Gibson and Bentham, 2003	Y		Y	Y	Y	Y						
Harris et al., 2002	Y		Y		Y	Y						
Jev et al., 1993	Y		Y		Y	Y						
Jolley et al., 2007*	Y	y	Y		Y					Y	Y	Y
Knai and Knipe, 1998*	Y		Y				Y			Y	y	Y

Koledoye et al., 2003	Y		Y			Y						
Myers et al., 2007*	Y	y	Y		Y	y				Y	Y	Y
Nybakken, 1991*	y		y		y	Y						
Sassi et al., 1992*	Y	y			Y		Y	y				
Smith, 1980	Y				Y					Y		
Speksnijder, 1987			Y									
Sverdrup et al., 2003*	Y		Y			Y	Y		y		y	y
Weber et al., 1978	y				Y							
Wehr et al., 2000	Y	y	Y		y		Y				y	Y
Welbon et al., 1997	y		y		y	y						

*In these papers the 'y' designation indicates that this data type likely exists based on figures presented (e.g, a cross-section suggests that 3D seismic data may have been available), but those original data remain unverified in the publication (i.e. unable to independently reproduce the result).

[a] Stratigraphic model derived from petrophysical data and cores collected in a vertical borehole.

[b] Stratigraphic model based on petrophysical data and seismic attributes, for example, used to condition interpretations away from wellbores.

[c] Primary issue is data resolution, which depends on rock physics, data acquisition parameters, and data processing steps; latter two improve with time.

[d] Height between crest of column and fluid contact; depth uncertainties usually unspecified, and fluid types sometimes unrecorded.

[e] Petrophysical data necessary to define gas, oil, and water saturations at all available well depths; fluid contacts included if encountered in well.

[f] Down-hole pressure profiles defined in newly drilled well; critical issues is vintage of gauge, which affects pressure precision and accuracy.

[g] In best case includes cores of small-scale and large-scale (seismically imaged) faults; complementary wellbore dipmeter and image logs.

[h] For example, Drill Stem Test (DST) in which pressure recovery is evaluated in same wellbore as flow has occurred (extraction or injection).

[i] Requires two wells, preferably located nearby across fault; one well monitors pressure while other well produces or injects fluid.

[j] In best case pressure gauges installed near reservoir interval monitor pressure (Flowing Bottom-Hole Pressure) during production and injection, especially when wells shut-in (Shut-in Pressure); otherwise derived from well-head pressures.

[k] Record of volumes of fluid produced from or injected into a well (daily basis); in best case well section open to flow is limited; may include 4D seismic results of changing sub-surface saturations or well tracer studies.

[l] Model used to evaluate effects of produced or injected volumes on resulting pressure history; models can be simple box-models or 3D simulations.

Flow studies that include cross-fault flow provide information on the aggregate permeability of a fault zone for Darcy flow (Fig. 28), a potentially valuable addition to the information provided through the capillary seal analysis. However, flow solutions are non-unique (e.g., Jolley et al., 2007), and it is difficult to find well-conceived cross-fault flow experiments that reduce the non-uniqueness of the solution. Although studies that incorporate cross-fault flow add further useful data (e.g., single and multi-well tests and downhole pressure records; Table 4), the resulting clay smear inferences are also dependent on the analytical methods used to analyze the data, ranging from analytical models used to interpret individual well tests to box-based mass-balance calculations to full 3D flow simulations (Table 4).

In this section we use the framework of observations and analyses defined in Table 4 to evaluate the literature devoted to field studies, and in particular clay smear calibrations based on this literature. An important perspective to maintain through this section is the sometimes competing interests and demands of commercial application of the clay smear concept and the scientific prerequisites that give us deeper insight into fault processes.

3.3.3. Clay smear interpreted as a capillary seal

Clay smear was originally applied to problems of capillary or membrane seals in which a continuous veneer of gouge maintains oil or gas saturation below a critical saturation threshold that is necessary for percolation (Fig. 28). Oil and gas on either side of a fault exist at different pressures at the same elevation, and if water is in pressure communication across the fault, contacts between different fluid phases will form at different elevations. This was the primary application considered by Weber et al. (1978)for trap considerations in the Niger Delta and Smith (1980) in the U. S. Gulf Coast. We infer that in both of these studies the stratigraphic framework applied at the analyzed faults was based on 1D models derived from well logs and that the fault framework was based on 2D seismic data of unknown spacing combined with fault interpretations based on missing section in wells (Table 4). An important component of these early studies is that they relied to a large extent on fluid contacts logged by petrophysical measurements because this was the most ubiquitous data type available. The value of pressure data was likely well appreciated but limited to rare drill-stem and production tests; no pressure data are reported in the early works.

Bouvier et al. (1989) offered the next advancement in the application of clay smear to subsurface field studies by introducing 3D seismic data into the analysis. Using newly developed interpretation and visualization techniques, the critical issue of fault continuity and cross-fault reservoir connectivity emerged, at least as expressed in seismic time slices. Fault juxtaposition diagrams were constructed using seismic sections extracted in the fault footwall and tied to well-log stratigraphic profiles. The effects of clay smear were overlain on this geometric representation by lumping calculated CSP values into high, medium, and low categories. Interestingly, higher CSP values are calculated where the proportion of shales in the stratigraphic section increases, and as the authors admit the chance for reservoir juxtaposition against shale increases (Bouvier et al., 1989). This is a conundrum that occurs in many, if not most, clay smear field studies.

Berg and Avery (1995) presented an interesting additional perspective in their discussion of complex fluid contact relations in the U. S. Gulf Coast. They used dip-meter data to document significant dip deviations in shale in fault zones. Although they describe this behavior as clay smear, the consistency of dips within limited intervals suggests tilted but coherent intervals of shale (lithology based on SP and resistivity curves). In a current context it might be more appropriate to think of these tilted, shale-dominated intervals as fault relays or lenses (e.g., Childs et al., 1997), a particular component of clay smear products.

Gibson (1994) used core, outcrop, and fluid contact data (defined as column heights) to infer clay smear processes for a number of gas and oil accumulations in the Columbus Basin of Trinidad. This paper is a good example of the part of the literature that seeks to apply clay smear models in order to reduce oil and gas exploration risk, but the insights it adds to clay smear processes are limited. Notwithstanding these comments, Gibson (1994) contributed to and further supported the idea that clay smear continuity is the single most critical aspect when applying a clay smear model. Moreover, Gibson's contribution is unique in that it introduced the idea of clay smear variability in a fault zone and explored that variability in the context of explaining traps that contain both oil and gas. Gibson (1994) realized that in order to trap both oil and gas in a fault-bounded trap, gas fill must be limited by capillary leak at the highest elevation of the fault-dependent trap component, precisely defining a critical pore throat diameter to limit the amount of gas trapped, while oil spills across the fault at the oil-water contact where capillary resistance is negligible. It is interesting that this insight appears to have been largely overlooked in subsequent field studies.

Similarly, Ottesen Ellevset et al. (1998) made inferences about fault properties based on core observations and offset gas-water contacts in the Sleipner Vest Field, Norway. Without incorporating pressure data to evaluate whether different contacts were the result of fluid segregation in the gas or water phases, these authors deduced fault compartmentalization in the gas column and inferred capillary fault properties, including clay smear, based on this assumption. However, in this field it is also plausible to interpret the offset contacts as a result of disconnected water bodies (Paul Hicks, personal communication to P. Vrolijk, 1997), an interpretation that alters the inferences of fault rock properties and continuity.

3.3.3.1. Clay smear interpretations including pressure data

In the 1990's the more common availability of wireline pressure data found its way into clay smear interpretations for subsurface studies. As described above, this development was important for helping to uniquely establish which fluid phases are connected and which are segmented across a fault. For example, Nybakken (1991) identified pressure differences in the Tampen Spur area of the North Sea, which he attributed to clay smear and injection as the most effective processes that create those pressure differences. However, Nybakken (1991) only inferred the effects of clay smearing in a qualitative sense; there is no attempt in this paper to select a clay smear model, apply it to all faults, and check for the internal consistency of all resulting interpretations (i.e. are capillary leaks and seals interpreted for the same value of clay smear quantity, like SGR?).

Welbon et al. (1997) combine the effects of shale smear with other deformation mechanisms like cataclasis and a generalized reservoir connectivity probability based on a paper by Knott (1993) into a 'Fault Seal Probability' product. Although the appeal for combining the results of so many complex processes into a single probability parameter is apparent, it also makes it difficult to isolate which processes are most important along any particular fault segment. Welbon et al. (1997) attempt to overcome this limitation by independently producing a shale smear map for one fault in their study. In constructing this map they appear to use a limiting SSF of 7 as defined by Lindsay et al. (1993) but for which other authors at this point had begun to suggest that this critical value is actually much lower (e.g., Gibson, 1994).

Pressure data were first used in a systematic way to evaluate cross-fault pressure differences by Fristad et al. (1997) and Yielding et al. (1997). In this approach wireline pressure data are used to define pressures in permeable sandstone intervals on both sides of a fault (in the context of a defined fluid phase), and those pressure differences compared with a spectrum of SGR values calculated on the entire fault surface (Fig. 30). Note that this approach relies on the definition of fluid pressure models that may be either loosely constrained with a single pressure measurement and an independent definition of fluid density (i.e. $\Delta p/\Delta z$), or well constrained with many pressure measurements. The resulting cross-plots of SGR versus pressure difference (Fig. 30) is then used to infer a capillary leak point and its capillary properties. A bounding envelop, typically drawn as a straight

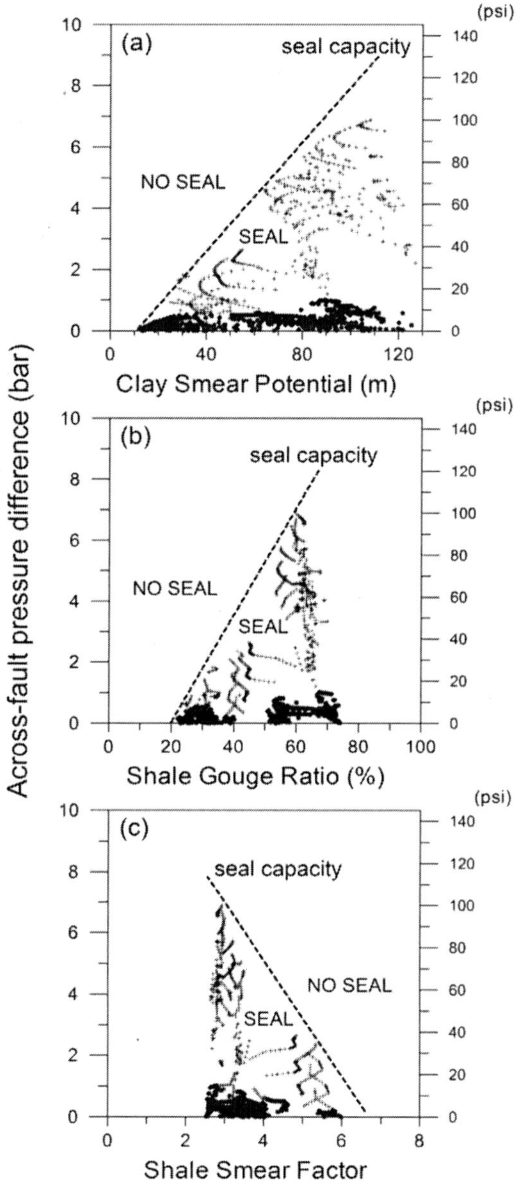

Figure 30. Calculated cross-fault fluid pressure differences for various representations of clay smear, including (a) CSP, (b) SGR, and (c) SSF. Plots developed to provide empirical definition of fault capillary seal for modeled clay-smear parameters. Reproduced *with permission* from Yielding et al. (1997).

line in SGR-pressure difference space, defines a calibrated leak envelope. In both of these publications, this approach was initially applied to the complexly faulted Oseberg Syd field in the Norwegian North Sea, and Yielding et al. (1997) extended the analysis to the Nigerian Nun River field (Bouvier et al., 1989) and accumulations in Trinidad described by Gibson (1994).

While this approach appears attractive in that it attempts to incorporate information across an entire fault zone, several questions arise. First, there is no *a priori* reason why entry pressure should increase with increasing SGR given experimental (Crawford et al., 2002) and theoretical information (Revil and Cathles, 1999), both of which achieve the greatest permeability reduction with incorporation of up to 30–40% clay and little or no further permeability reduction with additional clay. Indeed, debate on whether this effect exists is ongoing (e.g., Childs et al., 2009b and Davies et al., 2003). Second, threshold SGR values (i.e. the lowest SGR values that support a cross-fault pressure difference) tend to occur at small pressure differences, as it does in Yielding et al. (1997). Small pressure differences are difficult to evaluate out of context, especially in light of the evolution of pressure gauge precision and accuracy that occurred over the period of these studies. Remarkably, the issue of accuracy continues today even with high precision and accuracy pressure gauges when considering pressure differences of 1–20 kPa because depth uncertainty yields that much error (e.g., an apparent 10 kPa pressure difference results from a 10 m depth error applied to a water gradient of 10 kPa/m). Caution is warranted when seeking significance of SGR points defined for small pressure differences (Fig. 30). Fewer data define the bounding envelope at higher SGR values, and it is valid to ask whether alternative geologic interpretations might account for the highest values (e.g., juxtaposition seal), thereby shifting the bounding curve to higher SGR values.

The development of this technique by Fristad et al. (1997) and Yielding et al. (1997)remains a significant contribution to the application of clay smear effects in the subsurface, and in spite of the reservations expressed above, this comprehensive approach of mapping SGR and pressure difference values across an entire fault zone succeeds in testing the logical, internal self-consistency of the interpretation – there can be only be one unique leak point for each hydrocarbon phase trapped – a check that other studies have failed to complete. This fault mapping approach is a step in the right direction toward building an internally consistent interpretation, but it should be expanded to the entire field using a compartment approach like that described by Vrolijk et al. (2005a) in which the fate of any oil or gas that leaks out of any single compartment must be traced to the next downstream compartment until it exits the field completely. Does the calibration of SGR on one fault and applied to another fault in the field cause the trapped oil to back up and fail to reach the field spill point? Sometimes the interpretation of leakage is as critical as the interpretation of a capillary seal.

This discussion helps illustrate the point made at the outset of this section that in the absence of comprehensive field studies, there are always questions that arise that limit the value of a calibration formed on a subset of the field data. Critical questions invariably arise about the input data for the analysis that render at least parts of that analysis irreproducible.

3.3.4. Clay smear defined by core analyses

While broad-scale field studies were being undertaken in the 1980's and '90's, another line of investigation opened up based on the examination of faults encountered in cores (e.g., Knipe, 1992). The basic methodology and resulting fault classification scheme was defined early on (Fisher and Knipe, 1998) so there is a reasonable uniformity found in the literature. Although it is difficult to discern from individual papers, it is likely that most faults encountered in core were sampled accidently rather than planned in the context of a scientific experiment (i.e. a specific, testable hypothesis identified before drilling and tested by the sampled core), and Hesthammer and Fossen (2000) discuss potential sampling biases encountered in a detailed study of the Gullfaks Field.

The analysis of various fault samples from core has resulted in vast databases of fault properties and characterization data, most of which are proprietary and unavailable to the scientific community. However, this characterization effort also brought an additional graphical tool for evaluating clay smear among other fault rock types: the triangle diagram of Knipe (1997) that evolved from Bentley and Barry (1991). This type of display attempts to allow consideration of a much richer geologic description of offset lithologies and resulting fault rocks than the simple, binary seal/non-seal classification. In particular, this approach defines a possible hybrid clay smear in a phyllosilicate framework fault rock (PFFR; Gibson, 1994 and Knipe, 1992), a rock type containing abundant sand grains but which may reflect the flow properties of a clay-dominated clay smear (Knipe, 1997). Although the PFFR type is conventionally interpreted as a product of faulting clay-bearing sandstones (Gibson, 1994 and Knipe, 1992), we think the experimental evidence of sand-clay mixing (Schmatz et al., 2010a and Schmatz et al., 2010b) offers an additional mechanism to achieve the same product, especially if sand has deformed by independent particulate flow, an issue broached by Freeman et al. (1998). Note that the limit of 40% clay used to define the PFFR type is at the boundary where the addition of more clay yields little or no additional permeability decrease (Crawford et al., 2002).

While the triangle diagram approach defines juxtaposition windows and gouge types across all possible fault offsets, the analysis relies on a deterministic stratigraphic section, often defined at an existing wellbore whose location was likely chosen for reasons other than its utility for defining a stratigraphic section for fault-seal analysis.

There are two principle limitations in the application of core sample results to the problem of cross-fault flow influenced by clay smear:

1. Core samples offer little insight into the critical aspect of clay smear – clay smear continuity. Although it is possible to measure in the laboratory the capillary and permeability properties across a clay smear, extrapolation of those properties away from the core requires a model that core observations lack the ability to test comprehensively.
2. Core samples come from proprietary cores, and results are reported in the literature as either 'typical' results without attribution to a specific fault or location in the Earth, or they are amalgamated into large databases of fault rock properties. Even when sample analyses are reported in the context of a field study (e.g., Ottesen Ellevset et al., 1998), the results are presented lacking the context of the fault from which the samples were derived. Gibson (1998) sought to overcome this issue by presenting data tables that report results on an individual sample basis.

The primary contribution of core studies to the clay smear problem is to define a trend of decreasing permeability with increasing clay content in the fault (e.g., Gibson, 1998,Manzocchi et al., 1999, Fisher and Knipe, 2001 and Sperrevik et al., 2002). While the issue of clay smear continuity is recognized in these studies (e.g., Fisher and Knipe, 2001), there are no unique observations made from core samples to address this issue.

3.3.5. Modern clay smear field studies – the rise of uncertainty

Since the late 1990's, the data types with which modern subsurface studies are performed became routine – 3D seismic data; abundant accurate and precise wireline pressure data; complete wireline log characterization (including gamma ray, neutron porosity, bulk density, resistivity, and corresponding derivative Vshale, saturation, and porosity curves); and a myriad of tools for the evaluation and integration of diverse data types. Complete integration of these diverse data in order to achieve a logical, comprehensive interpretation of multiple oil and gas columns in a complexly faulted stratigraphic section can prove daunting. In an example from the Exmouth basin of Australia, Bailey et al. (2006) consider clay smear (high SGR) in the context of other trap leak controls and attribute a spectrum of mechanisms for the observed fluid type and pressure distributions.

A pair of studies by Childs et al. (2002) and Davies et al. (2003) consider a similar case of gas and water at the same elevation across a fault combined with a large difference in water pressure across the fault. In the Tune Field (Norway) example discussed by Childs et al. (2002), gas occurs in the hanging wall and is connected to an aquifer with a much higher pressure than the aquifer in the footwall. In contrast, Davies et al. (2003) discuss a Gulf of Mexico example in which the higher aquifer pressure is observed in the footwall while gas is trapped in the lower pressure hanging wall. In both of these studies, however, an attempt is made to account for the gas distribution and water pressure offset across the fault by interpreting a cross-fault water flux. Childs et al. (2002) model the relative permeability of water through the gas cap where the water saturation is above the irreducible water saturation value. Davies et al. (2003) interpret water flow in the opposite direction (i.e. from the higher pressure, water-saturated footwall into the gas column in the hanging wall) but without the corresponding calculations to illustrate the feasibility of this interpretation. And although Davies et al. (2003) acknowledge data uncertainty in their analysis, they dismiss the impact of that uncertainty on their conclusions without reporting the work done to reach that conclusion.

This interpretation of faults that form a capillary seal while allowing the flux of water through the gas column is interesting in that it has potentially significant implications for pressure histories during production. For example, the Tune Field example by Childs et al. (2002) suggests that pressure depletion during gas production will be offset to some degree by aquifer support, an interpretation that should be apparent in plots of pressure normalized for gas compressibility (P/z) as a function of time and through the early occurrence of water break-through. Moreover, as contacts rise, water saturations and the relative permeability of water will also increase, resulting in an increasing flux of water across the fault and a rapid pressure decline that offsets the aquifer support.

Other authors more directly addressed two types of uncertainties in predicting subsurface clay smear development: (1) log definition of the clay fraction in a faulted stratigraphic section (e.g., Bretan et al., 2003) and the resulting threshold value of a clay smear parameter like SGR that results in a capillary seal; and (2) detailed fault zone definition using core data and high resolution seismic reflection data (e.g., Childs et al., 1997, Faerseth et al., 2007, Hesthammer and Fossen, 2000 and Koledoye et al., 2003). In certain settings challenges arise in defining the stratigraphic section. For example,Bretan et al. (2003) claim that: 'Different vintages of V_{shale} analysis of the same well by different petrophysicists working in the same company can be alarmingly different.'Bretan et al. (2003) then go on to apply a revised method for determining the input V_{shale}value (with respect to the original work reported by Fristad et al., 1997), derive different SGR values from a V_{shale} curve that includes 'mica and kaolin' (Bretan et al., 2003), and define a higher SGR value necessary to achieve a capillary seal. Childs et al. (1997)illustrate how various natural fault zone complexities, including multiple slip surfaces that define fault relays and lenses, result in different interpretations of cross-fault connectivity, the fundamental basis necessary for considering the effects of clay smear. Hesthammer and Fossen (2000) discuss additional detailed fault complexities that affect cross-fault connectivity based on core description studies in the Gullfaks Field, and Koledoye et al. (2003) document how paired slip surfaces identified in high resolution seismic data from Nigeria promote the incorporation of shale in the fault zone, perhaps in the manner of fault-bounded, coherent shale lenses in the context of Berg and Avery (1995). These ideas are further developed by Faerseth et al. (2007) in which they attempt to define a methodology for accounting for these uncertainties within a risk framework.

The problem with recognizing these uncertainties is that solutions for addressing them are complex. Attempts to generalize complexity (e.g., Knott, 1993) lack the necessary detail for addressing questions related to individual faults. Specifically, Childs et al. (1997) suggest that: 'The distribution of fault rocks cannot be characterized from well data, raising the question of whether purely deterministic methods for fault seal prediction can ever be successful.' This kind of doubt ultimately led to efforts to systematically explore likely uncertainties (e.g., Childs et al., 2009b, Manzocchi et al., 2008a, Manzocchi et al., 2008b and Manzocchi et al., 2008c).

On the one hand this emerging realization of the impact and importance of uncertainty in the input parameters and its impact on the resulting clay smear calculations presents a sobering picture. On the other hand this period is also marked by studies attempting to 'calibrate' clay smear calculations against sub-surface observations using the methods of Yielding et al. (1997). From papers like Bretan et al., 2003, Bretan and Yielding, 2005 and Harris et al., 2002, and Yielding (2002) there arises a definition of SGR values above 15–20% necessary to create a capillary seal. Specifically, there are examples of low SGR where there is little evidence of a capillary seal and others with high SGR values that are associated with a capillary seal. This conclusion is further supported by the uncertainty study undertaken by Childs et al. (2009b). The transition region around an SGR value of 20% remains problematic, however. Does the resulting threshold entry pressure increase with increasing SGR? Over what range of SGR values does this transition occur? While exploring some uncertainty parameters (i.e. it remains challenging to incorporate alternative fault geometry interpretations), Childs et al. (2009b) conclude the following: 'The data also demonstrate that the level of fill in individual traps can be extremely sensitive to minor variations in the seal capacity envelope.'

Further complicating matters, Dee et al. (2007) used a clay smear approach to evaluate a field study described by James et al. (2004) in which a stochastic approach to evaluating juxtaposition geometries was employed. Dee et al. (2007) produced a similar gas column distribution to the one originally described by James et al. (2004). A logical conclusion from this result is that both approaches are an imperfect description of subsurface geology and that both predict faults that form capillary seals when shale is relatively abundant in the faulted stratigraphic section and the absence of seals in shale-poor sections. Indeed, when more comprehensive analyses of stratigraphic uncertainty are undertaken (e.g., Clarke et al., 2006 and Hovadik and Larue, 2010), it seems likely that at least some documented examples of capillary seal from clay smear might better be attributed to uninterpreted stratigraphic complexities. Consequently, it is difficult to glean deep physical insights into clay smear processes from the existing field studies literature that considers clay smear as a capillary seal.

3.3.6. Clay smear under transient, production-scale flow conditions

An alternative application of clay smear predictions is for the interpretation of cross-fault flow during the hours to years time-scales common to fluid production or injection. In these studies the effects of pressure transients introduced into the subsurface by wells producing or injecting gas, oil, or water fluids is monitored and the results interpreted in terms of aggregate permeability properties of intervening faults. With the increased number of parameters needed to define such a system (Fig. 28), an additional strategy is required to model the effects of faults in flow simulations (e.g., Manzocchi et al., 2002). AsManzocchi et al. (2010) point out, evaluation of the flow effects of clay smear are complementary to those of capillary membrane seals because: 'fault seal capacity relies on outliers of fault property distributions (the sealing capacity of a fault is controlled by the weakest point) while across-fault Darcy flow is controlled by average properties.' Interestingly, this facet of the literature is dominated by field studies from Mesozoic reservoirs in the North Sea with a minor, secondary contribution of older literature from the Niger Delta.

Speksnijder (1987) commented for the first time on the effectiveness of clay smear for cross-fault flow in the Cormorant Field of the North Sea. Speksnijder (1987) focused on the complex fault interpretation revealed by new 3D seismic data, and the impact of that interpretation on cross-fault flow is only briefly defined in the conclusion of the paper for CSP values that are apparently associated with production performance data, none of which is presented.

Fault patterns in the Cormorant Block IV field were evaluated by Sassi et al. (1992) in order to define the impact of various size groups of normal faults (defined in terms of throw) on flow within the Brent Group. Note that in their interpretation, cross-fault flow restriction only becomes important on medium-scale faults (throw 10–50 ft; 3–15 m), and only in the thin-bedded Ness Formation where the effects of juxtaposition seal and clay smear are hard to differentiate.

In a detailed study of a trap-bounding fault in the Niger Delta, Jev et al. (1993) reported multiple independent oil columns accumulated in a stacked deltaic reservoir sequence. In the younger, more shale-prone interval of the stratigraphic section, higher CSP values are calculated than in the deeper, older, more sandstone-rich part of the section. Based on this geologic description, Jev et al. (1993) were able to rationalize a capillary seal for the oil accumulations in the upper intervals while attributing water flow across the fault in the deeper sections resulting from oil production and pressure depletion in the neighboring Cawthorne Channel Field; water flows from deeper unproduced intervals (Akaso G reservoirs) across the fault to replenish the fluid withdrawn by production (Cawthorne Channel E reservoirs). Jev et al. (1993) also recognize the value of detailed reservoir simulation studies for unraveling the details of this kind of fluid pressure depletion response but show the value of initial geologically comprehensive analytical evaluations.

While the pressure history that arises from production fluid extraction yields additional insight into the permeability properties of a fault zone, those properties can only be inferred with additional information about production rates, connected pore volumes, the properties of the fluids contained within those pores (e.g., compressibility), the timing and history of all wells either producing or injecting fluid, and additional case-specific parameters. The calculated pressure histories introduced by Fulljames et al. (1997)begin to illustrate the complexity of even a simple pressure response. Rivenaes and Dart (2002) suggest that capillary effects maintaining higher water saturation in fault zones further reduce the relative permeability of a fault, adding another dimension to the problem. Fisher and Jolley (2007) further developed the idea of relative permeability in a fault zone. Ultimately the use of a geocellular reservoir simulation model was introduced to help keep track of these various parameters and dependencies.

Bentley and Barry (1991) provide the first application of a reservoir simulation study of a faulted reservoir, also basing their work on Cormorant Block IV. They describe different intervals of CSP into clay smear 'type panels,' which they used to classify different types of cross-fault flow. Interestingly, they comment that 'the complexity of the structural pattern and production history from Block IV rule out the option of simple calibration using production data' (Bentley and Barry, 1991), although they did iterate on leak paths within the model in the context of a well history match to define a threshold CSP = 5 for fault 'sealing' (which is interpreted to be negligible cross-fault flow rather than no-flow). Subsequent work by Wehr et al. (2000) on the same field investigated the flow effects of various fault gouge scenarios, including clay smear (CSP), on reservoir flow simulation models compared against water-cut data from production wells. Wehr et al. (2000)advanced the simulation approach toward the current practice of applying different fault transmissibility values to different parts of a fault guided by an independent geologic interpretation and illustrated the efficiencies gained in the reservoir simulation and history-matching process given *a priori* definition of fault gouge properties and distribution.

As reservoir simulations improved through evolution of raw computing power, it ultimately became possible to treat and evaluate a full continuum of fault properties in a comprehensive reservoir simulation. Jolley et al. (2007) provided one of the first examples in an evaluation of the Brent Group fields in the North Sea. They recognized how hard it is to construct the faulted reservoir framework needed to evaluate the flow effects of fault gouge, including clay smear, and identified the definition of the fault framework as a 1st-order sensitivity for the modeling process. Jolley et al. (2007) also describe the application of fault properties via the assignment of transmissibility multipliers; the methodologies for defining transmissibility multipliers are described in other literature (e.g., Fisher and Jolley, 2007 and Manzocchi et al., 1999). An important outcome of this study is the conclusion that with the time and care invested in constructing a viable reservoir framework and the systematic application of transmissibility multipliers based on geologic principles and clay smear models, the path to achieving an acceptable history match is far more efficient.

Many similar conclusions were reached by Myers et al. (2007) in a reservoir simulation study of the Ringhorne Field, Norway: the fundamental necessity of accurately constructing the faulted reservoir framework in the simulation model and the vast advantage in the history-matching exercise of starting with an array of geologically-defined fault transmissibilities that reflect clay smear properties. An important distinction in the approach used by Jolley et al. (2007) and Myers et al. (2007) is the way in which the clay smear product (e.g., SGR) is transformed to permeability. Jolley et al. (2007) used an empirical transform that relates fault permeability to shale fraction as expressed by SGR. In contrast, Myers et al. (2007) implemented a permeability calculation based on the geometry of the clay component in an aggregate fault zone consisting of multiple gouge types. Myers et al. (2007) also used this approach to explore alternative fault permeability scenarios for different material property definitions arising from different burial and faulting histories (Fig. 31). This approach represents a significant development in that it for the first time attempts to link a fault-process model with a resulting cross-fault flow calculation. Further information about how these two approaches (empirical permeability definition versus process-based evaluation) relate to the flow problem is outlined in the text inset.

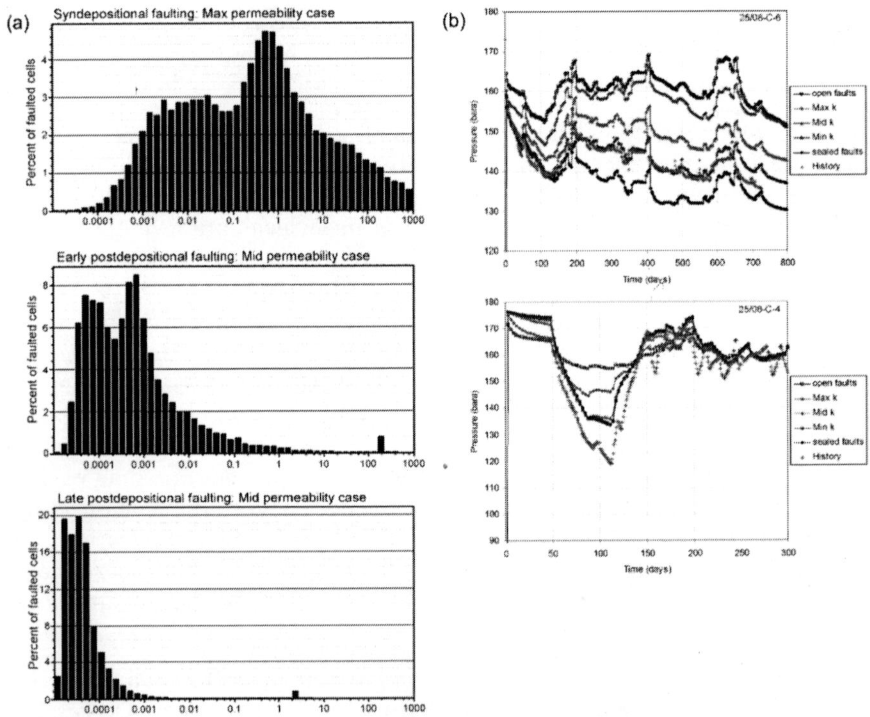

Figure 31. Illustration of how modeled effects of different failure modes, expressed here as burial depth of faulting, result in variable fault permeabilities and resulting producing well pressure histories. Note that history match was only possible after various other geologic problems in model were identified and rectified. (a) Histogram of fault permeabilities (milliDarcies) applied across entire model for three faulting scenarios: syndepositional, early post-depositional (shallow), and late post-depositional (deep) faulting. (b) Flowing bottom-hole pressure records for two producing oil wells in model compared against calculated pressures for three fault histories described in (a) as well as fault-open and fault-closed end-member scenarios. Reproduced *with permission* from Myers et al. (2007).

Even when reservoir simulation is used to test models of fault properties, the results are non-unique given the large number of degrees of freedom that remain in the model (i.e. multiple combinations of parameters consistent with measured data produce the same fluid pressure and rate histories). One way to further reduce the degrees of freedom in a model is with an independent data type, and Sverdrup et al. (2003) have shown how the application of 4D (time-lapse) seismic data are used to track the spatial movement of gas and water fluids and the corresponding pressure changes between repeat surveys. In particular, the amplitude difference map in Sverdrup et al. (2003; their Fig. 11) is a beautiful illustration of gas pooling against a fault and begs the questions of whether a sandstone connection exists at that location to support potential gas leakage across the fault, and if so, whether the gas is pooling in a small attic above the juxtaposition connection. If both of those geometric criteria are met, then cross-fault flow is restricted by fault gouge, possibly a clay smear.

Clay smear algorithms have the following units basis:

- $$SGR = \frac{\sum Vcl \times \Delta z}{throw} * 100\% \quad \textit{dimensionless length} \tag{2}$$
- $$CSP = \sum \frac{thickness^2}{distance} \quad \textit{length} \tag{3}$$
- $$SSF = \frac{throw}{thickness} \quad \textit{dimensionless length} \tag{4}$$

Insofar as these parameters are used to evaluate cross-fault flow, we compare them via Darcy's Law to properties that resist flow (permeability or hydraulic conductivity) which have units of L^2 and L/t, respectively. In this context it is possible to see SGR and SSF as proportionality constants that could be applied to the permeability definition (Hubbert, 1940):

$$k = Cd^2 \tag{5}$$

where C is the proportionality constant that further modifies *SGR* or *SSF* parameters (*d* is the diameter of uniform glass bead particles in Hubbert's formulation); both define properties of the medium through which flow occurs. *CSP* values with a length dimension may require additional information from the taper geometry to become dimensionless.

Attempts to relate some measure of clay smear to flow have used *SGR* as a non-linear proportionality constant in a permeability definition:

$$\log k_f = 0.4 - 4SGR - 0.25 \log(D)*(1 - SGR)^5 \tag{6}$$

$$k_f = a*SGR^{-b} \tag{7}$$

$$k_f = \left(F_{vsh}*k_{sh}^{(1-2Cf)} + (1 - F_{vsh})*(k_{ss}*k_{cat})^{(1-2Cf)}\right)^{-(1-2Cf)} \tag{8}$$

In all cases k_f = fault permeability. Equation (6) is from Manzocchi et al. (1999; arithmetic average permeability); *D* = fault throw (m). Equation (7) is from Jolley et al. (2007) where a and b are constants from k_f – clay % transforms. Equation (8) comes from Myers et al. (2014), and F_{vsh} = volume fraction of shale in the fault, k_{sh} and k_{ss} represent shale and sandstone permeabilities, respectively, and Cf represents a clay smear continuity factor.

Notwithstanding the success of these history-matching exercises, the issues of geologic uncertainty are still prominent for reservoir simulation problems, not least for models constructed to predict reservoir performance before production wells are drilled. In particular, some of the geometric uncertainties explored by Manzocchi et al. (2010)warrant careful thought when considering cross-fault flow.

3.3.6.1. Aquifer flow studies

Additional information about cross-fault flow is available from aquifer studies. Darcy flow is simpler in aquifer studies with a single fluid phase in the fully saturated zone (Fig. 28). However, the subsurface geologic structure, which may be constrained by surface outcrop exposures, is often less well defined in the absence of seismic reflection data that prove so useful in petroleum field studies. For example, Bense et al. (2003) offer detailed descriptions of faults and clay smear in shallow, unlithified sediments studied in trenches in the Netherlands with detail well beyond what is possible in petroleum field studies. They infer flow restriction across the fault based on an observed 5 m difference in the phreatic surface over a 25 m distance.

In a more comprehensive study in the Lower Rhine Embayment, Bense and Van Balen (2004) document differences in hydraulic head between different fault blocks. The area studied encompasses the lignite mines reported by Weber et al. (1978) where the original detailed observations of clay smear were made. The geologic and fault framework is derived from exposures in the open pit mines and numerous wellbores used to define coal distribution so some level of detail in the fault structure is expected to be missing from this extrapolation. In spite of this limitation, hydraulic head differences >100 m are interpreted across faults near open-pit mines where wells locally drop the local groundwater level in order to keep the mines dry. In some areas, though, the effects of the prominent clay smear are defeated by unbreached fault relays that enable fault-unhindered groundwater flow between fault blocks, albeit through a more restricted cross-section.

Results from the Bense and Van Balen (2004) work illustrate one difficulty in deducing fault flow properties from subsurface flow studies. In that work, they plot SGR-value against aquifer potential difference, from which they infer that SGR is related to fault permeability. However, this approach assumes that there are no holes in the clay gouge whose presence would dominate flow. If holes were present, the same pattern of hydraulic potential contours could be possible, even for the case of limited cross-fault flux (Fig. 32); in this example, fault flow (i.e. SGR) properties are constant, and the range in hydraulic potential difference is large.

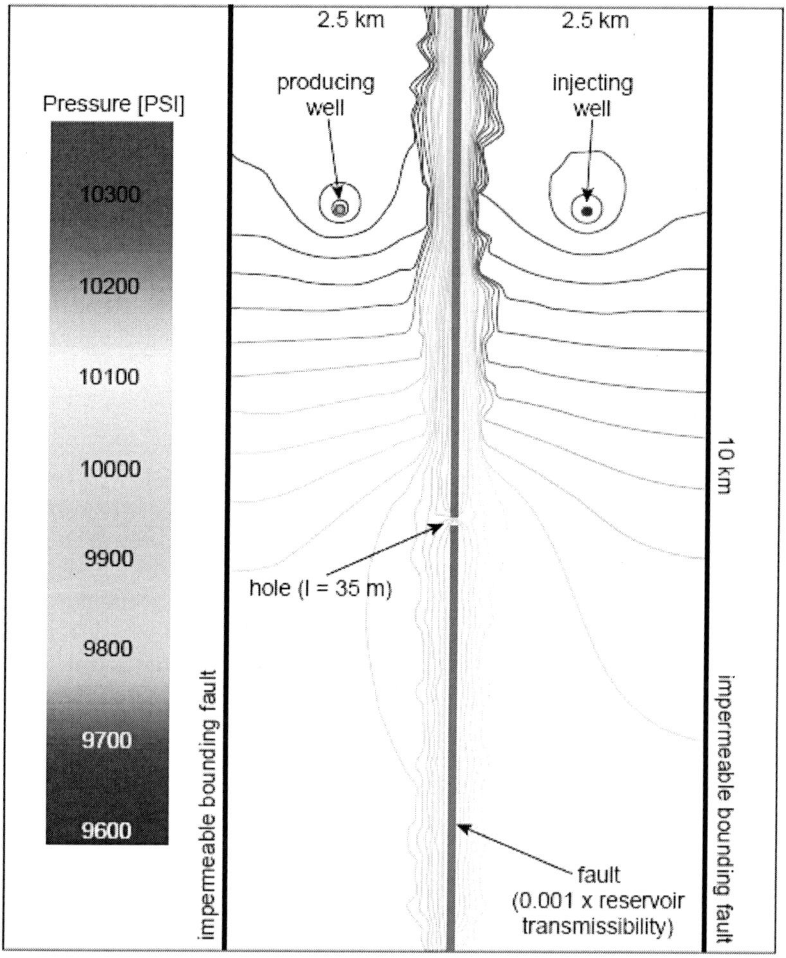

Figure 32. Cross-fault flow simulation to illustrate complexities in deducing fault properties from general hydraulic data. Model includes closed, limited area (area of figure), bounding faults (vertical black lines extending vertically across figure representing no-flow boundaries), middle fault (middle line) with constant, limited transmissibility (0.001 × reservoir transmissibility) and 35 m long hole in middle of model, uniform reservoir properties (1000 mD and 0.79 m thick), producing well (left fault block) and injecting well (right fault block) maintaining mass balance (5000 bbl/day; 795 m^3/day). Resulting steady-state flow defined after 1 year of flow, and pressure contours (20 psi contour interval; 138 kPa interval). Small flux across fault results in tight clustering of contours along fault, but majority of flow along fault block and through hole in middle. Note that biggest pressure difference across fault defined between wells, and smallest pressure difference immediately adjacent hole, even though gouge permeability is constant.

In subsequent work in the Lower Rhine Embayment, Bense et al. (2008) document thermal anomalies, which they ascribe to vertical water flux. Spiller et al. (2004) similarly address vertical aquifer flux in the open-pit mines in the area and interpret a deep aquifer source based on temperature and water chemistry results. Although the geologic cross-section presented by Spiller et al. (2004) indicates potential flow paths back-and-forth across faults through juxtaposed permeable intervals, this kind of pathway is probably inadequate for producing thermal anomalies. Bense et al. (2008) appeal to high permeability along the fault dip, possibly resulting from sand lenses entrained within the fault zone. However, Spiller et al. (2004) infer erosion processes along the fault that create high permeability conduits, erosion that may be specific to the local environment of unlithified sediments and low stresses.

Ironically then, in the area where the clay smear process was first identified and documented, groundwater flow is influenced by an array of geologic processes and conditions, including fault relays and groundwater-induced erosion processes. These examples illustrate the difficulty in inverting flow responses for clay-smear properties alone without considering the potential impact of other processes that may also affect cross-fault flow. However, combining fluid potential data with independent temperature (Bense et al., 2008) and chemical tracer (Cilona et al., 2015) data help reduce the number of degrees of freedom in solutions, much as 4D seismic images help define the distribution and movement of different fluid types during gas, oil, and water production and injection.

3.3.7. Clay smear dilemma in subsurface studies

The problem with subsurface studies is that evidence of the presence of clay smear fails to establish that clay smear has a material impact on subsurface fluid flow; the observation of clay smear provides necessary but not sufficient information to reach this conclusion. Moreover, the failure to account for subsurface fluid distributions with simple juxtapositions (i.e. offsets along a single fault surface) is insufficient to deduce the presence and effect of clay smear. To illustrate this point further by specific example, two fault interpretation examples are presented from the Hibernia Field, Canada.

3.3.7.1. Hibernia background

The Hibernia Field is a large oil and gas field on the Grand Banks offshore Eastern Canada (Mackay and Tankard, 1990). It is well suited for evaluating problems of cross-fault fluid flow because the field is heavily faulted (two near-orthogonal fault strike orientations interpreted to have formed in the

Cretaceous; Sinclair et al., 1999), and the Cretaceous braided fluvial reservoir varies on a scale that is much larger than a fault block (i.e. local fault problems are evaluated with high confidence in the stratigraphic distribution because of the interpreted stratigraphic continuity). Faults with offsets larger than the reservoir thickness result in separation of gas, oil, and water fluids, and faults with smaller offsets result in fluid continuity across the fault. Even though the reservoir is heavily faulted, gas, oil, and water fluids communicate where the geometric juxtaposition of permeable reservoir intervals permit that communication. Faults between permeable intervals rarely prevent cross-fault flow but rather serve to hinder or re-direct it.

Cross-fault fluid continuity is defined as the same original, pre-production fluid pressure in the same fluid type at the same depth on either side of a fault (i.e. no capillary membrane seal), even though that connectivity may be established via a long meandering pathway through multiple fault blocks. For fault blocks sharing the same initial conditions, the question becomes whether those juxtaposition connections are sufficient for allowing pressure communication during production (i.e. are rates of cross-fault flow significant?). There is substantial evidence for cross-fault fluid connectivity during production in Hibernia based on: (1) deviations of fluid pressure from original pressure conditions in undeveloped fault blocks as a result of production or injection of fluids in adjacent fault blocks; and (2) interpretation of cross-fault fluid flux necessary to account for mass balance constraints invoked in reservoir simulation. The production data suggest that where porous and permeable reservoir sandstones exist, cross-fault flow is common.

However, two significant exceptions to these generalizations are known, and they are instructive for considering the uniqueness of geologic interpretations for clay smear calibrations. Here we can address the following questions: Does clay smear form in these locations? Is it continuous? Are there other mechanisms for separating permeable sandstones?

3.3.7.2. Oil column separation – clay-smear or complex faulting?

In 2011 a well was drilled into an untested fault block in the southern part of the field. This fault block is at the junction of two independent oil columns (i.e. oil columns with different original pressures at the same elevation) – a higher pressure oil column documented to the east, and a lower pressure oil column to the west (Fig. 33). The question was which oil column defined the pressures in the untested fault block, and what geologic interpretation would yield a successful prediction? A range of pre-drill scenarios were defined that separated the oils along a fault on the western side and eastern side of the fault block to be drilled, but here we focus on the interpretations on the W-bounding fault.

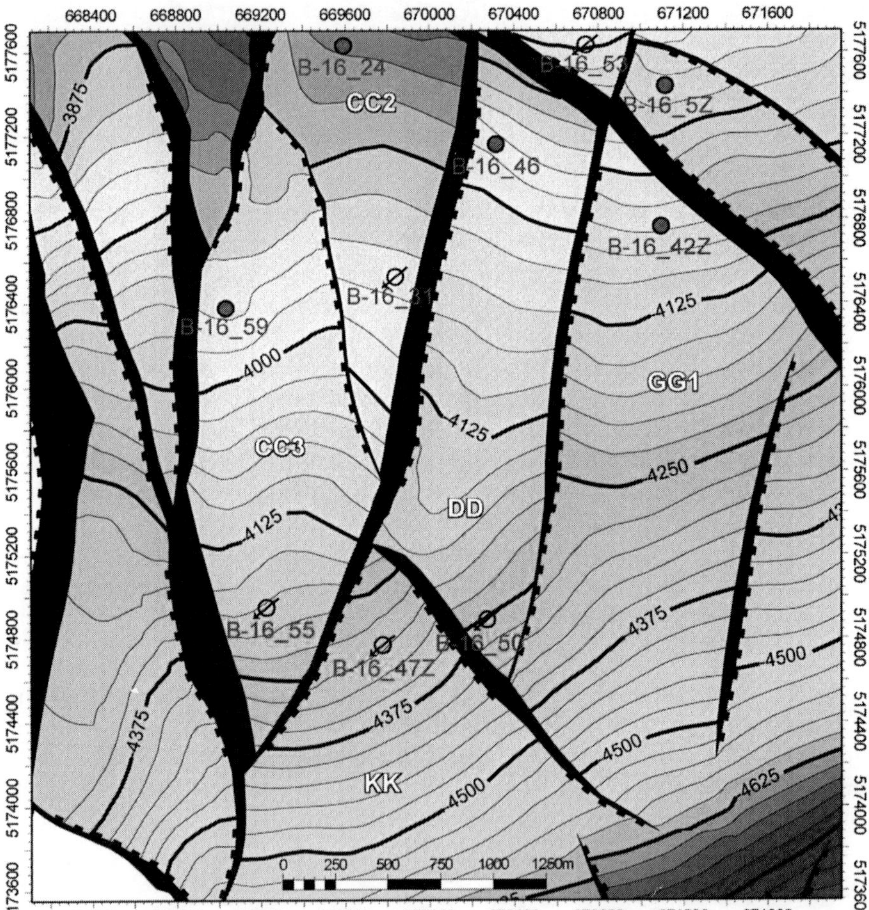

Figure 33. Depth-structure map of southern portion of Hibernia Field, offshore Eastern Canada. Map shows fault blocks on west (CC 2 & CC3) that define lower pressure oil column (at specified depth) than fault blocks further to east (DD). Geologic problem involves accounting for oil pressure properties observed by drilling into KK-fault block. Note that hanging wall (downthrown) block on each fault depicted by fault-teeth symbols. Warm colors are shallower than cool colors, and contour labels in meters below sea level; filled circles are oil producing wells, and open circles with arrows are water injection wells.

Seismic profiles across down-dip portions of the W-bounding fault suggest there should be a geometric connection between permeable intervals in this reservoir section that should allow the western, lower pressure oil to communicate across this fault (Fig. 34). The sandstone-rich nature of the faulted section results in low SGR values for all plausible offsets up to full reservoir separation (Fig. 35) with calculated SGR values < 0.2. Such low SGR values occur elsewhere in the field for the same Layer 3 juxtapositions and are associated with cross-fault flow during production, so it is unlikely that a similar SGR value should produce a capillary seal here (Table 5).

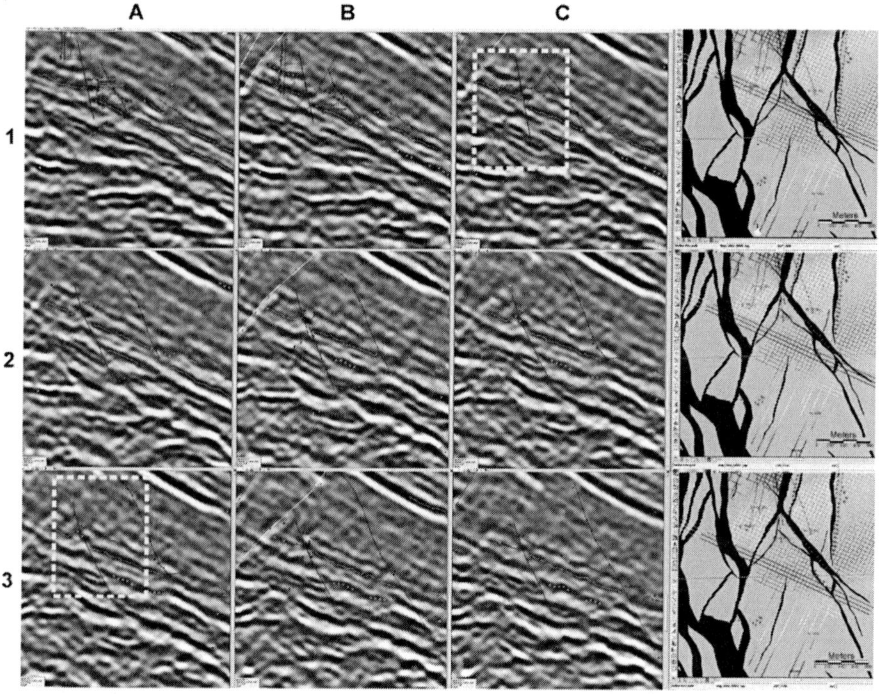

Figure 34. Serial seismic profiles across fault that separates CC3 and KK fault blocks. Seismic profiles progress from north (upper left corner: A1) to south (lower right corner: C3). Each row of 3 profiles correspond to blue and red lines (oriented WNW-ESE) in map at end of row (A to C = N to S). Note how E-dipping purple fault has little displacement, especially in the last row of profiles where imaging is compromised, in part, by fault shadowing effects beneath white fault. Reservoir interval contained between purple and red/pink horizons. Seismic data outlined in yellow boxes shown in Fig. 36.

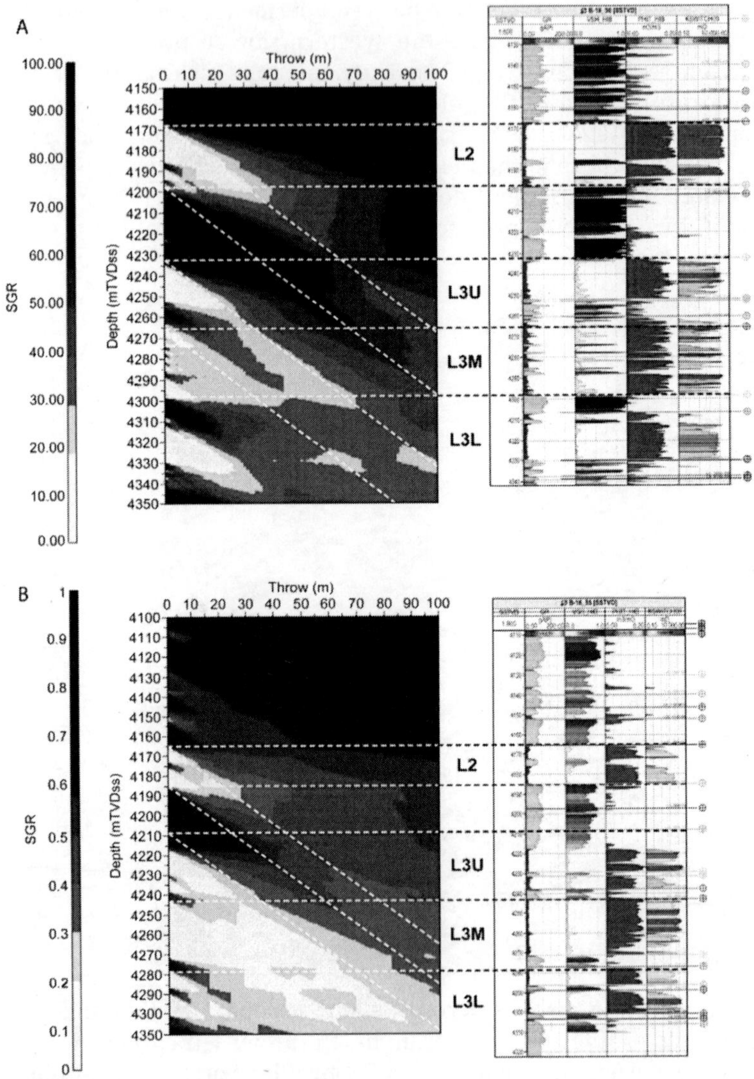

Figure 35. Well log representation of offset reservoir stratigraphy, including calculated SGR values for fault throws up to full reservoir offset. Note that sandstone-rich nature of Layer 3 limits value of SGR calculated, resulting in inference of SGR value no greater than 0.35–0.38 in (A) and <0.2 for the same L3U/L3L juxtapositions in (B); L2 = Layer 2, L3U = Layer 3 Upper, L3M = Layer 3 Middle, L3L = Layer 3 Lower. Although these wells are bound and are separated by <500 m from fault of interest (Fig. 33), subtle differences in stratigraphy yield some fairly significant differences in SGR calculation and resulting inference of capillary-sealing versus leaking for same critical SGR value (e.g., 0.2) depending upon precise stratigraphic development of L3M.

Table 5. Summary of layer 3 juxtaposition properties.

Layer 3 reservoir connections	Pre-production pressure connection	Neighboring pressure disturbance	Significant mass flow to neighbor
KK-CC blocks	No	No	No
Rest of field	Yes	Yes	Yes

Further complicating matters is the observation along the updip section of the W-bounding fault where reservoir separation is interpreted by a complex intersection fault geometry imaged in the seismic data (Fig. 36). In these sections a W-dipping fault is cut by the primary E-dipping fault with the result that the apparent offset across the entire fault zone appears minimal while in fact the offset on the E-dipping fault creates a juxtaposition seal with younger shales. The offset on the E-dipping fault is compensated by the older W-dipping fault with the result that an additional juxtaposition seal is created by insertion of an older shale interval.

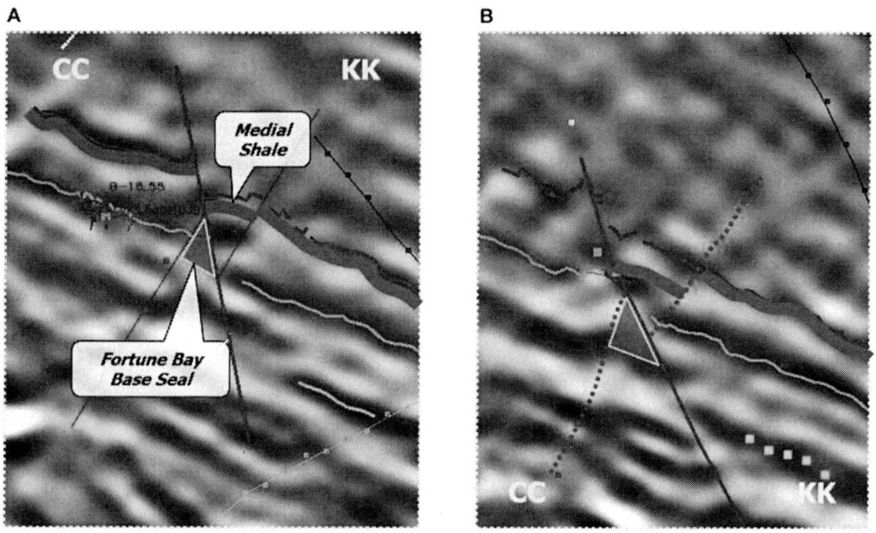

Figure 36. Close-up views of up-dip and down-dip seismic profiles highlighted in yellow boxes in Fig. 34 (width of view approximately 1 km). In up-dip profile (A), W-dipping fault intersects E-dipping fault within reservoir sequence and creates downward displacement of top-seal wedge and upward displacement of base-seal wedge relative to reservoir interval on both sides of fault system, which appears to have minimal offset across this system, although fault throw on each segment is significant. In down-dip section (B), W-dipping fault is no longer evident in seismic data even though horizon offset across entire fault system is minimal. Speculative extrapolation of W-dipping fault is shown with dotted fault interpretation and maintains juxtaposition seal geometries interpreted in up-dip profile. Medial shale is marine shale associated with Maximum Flooding Surface below Layer 2 and above thick shale-rich interval at top of Layer 3U (Fig. 35). Fortune Bay Base Seal is shale interval below oldest reservoir interval that supports large pressure difference with underlying Jeanne d'Arc units.

In down-dip sections only the E-dipping fault is recognizable in the seismic data, and the apparent offset remains minimal. However, given the complexity of the geology and the seismic resolution available at a depth >4500 m, an additional plausible interpretation involves maintaining the W-dipping fault geometry observed up-dip, even though the seismic data fail to require that interpretation, with the result that same top- and base-seal juxtapositions are maintained (Fig. 36). Note that the interpretation presented inFig. 36b is non-unique, which it must be for an interpretation under-constrained by data. Indeed the seismic data also allow a more complicated interpretation than the one shown, especially for faults that are closer to being synchronous (e.g., mutually offsetting synthetic and anthithetic faults).

Oil pressures measured in the drilled well were the same as those in fault blocks to the east, indicating that the W-bounding fault was responsible for separating oil columns. In this example we conclude that the observed oil separation is unlikely for juxtaposition seal on a simple fault surface, unlikely for a capillary clay smear, and thus more likely for the complex cross-cutting fault pattern observed updip in the seismic data. Note that if we had failed to observe this fault complexity, we might have been tempted to interpret a clay smear capillary-seal at an anomalously low SGR value.

3.3.7.3. Water trapped above oil – fault lens-enclosed shale?

In our clay smear definition, we include both the high shear-strain clay that accommodates significant fault offset and stratigraphically coherent lenses of shale that exist as part of the clay smear fault structure. This example further develops the idea that shale lenses may contribute to capillary seals in the sense of Berg and Avery (1995).

In 2011 another well was drilled into the Hibernia reservoir and encountered water in the midst of a thick oil column; the well documented an oil-water contact, and another well structurally down-dip in the next fault block previously documented oil (Fig. 37). Both oil occurrences are interpreted to be parts of the same continuous oil column in original pressure communication (pressures in the 2011 well were perturbed by production, which is why this is an interpretation rather than an observation), and the seismic data (Fig. 38) indicate a Layer 2–3 juxtaposition connection for a simple fault surface interpretation with no evidence for a complex antithetic fault as in the previous example (notwithstanding the fact that an antithetic fault would like be unable to create a juxtaposition seal in this part of the stratigraphic section). As in the previous example, the reservoir Layer 2–3 fault juxtaposition connection that traps water in the up-dip fault block (Fig. 38) is similarly associated with extensive pressure and fluid communication elsewhere in the field

(Table 6), and so the automatic deduction of a clay smear capillary seal is difficult based on conflicting evidence.

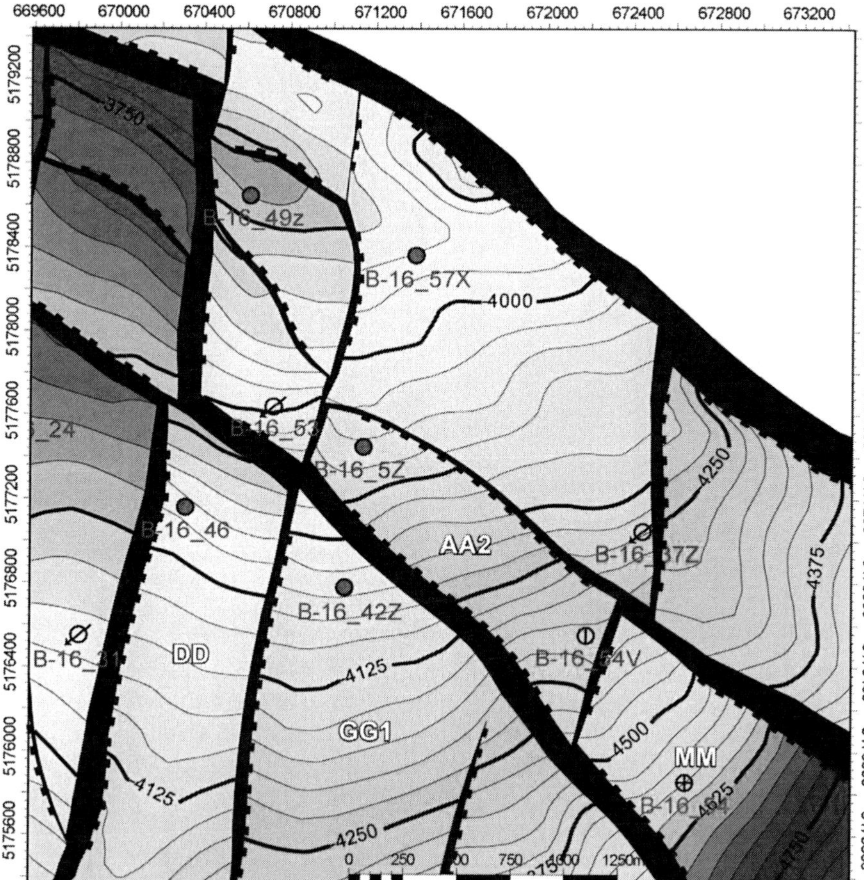

Figure 37. Depth-structure map of southern portion of Hibernia Field, offshore Eastern Canada. Map shows up-dip fault block (AA2) that contains thick interval of water above oil observed down-dip in MM fault block (interpreted original pressure communication). Fault offset with GG1-block is enough to create full separation of reservoir intervals, but AA2-MM fault offset is too small.

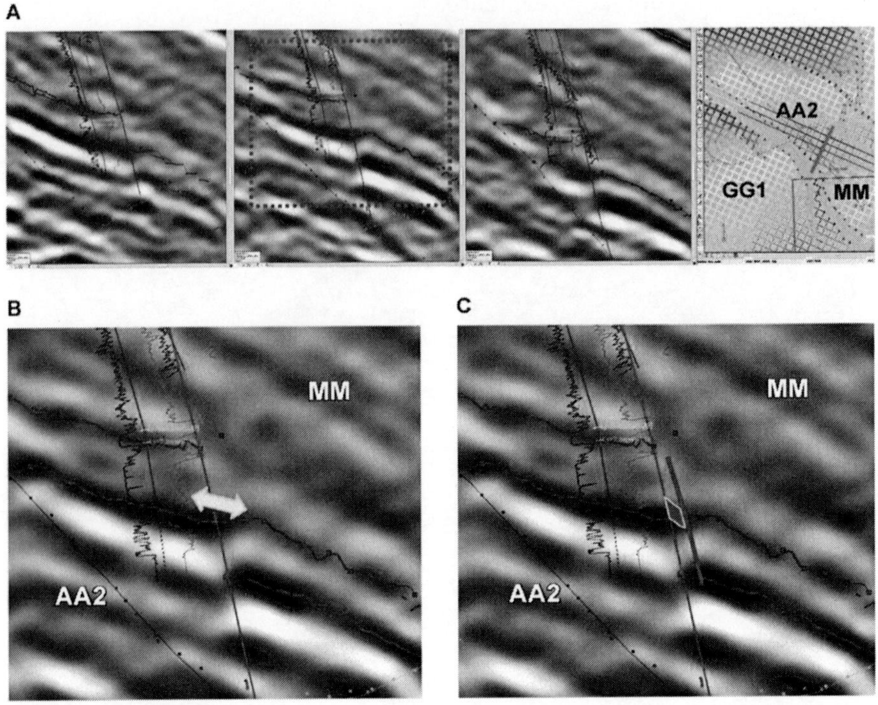

Figure 38. (A) Series of 3 seismic profiles crossing fault separating AA2 (up-dip) and MM (down-dip) fault blocks. Position of lines shown on map at end of row with profiles progressing from N (left) to S (right). Width of profiles approximately 2.4 km; no vertical exaggeration. (B) Blow-up of middle profile (inside dotted red box). Red curve that parallels pink fault interpretation is path of well bore in down-dip portion of AA2 fault block. Adjacent blue curve is gamma-ray curve taken in well; shale intervals identified by parts of curve closest to wellbore curve, and sandstone intervals by those farthest from wellbore curve. Green curve on right side of wellbore is resistivity curve; in sandstone intervals high resistivity (green curve farthest from wellbore curve) intervals represent oil and low resistivity (green curve close to wellbore curve) represents water. Note that for thick section of sandstone (blue curve) between purple and red horizons, resistivity indicates water with minor interpretations that a bit of oil exists at top but much of interval contains water in spite of Layer 2–3 juxtaposition (yellow arrow). Layer 2 reservoir interval lies above purple horizon interpretation. (C) Speculative fault relay or lens introduces wedge of Medial Shale (brown polygon outlined in yellow) displaced short distance from hanging wall but enough to occlude Layer 2–3 juxtaposition.

Table 6. Summary of layer 2–3 juxtaposition properties.

Layer 2–3 reservoir connections	Pre-production pressure connection	Neighboring pressure disturbance	Significant mass flow to neighbor
AA2-MM blocks	No – water[a]	Unknown	No – water[a]
Rest of field	Yes	Yes	Yes

[a] Can only be established for water phase encountered in well; no compelling evidence to conclude that original oil pressure communication or subsequent oil flux occurs along upper part of fault where oil is found on both sides of fault or via some alternative path of fault connections.

Although faulting of the Medial Shale unit, a continuous marine shale interval that separates the Layer 2 and 3 reservoirs, results in higher calculated SGR values than in the Layer 3 self-juxtaposition example above (Fig. 39), most other Layer 2–3 juxtapositions allow production time-scale pressure communication. However, the Medial Shale unit is comparable in thickness to the Layer 2 reservoir interval, and it is feasible to interpret a fault lens containing the Medial Shale between the Layer 2 and 3 reservoirs as a juxtaposition seal based on incorporation of a structurally coherent shale lens in the fault zone (Fig. 38), but a shale lens incorporated as part of clay smear processes.

How does a fault lens component of a clay smear interpretation satisfy the argument of conflicting observations for the same SGR? For this we assert that SGR is a poor predictor of the development of fault lenses (Noorsalehi-Garakani et al., 2013). Next, we conclude that fault lenses have limited lateral extent (e.g., Noorsalehi-Garakani et al., 2013). Then we observe that the fault separating water and oil is among the shortest fault-block segments in the field and deduce that all other fault segments are longer than any similar fault lenses that may be developed on those faults. Accepting that this interpretation is correct because alternative interpretations are excluded based on lack of evidence or conflicting observations, we then find evidence that the mechanics of clay smear processes are critical for predicting when shale-filled lenses will form. It also becomes crucial to know how long lenses can be because this defines which fault segments are candidates for this prediction.

These two examples illustrate the fact that complex fault geometries are a third alternative to simple juxtaposition and SGR-based clay smear predictions. Any clay smear calibration studies that have been undertaken without considering the potential for complex fault geometries to contribute to cross-fault flow must be viewed with caution. The clay-smear, fault-lens interpretation is a strong motivator for including a clay-smear mechanics component in prediction models.

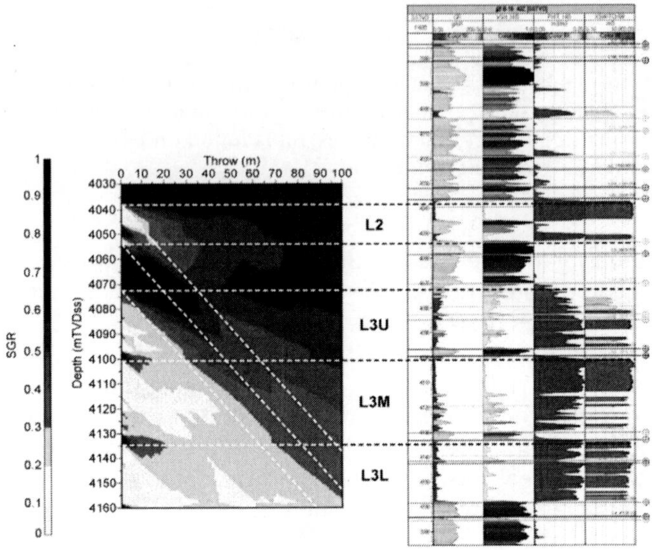

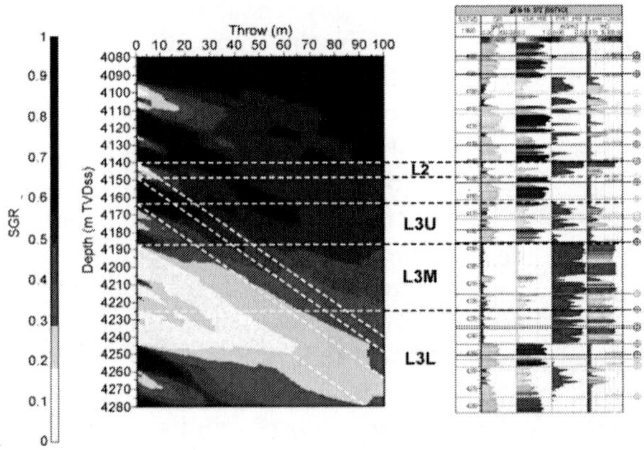

Figure 39. Well log representation of offset reservoir stratigraphy, including calculated SGR values for fault throws up to full reservoir offset. Note that SGR values increase as L2 is increasingly offset to progressively juxtapose L3U, L3M, and L3L, with SGR values decreasing from 0.55 to 0.36; L2 = Layer 2, L3U = Layer 3 Upper, L3M = Layer 3 Middle, L3L = Layer 3 Lower. In this instance there is little difference between SGR values calculated for two closest wells; both well results are dominated by relatively shaley development of L3U interval. Note that for capillary seal required by seismic data in Fig. 38, SGR values are lowest. Applying 'typical' threshold SGR value of 0.2 for creation of capillary seal, though, contradicts observations of production-scale pressure depletion elsewhere in field across same kind of juxtaposition connection and for similar stratigraphic development (NB both cross-fault pressure depletion and build-up above original pressure observed for pressure-maintained oil production via both gas and water injection).

3.3.8. Summary of subsurface field studies

Clay smear is interpreted to hinder cross-fault flow in many different field studies. Although clay smear was originally conceived of as a capillary, membrane seal, its impact on throttling the flux of fluids across faults in the presence of a fluid potential difference (i.e. production flow) was also appreciated early on. Much of the literature on the application of clay smear processes is derived from Tertiary deltaic sedimentary systems where rocks are often underconsolidated and the stratigraphic continuity of intervals up to faults where fault juxtaposition diagrams are created may be open to alternative interpretations. Moreover, although disturbed juxtaposition relations that arise from non-planar fault structures have been recognized, there is no coherent strategy for accounting for this complexity.

We think it is hard to gain meaningful insights into clay smear processes based on the types of field studies to date (Table 7). Although there is a considerable literature that addresses clay smear as a capillary seal, the interpretation of a continuous clay smear without the sparse holes described by Wehr et al. (2000), especially in light of data and interpretation uncertainties, seems difficult to achieve. Although there are no doubt examples in nature where clay smear has created a capillary seal to oil and gas, it is difficult to specify which faults achieve a continuous clay smear from the perspective of reproducibility embodied in the scientific method.

Table 7. Rating summary of subsurface studies and clay smear.

Contribution	Understanding	Comments
Capillary seal	☆☆☆☆☆	Although claimed in the literature, uncertainty evaluation and consideration of alternative interpretations left largely under-explored; interpretation reproducibility limited; presence of clay smear largely untested by independent core samples
Permeability reduction in oil & gas flow	☆☆☆☆☆	Same limitations as for capillary seal; production data used in history-match analysis offers more definitive but non-unique data constraints; tested in full 3D models that help enforce internal interpretation consistency
Permeability reduction – aquifer flow	☆☆☆☆☆	Same limitations as for capillary seal; less sub-surface definition than for oil & gas flow and more restricted data constraints (changes in potential surface); better integration of surface outcrop data possible
Fault mechanisms responsible	☆☆☆☆☆	Subsurface studies focus on the flow effects of clay smear and offer few insights into the processes that generate clay smear; effects of geometric ambiguities recognized, but data resolution often prevents independent evaluation of those geometries

Note: Star rating reflects score (red stars) out of maximum potential score (blue stars).

Most studies remain under-constrained (Table 4), but the integrated simulation studies (Jolley et al., 2007 and Myers et al., 2007) begin to define more detailed information about fault zone properties (Table 4), especially when constructed from first-principles geologic models (i.e. models in which various fault geometries and properties are defined as an input). The logical continuation of this approach is to make specific geologic and flow predictions and then drill wells, collect and analyze samples, and perform specific flow experiments to test those models. Aquifer studies (Table 7) offer additional insights on cross-fault flow if only because nearby outcrops offer relevant information about aquifer properties in the sub-surface.

However, the subsurface aquifer definition is poorer than in oil and gas fields, and the history-matching approach seems less well developed. The aquifer environment offers the opportunity for inexpensive diagnostic cross-fault flow tests because well costs are comparatively cheap.

Subsurface field studies offer little about the mechanics of the clay smear process – how clays are incorporated into a fault zone, how they are deformed once between the shear zone boundaries, and how clay smears develop holes. Core analyses in subsurface studies offer confirmation for the existence of clay smear but the scale of these samples hinders further interpretation. In the end subsurface studies are essential for testing clay smear models developed elsewhere, but to do so requires public disclosure of a comprehensive dataset, including all raw data to insure interpretation reproducibility, and the application of well-conceived cross-fault flow tests designed to test and differentiate clay smear models. There is no more effective means to interrogate the geometry and continuity of clay smear products in a fault zone than with subsurface flow studies, and current approaches can be useful for addressing questions of reservoir compartmentalization, but advances in understanding will require more definitive subsurface tests.

3.4. Discussion and synthesis from laboratory to subsurface

We have shown in this review that clay smear is much more than the weighted average of a faulted lithologic section. Rather, it is the ensemble of fault structures that develop during fault evolution. That ensemble is the result of the deformation regime during faulting, the mechanical properties of the faulted rocks, the effective stress, the geometric evolution of the fault network, and the lithology of the faulted section. In addition, mineral reaction and recrystallization processes are possible but have remained outside the clay smear context. When considering the effects of clay smear on subsurface flow problems, the single most important fault structures are the holes in the gouge. Therefore, even if all other aspects of the clay smear process are fully understood, predicted flow properties will remain uncertain if our understanding of how holes form remains incomplete.

Although we consider faults in this review that rarely have offsets greater than a hundred meters, the complexity is already enormous and diverse. Thus research into the clay smear problem for the purpose of improving subsurface flow predictions has also yielded important insights into fault localization phenomena. Exchange with other applications like earthquake studies is seen as promising, but those associations have yet to be fully developed.

In this section we synthesize insights across the three areas we developed in this paper: Outcrop Studies, Laboratory and Numerical Modeling Studies, and Subsurface Studies. We seek to better define what 'clay smear' is, trying to remain faithful to the history of the subject, providing a terminology framework useful for future studies, and thereby reducing potential confusion. We then step back and evaluate how deep we think our understanding has reached into the various processes that create clay smear. In a complementary evaluation, we summarize our knowledge of how clay smear has affected subsurface fluid flow, leading to the question of how effective current predictive models are. On the basis of these evaluations, we then conclude with a section that describes how we think future advancements will arise.

3.4.1. What is clay smear?

'Clay smear is a loosely defined term born in hydrocarbon geology; its usage differs between publications, and the definition of processes operating is often unclear. In the most general meaning, the term includes all processes, which somehow transform clay in the wall rock into clay that is part of the fault zone' (van der Zee et al., 2003). More than a decade after this perspective was given, we think this assessment is still correct.

Therefore clay smear, although sometimes referred to as a process, is better considered encompassing a multitude of fault processes that combine and interact in various fashions to create a myriad of fault structures, some of which collectively define a clay smear. Clay smear is an umbrella term for a series of fault processes, and similarly a term that includes many different fault structures. Thus we use the term 'clay smear' when it is efficient for referring to a collection of processes and structures but then rely upon the more detailed and explicit terms (e.g., abrasion, shear, and layered gouge) included beneath the umbrella term to achieve better insight into the problem.

Processes associated with clay smear reflect the evolution of a fault propagating through a mechanically heterogeneous sedimentary section. There are many different paths available for this evolution, resulting in the rich diversity of structures described in the literature. At its root, incipient clay smear is a fault localization phenomenon combined with reworking in high shear strain in the fault zone.

Lastly, although there is a temptation to focus the attention of the clay smear processes onto the high shear strain component of the fault zone, the available outcrop and experimental observations illustrate the crucial importance and ephemeral nature of structurally coherent lenses of sandstone, limestone, and in particular shale for the overall fault zone structure. The creation of shale-rich fault lenses is an integral part of the clay smear process, especially for considering the cross-fault flow problem and as a source of clay-rich material available to promote localization with increasing fault offset.

In the final analysis, clay smear is a special case of multiple fault processes and structures that arise when a faulted section contains mudstones and shales. It encompasses the presence and distribution, geometry and continuity of the deformed and faulted lithologies. In outcrop studies without mudstones or shales similar fault structures are found, and in the experimental section arguments were developed to build a clay smear process model based on localization models developed in pure sand systems.

3.4.2. How deep is our understanding of fault processes that result in clay smear?

From our review of available outcrop studies and complementary laboratory and numerical experiments, we conclude that the development of clay smear depends on three components:

1. how much clay exists in the faulted section
2. the geomechanical properties of the faulted lithologies defined in relation to:
    a. in-situ effective stress (i.e. brittleness)
    b. geometric evolution of fault system

These components have many complicated feedbacks, and different domains of processes arise out of these interactions.

The results of experimental and numerical simulations of fault processes, outcrop observations, and subsurface studies are very different windows into the clay smear phenomenon, and we face hurdles in relating the results from one area of investigation to another. For example, in the experimental section we discussed the limitations of grain size in experiments and grid size in simulations and the issues associated with scaling processes to natural dimensions. The boundary conditions in existing experimental and numerical simulations of clay smear processes are strongly influenced by the need to produce a large shear in a small model. Many boundary conditions in nature are unattainable in simulation. Conversely, outcrops lack the height and width needed to completely examine a faulted layer and the incremental strain history necessary to relate a series of final structures to the initial boundary conditions and subsequent fault evolution; process is an interpretation rather than an observation in this environment. And subsurface studies, which through the fluid flow problem allow us to interrogate complete fault zones, lack the detailed resolution necessary to observe the critical fault structures.

We find it useful to think of the deformation domains defined in experimental studies because they provide the means to relate a group of distinguishing processes in a domain to a series of diagnostic fault structures (Fig. 27). What makes this approach appealing is that the deformation behavior of sand and shale components is a consequence of material properties, stress conditions during faulting, and the previous stress history of the materials. In other words, weak clay deformed at low stress can deform by the same processes and result in the same diagnostic structures as stronger clay deformed at higher effective stresses. We arrive at this understanding as a result of observing clay fragments formed by extensional fractures at a propagating fault tip and then 'flowing' in the fault zone as a result of that same clay deforming in critical state (degree of softening depending on initial cementation). Although these deformation domains provide a highly simplified framework, it is nonetheless useful as a starting point for incorporating the mechanics of deformation into clay smear predictions, which currently lack any input from this perspective.

Fault geometric evolution plays a crucial role in early fault zone development and is as important a consideration for clay smear evolution as the geomechanical aspects described above. In outcrops and experiments we observe fault zones with clay smear associated with multiple slip surfaces (zones of localization), and laboratory and numerical experiments demonstrate both a discontinuous (or continuous) migration of localized shearing through the material and simultaneous displacement on multiple localizations. Moreover, the evolution of fault relays and lenses has a profound effect on how clay is incorporated into the fault zone and how it deforms once it is within the fault zone (e.g., reworking of coherent fragments in fault lenses). Fundamental questions like what controls the migration of deformation into the footwall versus hanging wall as the fault zone evolves and what determines the total number of slip surfaces (deformation bands) that develop have an important effect on the final clay smear structure.

Unfortunately, our understanding of the geometry of early localization is limited. Some insights are provided by models of a basement fault that propagates into a sedimentary section (Nollet et al., 2012), and outcrop observations show fault relay and lens structures resulting from different stratigraphic stacking patterns (Vrolijk et al., 2005b). Moreover, there is information about geometric processes like tipline bifurcation (e.g.,Childs et al., 2007) at our disposal, but the means to know when to consider these geometric processes is lacking. There is a consensus in the literature that large contrasts in strength between adjacent intervals lead to wider and more complex fault zones, especially when one of the layers fails in extension. But understanding how a fault zone propagates in 3D through a mechanically layered package is still limited, even though this insight is required for understanding how clay smear develops in this fault.

Interestingly, perhaps our greatest deficiency in understanding clay smear processes is in the one aspect most important for evaluating subsurface flow problems – the creation of holes in the clay smear structure. Holes are made sometimes in experiments, but we rarely observe how they emerge. They are rarely described in outcrop, and when they are the structures associated with them are poorly described. Underlying this quandary is the model proposed by Weber et al. (1978) that holes form as the result of attenuation and only once the clay smear has been offset significantly from its source bed. However, we now recognize that holes form at the source clay bed if the clay is incorporated via a brittle fragmentation process (shear or tensile failure). This process is reflected in no current shale smear algorithm other than the probabilistic shale smear methods, but this realization introduces a whole new perspective on the flow problem – clay smears can improve with displacement if an initial shale fragment associated with a continuous sand pathway across the fault becomes reworked into a clay smear.

With existing observations and incorporation of geomechanical principles, there are some generalizations that help guide us to the conditions that favor hole formation. For example, a wider fault zone (for the same throw) has a higher chance for continuous clay gouge, (once the sand layers are disrupted by boudinage processes), especially in the low mechanical contrast case (failure of both lithologies in shear). Alternatively in the strong contrast case (i.e. clay failing in extension and sand in shear), a wider fault zone may cause the clay continuity to decrease because the ability of a clay fragment to become reworked in a critical state may be reduced in a wider fault zone.

The role of secondary fault structures like Riedel shears has also been associated with the potential for holes, and this effect is likely enhanced when the clay smear is thin relative to the shear zone thickness and Riedels are formed at high angles to the smear. Are there conditions that favor the development of Riedels, and if so, at what orientation? An interesting and consistent observation is the absence of taper in clay gouge away from the source layer, both in outcrops and in models. Rather, the persistence of clay smear thickness is associated with erosion of the clay source bed and continuous addition of clay.

It is also worth remembering that although clay and other mineral reactions are important in other fault settings, they are typically excluded from consideration in clay smear studies. Pressure solution of quartz, feldspar, and carbonate grains and clay recrystallization are common in some of the systems considered in this paper and are manifested by enhanced softening of fault zone materials. Moreover, these chemical and mineralogical processes can influence the amount of clay found in a fault zone.

Moreover, we think our understanding could benefit from greater collaboration with related fields. For example, there are opportunities for cross fertilization between clay smear research and the part of the rock mechanics community concerned with containment assurance and fault reactivation prediction. Brittleness, dilatancy, and resealing by reworking in faults has been recognized and is becoming more important in many different rock mechanics applications – a few examples of relevant papers areOlgaard et al., 1997, Ingram and Urai, 1999, Ishii et al., 2011 and Schultz et al., 2014, andBourne et al. (2014). There is a potential for incorporating better predictors for shale strength and brittleness that have been developed in recent years into improved mechanical clay smear calculations (e.g., van der Zee and Urai, 2005 and Kleine-Vennekate, 2013). On the other hand, containment and reactivation studies often use highly simplified models of the fault zones under consideration, and the concepts reviewed in this paper can provide improved fault zone models for these calculations.

Perhaps even more surprising is the potential for cross fertilization between the clay smear community and those concerned with studies of fault gouge interpreted to have formed by seismic faulting using observations from outcrop, underground observatories and from laboratory tests. Perhaps the best attempt to bridge these disciplines was accomplished by Faulkner et al. (2010) in which a clear and concise section on clay smear and fault sealing is presented. Bullock et al. (2015) recognize how modest amounts of clay, perhaps introduced by clay smear mechanisms, can alter the seismic behavior of a carbonate system, by evaluating the effects of clay on velocity weakening, dilatancy, permeability evolution, and other effects of earthquake faulting. Earthquake processes are related in complicated couplings, for example flash heating (Platt et al., 2015), pressure solution (Gratier, 2011), and mechanical effects of fabric evolution. Here we ask the question: did the clay smears discussed in our paper form as a result of seismic slip? The clay smear literature never approaches this question, although most papers seem to imply that faulting is aseismic. There have been no studies of velocity weakening in clay smears, and usually the faults for which clay smear is important have formed at effective stresses and areas which are too low to generate significant seismic moment. Perhaps one way to form clay smears by seismic slip is when the seismic event is triggered in asperities in hard layers above or below the zone with clay smear, and propagates into the clay smear. Alternatively, clay smear in a normal fault above basement could lead to a larger fault rupture area for an earthquake initiated in the basement. On the other hand, clay smear research has provided interesting data on 3D geometry of fault structures, and these have received little attention in the quake-related studies. Insights gained from how clay is incorporated into a fault zone and subsequently evolves provide input for the geometries in models of seismic faulting.

In summary, our physical understanding of clay smear processes is incomplete. Overall we might grade our knowledge with a C with some components like the geomechanical effects somewhat better understood and others like the processes that result in holes somewhat worse. In our opinion there is substantial opportunity for improving physical understanding through better interactions with complementary rock mechanics research, and in particular certain aspects of earthquake research. We now ask, however, whether and how this current, albeit incomplete understanding is currently used in subsurface flow evaluations.

3.4.3. How does clay smear affect subsurface fluid flow?

There is no question that clay smear has the ability to impede subsurface fluid flow, and each of the components of clay smear from the high shear strain, localized, clay-rich fault zone to the coherent, undeformed block of shale caught in a fault lens has the ability to contribute to that impediment. Perhaps what is more uncertain is the extent to which each of these components contributes to reduced flow in any particular fault and for all faults in general. Moreover, in the little bit of work that has been done to evaluate processes that create gouge holes, it appears that they might develop quite readily, although more easily in some environments than others. So holes are an integral part of the ensemble of clay smear structures for the subsurface flow problem, even though they may only occupy a small fraction of the fault zone and be of far lesser consequence for its mechanics.

Therefore, for a larger fault area, the probability of finding a hole somewhere on this fault is greater than on a small fault area. Based on the arguments above, we propose that fully continuous clay smear on large faults is rare, and therefore the creation of a capillary seal by clay smear on a large fault surface is rare. Wehr et al. (2000) illustrated the consequences of introducing a small fraction of holes in a fault zone, consequences that may well be realized through the application of probabilistic clay smear calculations (e.g.,Childs et al., 2007 and Yielding, 2012). Returning to the contrast between multi-fault geometric analyses (James et al., 2004) and clay smear calculations (Dee et al., 2007), it seems reasonable to conclude that a large part of the cross-fault capillary seal problem is dictated by the geometry of fault juxtapositions that form by faulting but that the development of clay smear distributed within the fault zone has the potential to modify that juxtaposition window. The question then becomes: is that contribution significant or negligible in the context of the geometric uncertainties arising from the fault juxtaposition evaluation?

The second example of the Hibernia field study in which a small perched water accumulation is trapped in a geometric configuration that requires a downdip barrier that prevents the draining of water, is intended to illustrate this conundrum. In this instance, the calculation of a clay smear parameter (SGR) yields a value that must be compatible with cross-fault oil flow and pressure depletion during production. For this reason, we explored the possible interpretation of a largely undeformed shale lens in the fault zone as the clay smear structure that could cause the trapping of water. Moreover, we identified this example as one occurring on a fault segment with limited extent, thereby acknowledging the idea that a larger fault would increase the chance of a hole developing at a depth that would have allowed more water to drain from the fault trap. This line of reasoning leads us to speculate on the characteristic length scales of fault lens and relay structures and whether these scales have an influence on the area of a fault where clay smear processes may supplement traps created by fault juxtaposition.

For the more general case of cross-fault water, oil, and gas flow, the effects of clay smear are more compelling. The evidence from comprehensive subsurface oil and gas field studies (e.g., Jolley et al., 2007 and Myers et al., 2007) and aquifer studies (e.g., Bense and Van Balen, 2004) show clear evidence of the effects of fault gouge, including and especially clay smear, on flow patterns resulting from fluid production. However, our understanding is plagued by similar issues to the capillary seal problem, including:

- How are the proportion and distribution of high shear strain clay found in the principle fault slip zones and less deformed shale lenses between those shear zones determined?
- What is the distribution and size of holes that form in the clay smear?
- How is flow (e.g., water) distributed between small, high permeability holes and large areas of thin, low permeability clay smear?

The effects of clay smear are present in subsurface flow problems, but the questions are how well the flow properties of clay smear are deduced from subsurface measurements, and how good the prediction of those effects can be. We pose these questions in the full knowledge that the fidelity of the answer often depends on the question posed. Although we strive to achieve a deep, simple, generalized understanding like Byerlee's Law and seek to define the means to know when we have found that insight, we recognize that there is utility in practical applications of less complete understanding.

In summary, improved understanding of clay smear processes and their effect on cross-fault fluid flow will only come about with more comprehensive, detailed, and reproducible experiments and analyses of subsurface data. In addition, improved understanding of clay smear processes will result in better predictions of deformation regimes where clay smear has more or less influence on cross-fault flow.

3.4.4. How effective are predictive models of clay smear?

Clay smear predictive algorithms (Eqs. , and) are incomplete because they only account for the amount of clay or shale in a stratigraphic section in the context of the fault throw and make no accommodation for the large variety of fault processes and structures that create that lithologic offset. CSP is the only model with a mechanical basis, but it is defined in terms of only one facet of the mechanical problem (i.e. the clay incorporation process).

Thus attempts to 'calibrate' or 'validate' must be viewed with suspicion. Are calibration datasets drawn from faults in the same deformation regime? We know of no fully convincing demonstration of subsurface fault seal (capillary membrane seal) by clay smear, although we accept that such cases may exist. These are strong, provocative statements intended to address the terms 'calibrate' and 'validate' used in the literature, which we also find to be strong claims. We think these terms can only be applied when studies incorporate the most rigorous standards of experimental testing which we think are largely lacking. Scientifically rigorous experiments are difficult to conduct because most subsurface tests are designed for other purposes. However, it is possible to design subsurface experiments to appropriately test clay smear models in the context of specific, testable hypotheses.

We accept that clay smear affects subsurface fluid flow, and we accept that current clay smear models and algorithmic approaches can prove useful in some cases to estimate the nature of those effects. Keeping in mind the four main parameters that control clay smear, it is difficult to accept that only one of these can be used to predict. However, the current state makes it difficult to understand when those models may be misleading or when they might be right for the wrong reasons. For example, could the development of a shale lens predictor be more useful than a clay smear predictive model?

So when we claim that we know of no fully convincing demonstration of subsurface fault seal by clay smear, we mean this in the context of the data used in the subsurface evaluation. Too many data interpretations are non-unique, fraught with undefined spatial uncertainty, or limited by unresolved geometric complexities. We find the balance of evidence brought forward in subsurface field studies to be insufficient to convict. However, we also acknowledge that application of the existing imperfect models has proved useful in contributing to practical subsurface applications when the geologic setting has had some resiliency and flexibility for imperfect models.

3.4.5. Ways forward

There are excellent opportunities for the advancement of clay smear research that will improve our understanding of fault processes in sedimentary rocks and the effects of the resulting fault structures on subsurface fluid flow. Those advancements will be realized via the development of advanced predictive models that will entail a forward deformation model based on an input stratigraphic section, effective stress conditions, and mechanical property characterizations for each faulted lithology. The resulting geometric description of faulted and deformed lithologies within a fault zone will then be used with a geologic model of the subsurface to better evaluate and predict cross-fault flow and for earthquake studies to predict whether any fault geometry prediction is more or less susceptible to seismicity for an imposed stress field.

We will achieve an advanced understanding of clay smear processes necessary for a predictive forward model by advancements in each of the contributing areas that have led to our current state of knowledge. In outcrop studies future work should identify which aspects of the clay smear evolution the observations relate to (shale incorporation, fault zone evolution, and clay smear disruption) with special attention dedicated to observations of clay smear disruption; this will likely involve a greater emphasis on outcrop observations in 3D. In addition, all relevant geometric aspects of the faulted outcrop should be recorded and documented to better inform numerical and laboratory modeling studies; a goal of outcrop studies should be a 3D retro-deformable fault interpretation. Lastly, outcrops need to be documented in a way that allows the faulted stratigraphy to be evaluated with current and future clay smear models and ultimately fault-deformation forward models.

Laboratory studies will help elucidate complex, nonlinear fault processes for given initial and boundary conditions. However, each contribution needs to be documented and qualified for the limitations imposed by the experimental design; we need to know which observations are independent of design attributes and which may be influenced by those conditions. Experimental material characterization must be documented in relation to mechanical standards like the Benchmark experiments in order to insure reproducibility. And like outcrop studies a geometric characterization of the resulting fault structure is required to compare against observations in nature.

Numerical models require many of the same qualifications as laboratory studies, including explicit statements about the limitations of the applied modeling approach, application of benchmarked, standard material properties, and geometric characterization of resulting fault structures. In our opinion this domain of research has been plagued by the lack of reproduced results and benchmarks. The path toward a fault-deformation forward model likely begins with models that reproduce laboratory experiments where initial and boundary conditions are known, expanding to models that reproduce complete outcrop observations (ideally a fault described tip-to-tip), and ultimately to prediction of fault structures where only the initial and boundary conditions are given. The ultimate fault forward-model may be a numerical simulation or it may be a simpler model based on numerical simulations. The most significant advancements, however, will come from research dedicated to defining processes that create holes in clay smear and that allows us to better evaluate the relative importance of holes compared with all other clay smear structures.

The single most important requirement of subsurface field studies is to conduct subsurface flow tests that interrogate limited areas of faults for which cross-fault flow predictions exist. In this instance and all other field evaluations that involve faults, all necessary information must be documented to allow others to reproduce the experiment and reach the same conclusion. Subsurface flow predictions must entail measurable quantities, like a prediction of fluid type, fluid pressure, and fluid contacts (Vrolijk et al., 2005b and Vrolijk et al., 2005a) in untested fault blocks. With these more rigorous analytical requirements, field studies of fluid flow can interrogate large fault areas, searching for the weak link that is difficult to find in outcrops, for example. Lastly, predictive models of fluid flow need to incorporate aspects of the mechanics of deformation to improve their accuracy.

We think that the existing research has laid a foundation for a more complete model of clay smear development in normal faults in clastic sedimentary sections. That model may build upon the model presented in Fig. 27 that emphasizes the geometric components of the resulting fault structures and lithologies contained within those structures. The resulting predictive model of fault deformation will nevertheless provide geometric attributes of clay smear components that will be used in an advanced clay smear flow predictive model, perhaps along the lines of Equations , and that involve existing fault zone predictions (e.g., SGR) and supplement them with geometric constraints (especially Eq. (8)).

The path to applying the results of a clay-smear predictive model for earthquake studies is more speculative, but *a priori* knowledge of heterogeneous fault structure may influence those who consider the effects of such heterogeneous structure on earthquakes. In addition, it is conceivable that different faults with different structure (e.g., amount and continuity of a clay component) may be more or less susceptible to induced seismicity caused by changes in stress state in an area.

4. SUMMARY

Significant advancements in understanding normal fault processes in clastic sedimentary sequences and the effects of those processes on cross-fault gas, oil, and water flow have come about as a result of clay smear studies over the past four decades. From the initial conception of clay distributed along the dip of a normal fault using a process model developed from outcrop observations, laboratory experiments, and geomechanical principles, our understanding has advanced from additional research in all these areas and from subsurface flow studies. An important part of this evolution includes technological advancements, which have resulted in better numerical simulation methods, better subsurface data characterization, and new laboratory capabilities.

Although current clay smear prediction algorithms are based on simple stratigraphic and fault offset parameters, our review of the work committed to exploring fault processes in outcrop and in laboratory and numerical models suggests that we stand on the precipice of a new type of model that incorporates the manner in which rocks deform during faulting taken in concert with models of fault geometric evolution. The evidence supporting the idea that these factors play a fundamental role in fault processes is unequivocal. We think our understanding of potential fault deformation regimes is sufficient to apply today, even though that understanding remains incomplete. However, better predictive capabilities for fault network evolution require additional work to be similarly useful.

Interestingly, although the importance of holes in clay smear was recognized at the beginning as a crucial part of the clay smear problem, and this realization has resurfaced repeatedly over the years, there has been surprisingly little work until recently to investigate the processes that create holes and the conditions that favor those processes. Ongoing work will continue to develop this understanding, and we think that this is one of the most exciting areas of current research in this area.

ACKNOWLEDGMENTS

The ideas in this paper have evolved for over twenty years and we would like to express our gratitude to the many colleagues who shared their ideas, helped with experiments, supported field work or discussed complex subsurface problems.

PJV wishes to thank a whole host of people who have influenced my thinking over the years, some of whom I acknowledge by citing their work in this review, but others who have worked on this problem without fully publishing their results, and others still who have participated in the publication process but gifted me their insights in informal discussions and debates. In particular, I acknowledge the contributions of Rod Myers, Brooks Clark, Bill Shea, Lee Fairchild, Bill James, Ellen Meurer, Michael Tsenn, Paul Hicks, Gretchen Nakayama, Martha Gerdes, B J McPherson, Dave Stern, Dave Chorneyko, Dave Braisted, Joyce Schmatz, Sohrab Noorsalehi-Garakani, Sofie Nollet, Gisa Kleine Vennekate, Conrad Childs, Tom Manzocchi, Martin Schoepfer, John Walsh, and Graham Yielding. Dana Butters helped me with the Hibernia Field example by making his seismic interpretation available to me. Lastly, I wish to thank Chuck Tautfest, who supported me in what would have been one of the most comprehensive and definitive cross-fault flow experiments had it not been for a critical mechanical failure.

JLU wishes to thank Georg Mandl and Florian Lehner, for initial inspiration, and Yves Leroy, Pascal Richard, John Walsh, Pieter Vermeer, Wouter van der Zee, Robrecht Schmitz, Joyce Schmatz, Heijn van Gent, Marc Holland, Steffen Abe, Sohrab Noorsalehi-Garakani, Alexander Raith, Martin Ziegler, Gisa Kleine Vennekate, Steffen Giese, and Werner Kraus for sharing their ideas and for important lessons learned in a wide variety of fields.

MK and JLU thank Sven Asmus, Peter Lokay and Ulrich Krüger from RWE Power AG for providing the possibility to investigate faults in the Hambach lignite mine and all students who helped in the field.

Thoughtful reviews by Brandon Dugan and Conrad Childs, both of whom demonstrated perseverance in their tasks, is gratefully acknowledged.

Finally, JLU and MK acknowledge financial support by the German Science Foundation (Projects UR 64-2, ZI 714/4-1 and UR 64/10-1); and by ExxonMobil Upstream Research Company.

REFERENCES

1. Abe, S., Latham, S., Mora, P., 2006. Dynamic rupture in a 3-D particle-based simulation of a rough planar fault. Pure Appl. Geophys. 163, 1881–1892.
2. Allan, U.S., 1989. Model for hydrocarbon migration and entrapment within faulted structures. AAPG Bull. 73, 803–811. Aydin, A., Eyal, Y., 2002. Anatomy of a normal fault with shale smear; implications for fault seal. AAPG Bull. 86, 1367–1381.
3. Bailey, W.R., Underschultz, J., Dewhurst, D.N., Kovack, G., Mildren, S., Raven, M., 2006. Multi-disciplinary approach to fault and top seal appraisal; PyreneesMacedon oil and gas fields, Exmouth Sub-basin, Australian Northwest Shelf. Mar. Pet. Geol. 23, 241–259.
4. Balthasar, K., Gudehus, G., Kulzer, M., Bertini, A., 2006. Thin layer shearing of a highly plastic clay. Nonlinear Process. Geophys. 13, 671–680.
5. Belytschko, T., Chian, H.-Y., Plaskacz, E., 1994. High resolution two-dimensional shear band computations: imperfections and mesh dependence. Comput. Methods Appl. Mech. Eng. 119, 1–15.
6. Bense, V.F., Person, M.A., Chaudhary, K., You, Y., Cremer, N., Simon, S., 2008. Thermal anomalies indicate preferential flow along faults in unconsolidated sedimentary aquifers. Geophys. Res. Lett. 35, L24406.
7. Bense, V.F., Van Balen, R., 2004. The effect of fault relay and clay smearing on groundwater flow patterns in the Lower Rhine Embayment. Basin Res. 16, 397–411.
8. Bense, V.F., Van den Berg, E.H., Van Balen, R.T., 2003. Deformation mechanisms and hydraulic properties of fault zones in unconsolidated sediments; the Roer Valley Rift System, The Netherlands. Hydrogeol. J. 11, 319–332.
9. Bentley, M.R., Barry, J.J., 1991. Representation of fault sealing in a reservoir simulation: Cormorant Block IV, UK North Sea. In: SPE Annual Technical Conference and Exhibition, 6–9 October, Dallas, Texas.
10. Berg, R.R., Avery, A.H., 1995. Sealing properties of tertiary growth faults, Texas Gulf Coast. AAPG Bull. 79, 375–393.
11. Bourne, S.J., Oates, S.J., van Elk, J., Doornhof, D., 2014. A seismological model for earthquakes induced by fluid extraction from a subsurface reservoir. J. Geophys. Res. Solid Earth 119, 8991–9015.

12. Bouvier, J.D., Kaars-Sijpesteijn, C.H., Kluesner, D.F., Onyejekwe, C.C., van der Pal, R.C., 1989. Three-dimensional seismic interpretation and fault sealing investigations, Nun River Field, Nigeria. AAPG Bull. 73, 1397–1414.

13. Bretan, P., Yielding, G., 2005. Using buoyancy pressure profiles to assess uncertainty in fault seal calibration. In: Boult, P., Kaldi, J. (Eds.), Evaluating Fault and Cap Rock Seals. The American Association of Petroleum Geologists, pp. 151–162.

14. Bretan, P., Yielding, G., Jones, H., 2003. Using calibrated shale gouge ratio to estimate hydrocarbon column heights. AAPG Bull. 87, 397–413.

15. Buiter, S., Babeyko, A., Ellis, S., Gerya, T., Kaus, B., Kellner, A., Schreurs, G., Yamada, Y., 2006. The numerical sandbox: comparison of model results for a shortening and an extension experiment. Analog. Numer. Model. Crustal-Scale Process. 253, 29–64.

16. Bullock, R.J., De Paola, N., Holdsworth, R.E., 2015. An experimental investigation into the role of phyllosilicate content on earthquake propagation during seismic slip in carbonate faults. J. Geophys. Res. Solid Earth 120, 3187–3207.

17. Burhannudinnur, M., Morley, C.K., 1997. Anatomy of growth fault zones in poorly lithified sandstones and shales: implications for reservoir studies and seismic interpretation: part 1, outcrop study. Pet. Geosci. 3, 211–224.

18. Byerlee, J., 1978. Friction of rocks. Pure Appl. Geophys. 116, 615–626.

19. Caine, J.S., Evans, J.P., Forster, C.B., 1996. Fault zone architecture and permeability structure. Geology 24, 1025–1028.

20. Caine, J.S., Minor, S.A., 2009. Structural and geochemical characteristics of faulted sediments and inferences on the role of water in deformation, Rio Grande Rift, New Mexico. Geol. Soc. Am. Bull. 121, 1325–1340.

21. Childs, C., Manzocchi, T., Walsh, J.J., Bonson, C.G., Nicol, A., Schopfer, M.P.J., 2009a. € A geometric model of fault zone and fault rock thickness variations. J. Struct. Geol. 31, 117–127.

22. Childs, C., Sylta, Ø., Moriya, S., 2002. A method for including the capillary properties of faults in hydrocarbon migration models. Nor. Pet. Soc. 11, 127–139.

23. Childs, C., Sylta, O., Moriya, S., Morewood, N., Manzocchi, T., Walsh, J.J., Hermanssen, D., 2009b. Calibrating fault seal using a hydrocarbon migration model of the Oseberg Syd area, Viking Graben. Mar. Pet. Geol. 26, 764–774.

24. Childs, C., Walsh, J.J., Manzocchi, T., Strand, J., Nicol, A., Tomasso, M., Schopfer, M.P.J., Aplin, A.C., 2007. Definition of a Fault Permeability Predictor from Outcrop Studies of a Faulted Turbidite Sequence, Taranaki, New Zealand, Structurally Complex Reservoirs. Geological Society of London, London, pp. 235–258.

25. Childs, C., Walsh, J.J., Watterson, J., 1997. Complexity in fault zone structure and implications for fault seal prediction. In: Møller-Pedersen, P., Koestler, A.G. (Eds.), Hydrocarbon Seals: Importance for Exploration and Production. Norwegian Petroleum Society, Trondheim, Norway, pp. 61–72.

26. Ciftçi, N.B., Giger, S.B., Clennell, M.B., 2012. Testing fault seal prediction algorithms using geomodels of experimentally produced fault zones. In: Fault and Top Seals 2012, Montpellier, France, pp. 1–3.

27. Çiftçi, N.B., Giger, S.B., Clennell, M.B., 2013. Three-dimensional structure of experimentally produced clay smears: implications for fault seal analysis. AAPG Bull. 97, 733–757.

28. Cilona, A., Aydin, A., Johnson, N., 2015. Permeability of a fault zone crosscutting a sequence of sandstones and shales and its influence on hydraulic head distribution in the Chatsworth Formation, California, USA. Hydrogeol. J. 23, 405–419.

29. Clarke, S.M., Littler, M., Burley, S.D., Williams, G.D., Hughes, D., Coogan, S., 2006. Modelling the effects of stratigraphical uncertainty on fault seal and trap-fill in faulted structures. Pet. Geosci. 12, 143–156.

30. Clausen, J.A., Gabrielsen, R.H., 2002. Parameters that control the development of clay smear at low stress states: an experimental study using ring-shear apparatus. J. Struct. Geol. 24, 1569–1586.

31. Clausen, J.A., Gabrielsen, R.H., Johnsen, E., Korstgård, J.A., 2003. Fault architecture and clay smear distribution. Examples from field studies and drained ring-shear experiments. Norw. J. Geol. 83, 131–146.

32. Cosserat, E., Cosserat, F., 1909. Theorie des corps deformables. A. Hermann et fils, Paris.

33. Crawford, B.R., Faulkner, D.R., Rutter, E.H., 2008. Strength, porosity, and permeability development during hydrostatic and shear loading of synthetic quartzclay fault gouge. J. Geophys. Res. 113, B03207.

34. Crawford, B.R., Myers, R.D., Woronow, A., Faulkner, D.R., Rutter, E.H., 2002. Porositypermeability relationships in clay-bearing fault gouge. In: SPE/ISRM Rock Mechanics Conference, 20–23 October. Society of Petroleum Engineers, Irving, Texas.

35. Crook, A.J.L., Willson, S.M., Yu, J.G., Owen, D.R.J., 2006. Predictive modelling of structure evolution in sandbox experiments. J. Struct. Geol. 28, 729–744.

36. Cuisiat, F., Skurtveit, E., 2010. An experimental investigation of the development and permeability of clay smears along faults in uncemented sediments. J. Struct. Geol. 32, 1850–1863.

37. Cundall, P.A., Drescher, A., Strack, O.D.L., 1982. Numerical experiments on granular assemblies; measurements and observations. In: Vermeer, P.A., Luger, H.J. (Eds.), IUTAM Conference on Deformation End Failure of Granular Materials. Delft, Balkema, Rotterdam, pp. 553–568.

38. Cundall, P.A., Strack, O.D.L., 1979. A discrete numerical model for granular assemblies. Geotechnique 29, 47–65.

39. Davatzes, N.C., Aydin, A., 2005. Distribution and nature of fault architecture in a layered sandstone and shale sequence; an example from the Moab Fault, Utah. AAPG Mem. 85, 153–180.

40. Davies, R.K., An, L., Jones, P., Mathis, A., Cornette, C., 2003. Fault-seal analysis south Marsh Island 36 field, Gulf of Mexico. AAPG Bull. 87, 479–491.

41. Dee, S.J., Yielding, G., Freeman, B., Bretan, P., 2007. A Comparison between Deterministic and Stochastic Fault Seal Techniques. In: Geological Society, London, Special Publications, vol. 292, pp. 259–270.

42. Doughty, P.T., 2003. Clay smear seals and fault sealing potential of an exhumed growth fault, Rio Grande rift, New Mexico. AAPG Bull. 87, 427–444.

43. Egholm, D.L., 2007. A new strategy for discrete element numerical models; 1, Theory. J. Geophys. Res. 112, B05203.

44. Egholm, D.L., Sandiford, M., Clausen, O.R., Nielsen, S.B., 2007. A new strategy for discrete element numerical models; 2, Sandbox applications. J. Geophys. Res. 112, B05204.

45. Egholm, D.L., Clausen, O.R., Sandiford, M., Kristensen, M.B., Korstgard, J.A., 2008. The mechanics of clay smearing along faults. Geology 36, 787–790.

46. Eichhubl, P., D'Onfro, P.S., Aydin, A., Waters, J., McCarty, D.K., 2005. Structure, petrophysics, and diagenesis of shale entrained along a normal fault at Black Diamond Mines, Californiaeimplications for fault seal. AAPG Bull. 89, 1113–1137.

47. Faerseth, R.B., 2006. Shale smear along large faults; continuity of smear and the fault seal capacity. J. Geol. Soc. Lond. 163, 741–751.

48. Faerseth, R.B., Johnsen, E., Sperrevik, S., 2007. Methodology for risking fault seal capacity; implications of fault zone architecture. AAPG Bull. 91, 1231–1246.

49. Faulkner, D.R., Jackson, C.A.L., Lunn, R.J., Schlische, R.W., Shipton, Z.K., Wibberley, C.A.J., Withjack, M.O., 2010. A review of recent developments concerning the structure, mechanics and fluid flow properties of fault zones. J. Struct. Geol. 32, 1557–1575.

50. Finch, E., Hardy, S., Gawthorpe, R., 2004. Discrete-element modelling of extensional fault-propagation folding above rigid basement fault blocks. Basin Res. 16, 489–506.

51. Fisher, Q.J., Jolley, S.J., 2007. Treatment of Faults in Production Simulation Models. In: Geological Society, London, Special Publications, vol. 292, pp. 219–233.

52. Fisher, Q.J., Knipe, R.J., 1998. Fault sealing processes in siliciclastic sediments. In: Jones, G., Fisher, Q.J., Knipe, R.J. (Eds.), Faulting, Fault Sealing and Fluid Flow in Hydrocarbon Reservoirs. Geological Society of London, London, pp. 117–134.

53. Fisher, Q.J., Knipe, R.J., 2001. The permeability of faults within siliciclastic petroleum reservoirs of the North Sea and Norwegian Continental Shelf. Mar. Pet. Geol. 18, 1063–1081.

54. Foxford, K.A., Walsh, J.J., Watterson, J., Garden, I.R., Guscott, S.C., Burley, S.D., 1998. Structure and content of the Moab Fault Zone, Utah, USA, and its implications for fault seal prediction. In: Jones, G., Fisher, Q.J., Knipe, R.J. (Eds.), Faulting, Fault Sealing and Fluid Flow in Hydrocarbon Reservoirs. Geological Society London, London, pp. 87–103.

55. Freeman, B., Yielding, G., Needham, D.T., Badley, M.E., 1998. Fault seal prediction: the gouge ratio method. In: Coward, M.P., Daltaban, T.S., Johnson, H. (Eds.), Structural Geology in Reservoir Characterization. Geological Society London, London, pp. 19–25.

56. Fristad, T., Groth, A., Yielding, G., Freeman, B., 1997. Quantitative fault seal prediction: a case study from Oseberg Syd. In: Møller-Pedersen, P., Koestler, A.G. (Eds.), Hydrocarbon Seals: Importance for Exploration and Production. Norwegian Petroleum Society, Trondheim, Norway, pp. 107–124.

57. Fulljames, J.R., Zijerveld, L.J.J., Franssen, R.C.M.W., 1997. Fault seal processes: systematic analysis of fault seals over geological and production time scales. In: Møller-Pedersen, P., Koestler, A.G. (Eds.), Hydrocarbon Seals: Importance for Exploration and Production. Norwegian Petroleum Society, pp. 51–59.

58. Gibson, R.G., 1994. Fault-zone seals in siliciclastic Strata of the Columbus Basin, Offshore Trinidad. AAPG Bull. 78, 1372–1385.

59. Gibson, R.G., 1998. Physical character and fluid-flow properties of sandstonederived fault zones. In: Coward, M.P., Daltaban, T.S., Johnson, H. (Eds.), Structural Geology in Reservoir Characterization. Geological Society London, London, pp. 83–97.

60. Gibson, R.G., Bentham, P.A., 2003. Use of fault-seal analysis in understanding petroleum migration in a complexly faulted anticlinal trap, Columbus Basin, offshore Trinidad. AAPG Bull. 87, 465–478.

61. Giger, S.B., Clennell, M.B., Ciftci, N.B., Harbers, C., Clark, P., Ricchetti, M., 2013. Fault transmissibility in clastic-argillaceous sequences controlled by clay smear evolution. AAPG Bull. 97, 705–731.

62. Giger, S.B., Clennell, M.B., Harbers, C., Clark, P., Ricchetti, M., Ter Heege, J.H., Wassing, B.B.T., Orlic, B., 2011. Design, operation and validation of a new fluidsealed direct shear apparatus capable of monitoring fault-related fluid flow to large displacements. Int. J. Rock Mech. Min. Sci. 48, 1160–1172.

63. Gratier, J.P., 2011. Fault permeability and strength evolution related to fracturing and healing episodic processes (years to millennia): the role of pressure solution. Oil Gas Sci. Technol. Rev. IFP Energies Nouv. 66, 491–506.

64. Gudehus, G., Karcher, C., 2007. Hypoplastic simulation of normal faults without and with clay smears. J. Struct. Geol. 29, 530–540.

65. Gudehus, G., Nubel, K., 2004. Evolution of shear bands in sand. Geotechnique 54, 187–201.

66. Haines, S.H., Kaproth, B., Marone, C., Saffer, D., van der Pluijm, B., 2013. Shear zones in clay-rich fault gouge: a laboratory study of fabric development and evolution. J. Struct. Geol. 51, 206–225.

67. Hall, S., Bornert, M., Desrues, J., Pannier, Y., Lenoir, N., Viggiani, G., Besuelle, P., 2010. Discrete and continuum analysis of localised deformation in sand using X-ray mCT and volumetric digital image correlation. Geotechnique 60, 315–322.

68. Hardy, S., Finch, E., 2005. Discrete-element modelling of detachment folding. Basin Res. 17, 507–520.

69. Hardy, S., Finch, E., 2006. Discrete element modelling of the influence of cover strength on basement-involved fault-propagation folding. Tectonophysics 415, 225–238.

70. Harris, D., Yielding, G., Levine, P., Maxwell, G., Rose, P.T., Nell, P., 2002. Using Shale Gouge Ratio (SGR) to model faults as transmissibility barriers in reservoirs; an example from the Strathspey Field, North Sea. Pet. Geosci. 8, 167–176.

71. Herle, I., Feda, J., 2001. Interaction of spread footings with sandy subsoil. Eng. Mech. 8, 1–12.

72. Herle, I., Feda, J., 2002. Interaction of spread footings with sandy subsoil. Part 2: plane strain FE modelling. Eng. Mech. 9, 259–272.

73. Hesthammer, J., Fossen, H., 2000. Uncertainties associated with fault sealing analysis. Pet. Geosci. 6, 37–45.

74. Heynekamp, M.R., Goodwin, L.B., Mozley, P.S., Haneberg, W.C., 1999. Controls on fault-zone architecture in poorly lithified sediments, Rio Grande Rift, New Mexico: implications for fault-zone permeability and fluid flow. In: Haneberg, W.C., Mozley, P.S., Moore, J.C., Goodwin, L.B. (Eds.), Faults and Subsurface Fluid Flow in the Shallow Crust. American Geophysical Union, Washington D.C, pp. 27–49.

75. Holland, M., Urai, J.L., van der Zee, W., Stanjek, H., Konstanty, J., 2006. Fault gouge evolution in highly overconsolidated claystones. J. Struct. Geol. 28, 323–332.

76. Hovadik, J.M., Larue, D.K., 2010. Stratigraphic and Structural Connectivity. In:

77. Geological Society, London, Special Publications, vol. 347, pp. 219–242.

78. Hubbert, M.K., 1940. The theory of ground-water motion. J. Geol. 785–944.

79. Hügel, H.M., 1995. Prognose von Bodenverformungen, Institut für Bodenmechanik und Felsmechanik. Fridericiana Universit€ at, Karlsruhe.

80. Ingram, G.M., Urai, J.L., 1999. Top-seal Leakage through Faults and Fractures: the Role of Mudrock Properties. In: Geological Society, London, Special Publications, vol. 158, pp. 125–135.

81. Ishii, E., Sanada, H., Funaki, H., Sugita, Y., Kurikami, H., 2011. The relationships among brittleness, deformation behavior, and transport properties in mudstones: an example from the Horonobe Underground Research Laboratory, Japan. J. Geophys. Res. Solid Earth 116.

82. James, W.R., Fairchild, L.H., Nakayama, G.P., Hippler, S.J., Vrolijk, P.J., 2004. Fault-seal analysis using a stochastic multifault approach. AAPG Bull. 88, 885–904.

83. Jev, B.I., Kaarssijpesteijn, C.H., Peters, M., Watts, N.L., Wilkie, J.T., 1993. Akaso field, Nigeria e use of integrated 3-D seismic, fault slicing, clay smearing, and RFT pressure data on fault trapping and dynamic leakage. AAPG Bull. 77, 1389–1404.

84. Jolley, S.J., Dijk, H., Lamens, J.H., Fisher, Q.J., Manzocchi, T., Eikmans, H., Huang, Y., 2007. Faulting and fault sealing in production simulation models: Brent Province, northern North Sea. Pet. Geosci. 13, 321–340.

85. Karakouzian, M., Hudyma, N., 2002. A new apparatus for analog modeling of clay smears. J. Struct. Geol. 24, 905–912.

86. Kettermann, M., Urai, J.L., 2015. Changes in structural style of normal faults due to failure mode transition: first results from excavated scale models. J. Struct. Geol. 74, 105–116.

87. Kleine-Vennekate, G., 2013. Numerische Simulationen zur Scherfugenentwicklung in Sand-Ton-Wechselfolgen/Numerical Simulations of Shear Zones in Layered Sand-clay Sequences, Fakultat für Bauingenieurwesen (PhD thesis). RWTH € Aachen University, Aachen.

88. Kleine-Vennekate, G., Noorsalehi-Garakani, S., Urai, J.L., Ziegler, M., 2014. Numerische simulationen des clay-smear-prozesses in sandboxversuchen. Geotechnik 37, 83–95.

89. Knai, T.A., Knipe, R.J., 1998. The Impact of Faults on Fluid Flow in the Heidrun Field. In: Geological Society, London, Special Publications, vol. 147, pp. 269–282.

90. Knipe, R.J., 1992. Faulting processes and fault seal. In: Larsen, R.M., Brekke, H., Larsen, B.T., Talleraas, E. (Eds.), Structural and Tectonic Modelling and its Application to Petroleum Geology, NPF Special Publication, vol. 1. Elsevier, Amsterdam, pp. 325–342.

91. Knipe, R.J., 1997. Juxtaposition and seal diagrams to help analyze fault seals in hydrocarbon reservoirs. AAPG Bull. 81, 187–195.

92. Knott, S.D., 1993. Fault seal analysis in the North Sea. AAPG Bull. 77, 778–792.

93. Koledoye, B.A., Aydin, A., May, E., 2003. A new process-based methodology for analysis of shale smear along normal faults in the Niger Delta. AAPG Bull. 87, 445–463

94. Kolymbas, D., 1988. Eine konstitutive Theorie für Boden und andere k € ornige Stoffe. € Institut für Bodenmechanik und Felsmechanik der Universit€ at Fridericiana, Karlsruhe. Kristensen, M.B., Childs, C., Olesen, N.O., Korstgard, J.A., 2013. The microstructure and internal

architecture of shear bands in sand-clay sequences. J. Struct. Geol. 46, 129–141.

95. Latham, S., Abe, S., Mora, P., 2006. Parallel 3-D simulation of a fault gouge using the lattice solid model. Pure Appl. Geophys. 163, 1949–1964.

96. Lehner, F.K., Pilaar, W.F., 1997. The emplacement of clay smears in synsedimentary normal faults: inferences from field observations near Frechen, Germany. In: Møller-Pedersen, P., Koestler, A.G. (Eds.), Norwegian Petroleum Society Special Publications: Hydrocarbon Seals: Importance for Exploration and Production. Elsevier Science, pp. 39–50.

97. Lewis, G., Knipe, R.J., Li, A., 2002. Fault seal analysis in unconsolidated sediments: a field study from Kentucky, USA. In: Koestler, A.G., Hunsdale, R. (Eds.), Hydrocarbon Seal Quantification, Norwegian Petroleum Society (NPF), Special Publications, vol. 11, pp. 243–253.

98. Lindsay, N.G., Walsh, J.J., Watterson, J., Murphy, F.C., 1993. Outcrop studies of shale smears on fault surfaces. In: Flint, S.S., Bryant, I.D. (Eds.), The Geological Modelling of Hydrocarbon Reservoirs and Outcrop Analogues. Blackwell Scientific Publications, Inc, Oxford, UK, pp. 113–123.

99. Loveless, S., Bense, V., Turner, J., 2011. Fault architecture and deformation processes within poorly lithified rift sediments, Central Greece. J. Struct. Geol. 33, 1554–1568.

100. Luding, S., Hermann, H.J., 2001. Micro-macro transition for cohesive granular media. In: Diebels, S. (Ed.), Zur Beschreibung Komplexen Materialverhaltens. Institut für Mechanik (Bauwesen), Stuttgart.

101. Lupini, J.F., Skinner, A.E., Vaughan, P.R., 1981. The drained residual strength of cohesive soils. Geotechnique 31, 181–213.

102. Mackay, A.H., Tankard, A.J., 1990. Hibernia Oil Field Canada, Jeanne D'Arc Basin, Grand Banks, Newfoundland. In: AAPG Treatise of Petroleum Geology, Atlas of Oil and Gas Fields A-019, pp. 145–175.

103. Maksimovic, M., 1989. Nonlinear failure envelope for soils. J. Geotech. Eng. 115, 581–586.

104. Mandl, G., 1988. Mechanics of tectonic faulting: models and basic concepts. In: Zwart, H.J. (Ed.), Developments in Structural Geology. Elsevier, Amsterdam. Mandl, G., Jong, L.N.J., Maltha, A., 1977. Shear zones in granular material. Rock Mech. 9, 95–144.

105. Manzocchi, T., Carter, J.N., Skorstad, A., Fjellvoll, B., Stephen, K.D., Howell, J.A., Matthews, J.D., Walsh, J.J., Nepveu, M., 2008a. Sensitivity of

the impact of geological uncertainty on production from faulted and unfaulted shallowmarine oil reservoirs: objectives and methods. Pet. Geosci. 14, 3–15.

106. Manzocchi, T., Childs, C., Walsh, J.J., 2010. Faults and fault properties in hydrocarbon flow models. Geofluids 10, 94–113.

107. Manzocchi, T., Heath, A.E., Palananthakumar, B., Childs, C., Walsh, J.J., 2008b. Faults in conventional flow simulation models: a consideration of representational assumptions and geological uncertainties. Pet. Geosci. 14, 91–110.

108. Manzocchi, T., Heath, A.E., Walsh, J.J., Childs, C., 2002. The representation of two phase fault-rock properties in flow simulation models. Pet. Geosci. 8, 119–132.

109. Manzocchi, T., Matthews, J.D., Strand, J.A., Carter, J.N., Skorstad, A., Howell, J.A., Stephen, K.D., Walsh, J.J., 2008c. A study of the structural controls on oil recovery from shallow-marine reservoirs. Pet. Geosci. 14, 55–70.

110. Manzocchi, T., Walsh, J.J., Nell, P., Yielding, G., 1999. Fault transmissibility multipliers for flow simulation models. Pet. Geosci. 5, 53–63.

111. Mühlhaus, H.B., 1986. Scherfugenanalyse bei granularem Material im Rahmen der Cosserat-Theorie. Ingenieur-Archiv 56, 389–399.

112. Myers, R.D., Allgood, A., Hjellbakk, A., Vrolijk, P., Briedis, N., 2007. Testing Fault Transmissibility Predictions in a Structurally Dominated Reservoir; Ringhorne Field, Norway. In: Geological Society Special Publications, vol. 292, pp. 271–294.

113. Myers, R.D., Vrolijk, P.J., Kiven, C.W., Tsenn, M., 2014. In: Patent, U.S. (Ed.), Method for Predicting Fluid Flow. ExxonMobil Upstream Research Company, Houston, TX, U. S., p. 21

114. Niemunis, A., 2003. Extended Hypoplastic Models for Soils. Th. Triantafyllidis, Bochum.

115. Nollet, S., Vennekate, G.J.K., Giese, S., Vrolijk, P., Urai, J.L., Ziegler, M., 2012. Localization patterns in sandbox-scale numerical experiments above a normal fault in basement. J. Struct. Geol. 39, 199–209.

116. Noorsalehi-Garakani, S., Vennekate, G.J.K., Vrolijk, P., Urai, J.L., 2013. Clay-smear continuity and normal fault zone geometry e first results from excavated sandbox models. J. Struct. Geol. 57, 58–80.

117. Nübel, K., 2002. Experimental and Numerical Investigation of Shear Localization in Granular Material. Inst. für Bodenmechanik und Felsmechanik. Fridericiana Universit€ at, Karlsruhe.

118. Nybakken, S., 1991. Sealing fault traps; an exploration concept in a mature petroleum province; Tampen Spur, northern North Sea. First Break 9, 209–222.

119. Olgaard, D.L., Urai, J.L., Dell'Angelo, L.N., Nüesch, R., Ingram, G., 1997. The influence of swelling clays on the deformation of mudrocks. Proceedings of the 36th U. S. Rock Mechanics Symposium, New York, NY, United States, June 29eJuly 2. Int. J. Rock Mech. Min. Sci. Geomech. 34 (3–4), 364–374.

120. Ottesen Ellevset, S., Knipe, R.J., Svava Olsen, T., Fisher, Q.J., 1998. Fault controlled communication in the Sleipner Vest field, Norwegian Continental Shelf; detailed, quantitative input for reservoir simulation and well planning. In: Jones, G., Fisher, Q.J., Knipe, R.J. (Eds.), Structure and Seal Analysis of Hydrocarbon Fields, Geological Society, London, Special Publications, pp. 283–297.

121. Peacock, D.C.P., Knipe, R.J., Sanderson, D.J., 2000. Glossary of normal faults. J. Struct. Geol. 22, 291–305.

122. Pevear, D.R., Vrolijk, P.J., Lomgstaffe, F.J., 1997. Timing of Moab Fault displacement and fluid movement integrated with burial history using radiogenic and stable isotopes. In: Carey, H.J., Parnell, J., Ruffell, A., Worden, R. (Eds.), Geofluids II '97: Contributions to the Second International Conference on Fluid Evolution, Migration & Interaction in Sedimentary Basins and Orogenic Belts. The Queen's University of Belfast, Belfast, pp. 42–45.

123. Platt, J.D., Brantut, N., Rice, J.R., 2015. Strain localization driven by thermal decomposition during seismic shear. J. Geophys. Res. Solid Earth 120, 4405–4433.

124. Raith, A., 2012. The Evolution of Fault Zones in Brittle Ductile-layered Rocks (MSc thesis). Department of Structural Geology, Tectonics and Geomechanics. RWTH Aachen University, Aachen, p. 76.

125. Rathbun, A.P., Marone, C., 2010. Effect of strain localization on frictional behavior of sheared granular materials. J. Geophys. Res. Solid Earth 115.

126. Revil, A., Cathles, L.M., 1999. Permeability of shaly sands. Water Resour. Res. 35, 651–662.

127. Richard, P.D., Urai, J.L., 2001. In: Nieuwland, D.A., Nijman, M. (Eds.), The Atlas of Structural Geometry: a Digital Collection of 25 Years of Analogue Modelling. Geol. Mijnb./Neth. J. Geosci. 80 (2), 59–60.

128. Rivenaes, J.C., Dart, C., 2002. Reservoir compartmentalization by water-saturated faults e is evaluation possible with today's tools? In: Koestler, A.G., Hunsdale, R. (Eds.), Hydrocarbon Seal Quantification, Norwegian Petroleum Society (NPF), Special Publications, vol. 11, pp. 173–186.

129. Rudnicki, J.W., Rice, J.R., 1975. Conditions for the localization of deformation in pressure-sensitive dilatant materials. J. Mech. Phys. Solids 23, 371–394.

130. Rycroft, C., Wong, Y., Bazant, M., 2010. Fast spot-based multiscale simulations of granular drainage. Powder Technol. 200, 1–11.

131. Sadrekarimi, A., Olson, S.M., 2010. Shear band formation observed in ring shear tests on sandy soils. J. Geotech. Geoenviron. Eng. 136, 366–375.

132. Santimano, T., Rosenau, M., Oncken, O., 2015. Intrinsic versus extrinsic variability of analogue sand-box experiments e insights from statistical analysis of repeated accretionary sand wedge experiments. J. Struct. Geol. 75, 80–100.

133. Sassi, W., Livera, S.E., Caline, B.P.R., 1992. Reservoir compartmentation by faults in Cormorant Block IV, UK Northern North Sea. In: Larsen, R.M., Brekke, H., Larsen, B.T., Talleras, E. (Eds.), Structural and Tectonic Modeling and its Application to Petroleum Geology, NPF Special Publication. Elsevier, Amsterdam.

134. Savage, J.C., Lockner, D.A., Byerlee, J.D., 1996. Failure in laboratory fault models in triaxial tests. J. Geophys. Res. Solid Earth 101, 22215–22224.

135. Schmatz, J., Holland, M., Giese, S., van der Zee, W., Urai, J.L., 2010a. Clay smear processes in mechanically layered sequences e results of water-saturated model experiments with free top surface. J. Geol. Soc. India 75, 74–88.

136. Schmatz, J., Vrolijk, P.J., Urai, J.L., 2010b. Clay smear in normal fault zones e the effect of multilayers and clay cementation in water-saturated model experiments. J. Struct. Geol. 32, 1834–1849.

137. Schopfer, M.P.J., Childs, C., Walsh, J.J., 2006. Localisation of normal faults in multi- € layer sequences. J. Struct. Geol. 28, 816–833.

138. Schopfer, M.P.J., Childs, C., Walsh, J.J., 2007a. Two-dimensional distinct element € modeling of the structure and growth of normal

faults in multilayer sequences: 1. Model calibration, boundary conditions, and selected results. J. Geophys. Res. 112.

139. Schopfer, M.P.J., Childs, C., Walsh, J.J., 2007b. Two-dimensional distinct element € modeling of the structure and growth of normal faults in multilayer sequences: 2. Impact of confining pressure and strength contrast on fault zone geometry and growth. J. Geophys. Res. 112.

140. Schultz, R.A., Summers, L.E., Lynch, K.W., Bouchard, A.J., 2014. Subsurface containment assurance program: key element overview and best practice examples. In: Offshore Technology Conference Asia, 25–28 March, Kuala Lumpur, Malaysia, OTC 24851.

141. Sinclair, I.K., Evand, J.E., Albrechtsons, E.A., Sydora, L.J., 1999. The Hibernia Oilfield e effects of episodic tectonism on structural character and reservoir compartmentalization. In: Fleet, A.J., Boldy, S.A.R. (Eds.), Petroleum Geology of Northwest Europe: Proceedings of the 5th Conference. The Geological Society, London, pp. 517–528.

142. Smith, D.A., 1966. Theoretical considerations of sealing and non-sealing faults. AAPG Bull. 50, 363–374.

143. Smith, D.A., 1980. Sealing and nonsealing faults in Louisiana Gulf Coast salt basin. AAPG Bull. 64, 145–172.

144. Solum, J.G., van der Pluijm, B.A., Peacor, D.R., 2005. Neocrystallization, fabrics and age of clay minerals from an exposure of the Moab Fault, Utah. J. Struct. Geol. 27, 1563–1576.

145. Speksnijder, A., 1987. The structural configuration of Cormorant Block IV in context of the northern Viking Graben structural framework. Geol. Mijnb. 65, 357–380.

146. Sperrevik, S., Faerseth, R.B., Gabrielsen, R.H., 2000. Experiments on clay smear formation along faults. Pet. Geosci. 6, 113–123.

147. Sperrevik, S., Gillespie, P.A., Fisher, Q.J., Knipe, R.J., Halvorsen, T., 2002. Empirical estimation of fault rock properties. In: Koestler, A.G., Hunsdale, R. (Eds.), Hydrocarbon Seal Quantification, Norwegian Petroleum Society (NPF), Special Publications, vol. 11, pp. 109–125.

148. Spiller, M., Forkel, C., Kongeter, J., 2004. Case study: inflow of groundwater into open-cast mine Hambach, Germany. J. Hydraul. Eng. ASCE 130, 608–615.

149. Sverdrup, E., Helgesen, J., Vold, J., 2003. Sealing properties of faults and their in- fluence on water-alternating-gas injection efficiency in the Snorre Field, northern North Sea. AAPG Bull. 87, 1437–1458.

150. Takahashi, M., 2003. Permeability change during experimental fault smearing. J. Geophys. Res. 108.

151. Tembe, S., Lockner, D.A., Wong, T.F., 2010. Effect of clay content and mineralogy on frictional sliding behavior of simulated gouges: binary and ternary mixtures of quartz, illite, and montmorillonite. J. Geophys. Res. Solid Earth 115.

152. TerHeege, J.H., Wassing, B.B.T., Orlic, B., Giger, S.B., Clennell, M.B., 2013. Constraints on the sealing capacity of faults with clay smears from discrete element models validated by laboratory experiments. Rock Mech. Rock Eng. 46, 465–478.

153. Thornton, C., 2000. Numerical simulations of deviatoric shear deformation of granular media. Geotechnique 50, 43 –53.

154. Thornton, C., Zhang, L., 2003. Numerical simulations of the direct shear test. Chem. Eng. Technol. 26, 153–156.

155. Urai, J.L., Schmitz, R.M., van der Zee, W., Holland, M., Vrolijk, P., 2003. Experimental investigation of clay smear processes in a geotechnical direct shear apparatus, faults and top seals: what do we know and where do we go?. In: EAGE, Montpellier, France, 8e11 September 2003, p. 43.

156. Urai, J.L., Nover, G., Zwach, C., Ondrak, R., Schoner, R., Kroos, B.M., 2008. Transport € processes. In: Littke, R., Bayer, U., Gajewski, D., Nelskamp, S. (Eds.), Dynamics of Complex Intracontinental Basins: the Central European Basin System. SpringerVerlag, Berlin, pp. 367–388.

157. van der Zee, W., Urai, J.L., 2005. Processes of normal fault evolution in a siliciclastic sequence: a case study from Miri, Sarawak, Malaysia. J. Struct. Geol. 27, 2281–2300.

158. van der Zee, W., Urai, J.L., Richard, P.D., 2003. Lateral clay injection into normal faults. GeoArabia 8, 501–522.

159. van Gent, H.W., Holland, M., Urai, J.L., Loosveld, R., 2010. Evolution of fault zones in carbonates with mechanical stratigraphy e insights from scale models using layered cohesive powder. J. Struct. Geol. 32, 1375–1391.

160. Vardoulakis, I., 1989. Shear-banding and liquefaction in granular materials on the basis of a Cosserat continuum theory. Ing. Arch. 59, 106–113.

161. Vardoulakis, I., Sulem, J., 1995. Bifurcation Analysis in Geomechanics. Blackie Academic and Professional, London.

162. Vrolijk, P., James, B., Maynard, J., Sumpter, L., Sweet, M., 2005a. Reservoir connectivity analysis e defining reservoir connections & plumbing. In: SPE Middle East Oil and Gas Show and Conference, 12e15 March. Society of Petroleum Engineers, SPE 93577, Kingdom of Bahrain.

163. Vrolijk, P., Myers, R., Sweet, M.L., Shipton, Z.K., Dockrill, B., Evans, J.P., Heath, J., Williams, A.P., 2005b. Anatomy of reservoir-scale normal faults in central Utah; stratigraphic controls and implications for fault zone evolution and fluid flow. GSA Field Guide 6, 261–282.

164. Vrolijk, P., van der Pluijm, B.A., 1999. Clay gouge. J. Struct. Geol. 21, 1039e1048. Wang, Y., Abe, S., Latham, S., Mora, P., 2006. Implementation of particle-scale rotation in the 3-d Lattice Solid Model. Pure Appl. Geophys. 163, 1769–1785.

165. Weber, K.J., Mandl, G., Pilaar, W.F., Lehner, F., Precious, R.G., 1978. The role of faults in hydrocarbon migration and trapping in Nigerian growth fault structures. In: Offshore Technology Conference, Dallas, TX, United States, pp. 2643–2653.

166. Wehr, F.L., Fairchild, L.H., Hudec, M.R., Shafto, R.K., Shea, W.T., White, J.P., 2000. Fault seal; contrasts between the exploration and production problem. AAPG Mem. 73, 121–132.

167. Welbon, A.I., Beach, A., Brockbank, P.J., 1997. Fault seal analysis in hydrocarbon exploration and appraisal: examples from offshore mid-Norway. Nor. Pet. Soc. Spec. Publ. 7, 125–138.

168. Welch, M.J., Knipe, R.J., Souque, C., Davies, R.K., 2009. A Quadshear kinematic model for folding and clay smear development in fault zones. Tectonophysics 471, 186–202.

169. Yielding, G., Freeman, B., Needham, D.T., 1997. Quantitative fault seal prediction. AAPG Bull. 81, 897–917.

170. Yielding, G., 2002. Shale Gouge Ratio e calibration by geohistory. In: Koestler, A.G., Hunsdale, R. (Eds.), Hydrocarbon Seal Quantification, Norwegian Petroleum Society (NPF), Special Publications, vol. 11, pp. 1–15.

171. Yielding, G., 2012. Using probabilistic shale smear modelling to relate SGR predictions of column height to fault-zone heterogeneity. Pet. Geosci. 18, 33–42.

172. Zhao, C., Hobbs, B., Ord, A., Robert, P., Hornby, P., Peng, S., 2007. Phenomenological modelling of crack generation in brittle crustal rocks using the particle simulation method. J. Struct. Geol. 29, 1034–1048.

173. Zhou, J., Chi, Y., 2002. Shear-band of sand simulated by Particle Flow Code (PFC). In: Konietzky, H. (Ed.), Numerical Modeling in Micromechanics via Particle Methods. Balkema, Rotterdam, Gelsenkirchen, Germany, pp. 205–210.

Index